T0249543

MATHEMATICS: A Second Start, 2nd edition

"Talking of education, people have now a-days" (said he) "got a strange opinion that every thing should be taught by lectures. Now, I cannot see that lectures can do so much good as reading the books from which the lectures are taken. I know nothing that can be best taught by lectures, except where experiments are to be shewn. You may teach chymestry by lectures — You might teach making of shoes by lectures!"

James Boswell: *Life of Samuel Johnson, 1766*

The direction in which education starts a man will determine his future life.

Plato (427-347 BC): *The Republic*

Mathematics possesses not only truth, but supreme beauty, cold and austere like that of sculpture, and capable of stern perfection, such as only great art can show.

Bertrand Russell (1872-1970): *The Principles of Mathematics*

Horwood Publishing Series: Mathematics and Applications

Teaching Practice in Mathematics
J.H. MASON, The Open University, Milton Keynes

Modelling and Mathematics Education (ICTMA 9)
J.F. MATOS *et al*, University of Lisbon, Portugal

Mathematical Modelling (ICTMA 8)
Editors: P. GALBRAITH *et al*, University of Queensland, Brisbane, Australia

Teaching & Learning Mathematical Modelling (ICTMA 7)
Editors: S.K. HOUSTON *et al*, University of Ulster, Northern Ireland

Statistical Mechanics: An Introduction
D.H. TREVENA, University of Aberystwyth

Manifold Theory
D. MARTIN, University of Glasgow

Ordinary Differential Equations and Applications
W.S. WEIGLHOFER & K.A. LINDSAY, University of Glasgow

Geometry with Trigonometry
P.D. BARRY, National University of Ireland, Cork

Cultural Diversity in Mathematics (Education) : CIEAM 51
Editors: A. AHMED *et al*, Universsity College, Chichester

Advanced Imaging: Theory, Computing, Practice
J.M. BLACKLEDGE & M.J.RYCROFT, De Montfort University, Leicester

Digital Signal Processing: Software Solutions and Applications
J.M. BLACKLEDGE & M.J. TURNER, De Montfort University, Leicester

Image Processing III: Mathematical Methods, Algorithms & Applications
J.M. BLACKLEDGE *et al*, De Montfort University, Leicester

Image Processing II: Mathematical Methods, Algorithms & Applications
J.M. BLACKLEDGE & M.J. TURNER, De Montfort University, Leicester

Electrical Engineering Mathematics
R. CLARKE, Imperial College, London

Experimental Design Techniques in Statistical Practice
W.P. GARDINER & G. GETTINBY, Glasgow Caledonian University

Delta Functions: Fundamental Introduction to Generalised Functions
R.F. HOSKINS, De Montfort University, Leicester

Calculus: Introduction to Theory and Applications in Physical and Life Sciences
R.M. JOHNSON, University of Paisley

Linear Differential and Difference Equations
R.M. JOHNSON, University of Paisley

Game Theory: Mathematical Models of Conflict
A.J. JONES, University College of Wales, Cardiff

Stochastic Differential Equations and Applications
X. MAO, University of Strathclyde

Fundamentals of University Mathematics, 2nd edition
C. McGREGOR *et al*, University of Glasgow

Mathematics for Earth Sciences
P. SHARKEY, University of Portsmouth

Decision and Discrete Mathematics
The SPODE GROUP, Truro School, Cornwall

Mathematical Analysis ad Proof
D.S.G. STIRLING, University of Reading

Geometry of Navigation
R. WILLIAMS, Master Mariner, Mersey Dock and Harbour Company, Liverpool

MATHEMATICS: A Second Start, 2nd edition

SHEILA PAGE
Lecturer in Mathematics, University of Bradford

JOHN BERRY
Professor of Mathematics Education, University of Plymouth

and

HOWARD HAMPSON
Department of Mathematics, Torquay College of Further Education

Horwood Publishing
Chichester

Published in 2002 by
HORWOOD PUBLISHING LIMITED
International Publishers
Coll House, Westergate, Chichester, West Sussex, PO20 6QL
England

First published in 1986 by Ellis Horwood Limited

British Library Cataloguing in Publication Data
A catalogue record of this book is available from the British Library

ISBN 1-898563-04-7

Printed in Great Britain by Antony Rowe Ltd, Eastbourne

Contents

Preface to Second Edition ... vii

Introduction to the First Edition ... vii

Chapter 0   Using a Scientific Calculator ... 1

Chapter 1   The Set of Real Numbers ... 9

Chapter 2   Number Skills ... 37

Chapter 3   Algebra: a Basic Toolkit ... 55

Chapter 4   Algebra: More Tools ... 75

Chapter 5   Products ... 90

Chapter 6   Factors ... 106

Chapter 7   Equations and Inequalities ... 121

Chapter 8   Manipulation of Formulae ... 153

Chapter 9   Quadratic Equations ... 170

Chapter 10   Simultaneous Equations ... 186

Chapter 11   Indices and Logarithms ... 196

Chapter 12   Functions and Graphs ... 211

Chapter 13   Linear Graphs and Their Use in Experimental Work ... 230

Chapter 14   Calculus – the Mathematics of Change ... 245

Chapter 15   Applications of Differentiation ... 266

Contents

Chapter 16	Integration	286
Chapter 17	Introduction to Trigonometry	302
Chapter 18	Calculus of Trigonometry	333
Chapter 19	The Law of Natural Growth	345
Chapter 20	Some Applications of Logarithms	362
Chapter 21	A First Look at Statistics	377
Chapter 22	Probability	400
Chapter 23	Probability Distributions	415
Appendix		439
Answers		440
Index		466

Preface to Second Edition

In revising this text we have attempted to retain the original aims of Mrs Page's book; that is to provide a second start at mathematics for those students who 'never could do maths'. The text provides the basic algebra, calculus and statistics for students in further and higher education who require mathematics to support their main study area.

Howard Hampson
John Berry

Introduction to the First Edition

This book is intended for the student 'who never could do maths', the student who for one reason or another missed his or her way at school, either by absence from some of the O-level course, or by having too many changes of masters or schools. Such a student has lost confidence in his own ability to tackle the subject. It is intended for the student whose knowledge extends only to O-level mathematics, probably obtained several years ago, but who would now like to be able to communicate with mathematicians, i.e. to know what is meant when one talks about 'an integral' or 'a differential equation'. It is based on a course of tutorials and discussions with students of this type, to whom I am indebted for their patient endeavours and encouragement.

So let us begin at the beginning. The student who does not require the elementary work, nevertheless, is advised to see that the examples at the end of the chapter are worked through conscientiously, as each step depends on the previous steps being fully understood. One must be completely competent in dealing with mathematical 'shorthand', i.e. the method of expressing an idea in as neat a way as possible, so that it can be handled easily and quickly.

0

Using a Scientific Calculator

INTRODUCTION

Few people attempt to study and use mathematics without taking the drudgery out of the arithmetical operation by using a calculator. The speed and accuracy of modern calculators can benefit everyone from scientists and engineers, to business people and to those in the home. Children are encouraged to use simple calculators from an early age in school. Students in colleges and universities use advanced scientific calculators, often with graph drawing facilities, in most courses.

You will need a calculator to study and use the mathematics in this book. A natural question to ask is 'which calculator should I buy?' There are many different calculators on the market. We have decided not to recommend a particular brand name or model because their features regularly change, but we do offer the following guidelines:

1. The calculator should be a scientific calculator containing trigonometric, logarithmic and exponential functions. Look for keys marked sin, cos, tan, e^x, ln, log. You may not understand their meaning yet; we will develop the theory in later chapters.

2. Look for simple statistics measures such as the mean and standard deviation.

3. Although a calculator with graph drawing facilities will help your understanding of functions and their properties, it is not essential to include this feature. Avoid a calculator with advanced programmable features.

When you have purchased a new calculator you should study the booklet provided with it very carefully. Explore the features of the different keys.

This chapter of the book is provided for two reasons:

1.   To attempt to help you develop good calculator skills and to prevent you developing bad habits.

2.   To provide exercises, with answers, so that you can familiarise yourself with your calculator.

(If your calculator uses Reverse Polish notation, methods of this chapter do not apply.)

GETTING TO KNOW YOUR CALCULATOR

An obvious start is to turn the calculator on!

1.   Arithmetic Operations $+, -, \times, \div$

     Press: 3, +, 5, =   The display should give 8

     Press: 8, ×, 6, =   The display should give 48

     Press: 8, ÷, 2, =   The display should give 4

On many calculators the = operation is the EXE key (for execute). On others it is the ENTER key.

         Press: 3, +, 5, ×, 6, =,

The calculator has evaluated $3 + 5 \times 6 =$. What answer does your calculator give?

If the answer on your calculator is 33 then your calculator performs the operations in a special order. The order of preference is

        multiplications and divisions   before   additions and subtractions

So the calculator does the following:

         multiplication: $5 \times 6 = 30$
         addition: $3 + 30 = 33$

If the answer on your calculator is 48 then your calculator does not exercise any hierarchy. It performs the calculations in the order that you type.

         addition: $3 + 5 = 8$
         multiplication: $8 \times 6 = 48$

Immediately you can see how easy it is to make errors. The calculator will do what you tell it which might not be what you want! At this early stage familiarise yourself with any built-in hierarchy of performing operations. As we shall see below, the use of brackets will often avoid errors caused by the order of doing operations.

The following order of preference is commonly used in scientific calculators:

expressions with brackets   (highest priority)
functions
powers
multiplications and divisions
additions and subtractions   (lowest priority)

Exercise 0.1

Carry out the following operations and explain the meaning of the answer given by your calculator.

1.   Press: $2, \times, 5, -, 3, =$

2.   Press: $1, +, 7, \times, 3, =$

3.   Press: $8, -, 4, \div, 2, =$

4.   Press: $8, \div, 2, -, 4, =$

5.   Press: $3, +, 11, \times, 5, -, 2, =$

6.   Press: $2, \times, 6, +, 9, \div, 3, =$

2.   The Mode Key

Calculators operate in several different ways or modes. Find out how to display the mode on your screen. (You may find a key marked MODE or M Disp.) You will find several different options. Don't worry about them all at this stage. The option that may cause the biggest problem in the trigonometric chapters is the choice of the units to measure angles.

Change from Degree Mode to Radian Mode

Press: sin, 30, =
You should obtain -0.9880316241

Change from Radian Mode to Degree Mode

Press: sin, 30, =
You should obtain 0.5

In the first operation 30 is measured in units called radians; in the second operation 30 is measured in degrees. In the trigonometry and calculus chapters you must ensure that your calculator is in the appropriate mode.

3.   2nd or SHIFT or INV Key

Most calculators have two (or more) functions for each key. For example, you may find the symbols Log and 10^x on the same key. To obtain the alternative function 10^x you will need to press two keys.

For example, to evaluate 10^3:

Press: SHIFT, log, 3, =   The display should give 1000

Your calculator may have an alternative key to SHIFT. The keys 2nd or INV are most common.

4.   The Fix Mode

This mode determines the number of decimal places shown on the calculator display. To change the number you may have to carry out one of the following operations:

(for Casio):   Press: MODE, Fix, followed by a number

For example, MODE, Fix, 5 will display five digits after the decimal point.

(for TEXAS):   Press: MODE, choose the number of decimal places from the various display options

Exercise 0.2

Calculate the following displaying the given number of decimal places.

1. 4.23×1.602 4 d.p.

2. 0.612×3.4712 6 d.p.

3. $6.02 \div 2.104$ 5 d.p.

4. $0.623 \div 8.172$ 8 d.p.

5.   The Negative Key (–)

This key allows you to enter a negatively signed number into the calculator. Sometimes, it may be marked '+/–'.

For example, to evaluate -2×-3

Press: (–), 2, ×, (–), 3, =   The display should give 6

For example, to evaluate -2×3

Press: (–), 2, ×, 3, =   The display should give –6

(If the meaning of these calculations is a mystery do not worry at this stage!)

CALCULATING WITH YOUR CALCULATOR

You have already seen the care that is needed when carrying out simple arithmetical operations. The calculator has its own built-in rules and so it is important to know that it calculates what you want. To avoid errors and develop good habits when using a calculator, here are some helpful hints:

1.   Always **check** the display when you have keyed in a number. It's easy to press the wrong key.

2.   Use **brackets** to ensure the required order of operations.

3.   Do a rough **mental calculation** to check that the computation looks correct. For example, when calculating 3.14×1.96 a display showing 61.544 means that the numbers have not been entered correctly; 3.14 is roughly 3 and 1.96 is roughly 2, so the correct answer should be nearer 6 than 60.

6.   Using Brackets (. . .)

The order of carrying out arithmetical operations affects the answer of a calculation. For example,

$$(2 + 3) \times 4 = 20 \qquad (6 \div 2) + 1 = 4$$
$$2 + (3 \times 4) = 14 \qquad 6 \div (2 + 1) = 2$$

The different positions of the brackets are important. Sometimes brackets are given in a problem for clarity; for example, $2 + (3 \times 4) - (4 \times 2)$ clearly shows the order of the calculation whereas $2 + 3 \times 4 - 4 \times 2$ is less obvious.

Now $\dfrac{1}{2+3}$ is clear to the mathematician, it means $\dfrac{1}{5} = 0.2$. But look at the effect of bad calculator habits of someone who keys in: $1 \div 2 + 3$. The answer is 3.5. The calculator has evaluated $(1 \div 2) + 3 = 0.5 + 3 = 3.5$. To ensure a correct answer, type in $1 \div (2 + 3)$. The use of brackets avoids errors.

Example 0.1   Calculate (a) $\dfrac{4.26 - 0.12}{5.10 + 2.32}$ (b) $\dfrac{\sqrt{11.01} - 1.2^2}{\sqrt{1.7 \times 2.4}}$.

Solution

(a)   To get an estimate for the answer we could approximate the numerator by 4 and the denominator by 7. So we expect an answer of roughly $\dfrac{4}{7}$ or 0.5.

Using the calculator we introduce brackets and calculate

$$(4.26 - 0.12) \div (5.10 + 2.32)$$

to give the answer 0.5580.

(b)   The calculation includes the use of the keys $\sqrt{}$ and x^2. Finding a rough estimate is more difficult now but is still an important activity so that we can check the answer.

$\sqrt{11.01}$ is roughly 3 1.2^2 is roughly 1

1.7×2.4 is roughly 4 $\sqrt{1.7 \times 2.4}$ is roughly 2

So a very approximate answer is $\dfrac{3-1}{2} =$.

Using the calculator we introduce brackets and calculate

$$\left(\sqrt{11.01} - 1.2^2\right) \div \left(\sqrt{(1.7 \times 2.4)}\right)$$

to give the answer 0.9298.

Note the importance of the brackets in the denominator:

$$\sqrt{}, (, 1.7, \times, 2.4,), \; = \; 2.0199 \text{ but } \sqrt{}, 1.7, \times, 2.4, = \; 3.1292$$

In the second operation the $\sqrt{}$ only applies to 1.7.

Exercise 0.3

Complete the following table:

Calculation	Estimate of the answer	Calculator display (to 4 decimal places)
4.23×5.71	$4 \times 6 = 24$	24.1533
$6.31 \div 1.72$	$6 \div 2 = 3$	3.6686
$\dfrac{81.23}{4.1 + 3.5}$		
$\dfrac{427}{11.2 - 3.14}$		
$\dfrac{6.71 - 2.04}{11.62 + 4.12}$		
$7.26 \times (6.91 - 1.32)$		
$\dfrac{1.42 + 6.17 - 0.32}{1.71 - 09.82}$		
$\dfrac{1}{4.86 - 2.49}$		
$\dfrac{3}{7} + \dfrac{2}{4}$		
$\dfrac{17.2}{1.4} - \dfrac{3.61}{0.42}$		
$\dfrac{17.7}{3.02 \times (9.61 - 4.26)}$		

Exercise 0.4

Use the $\sqrt{}$ and x^2 keys to calculate the following:

1. $\sqrt{25}$ 2. $\sqrt{196}$ 3. $\sqrt{14}$ 4. $\sqrt{16.8}$

5. $\sqrt{7.8}$ 6. $\sqrt{1.35}$ 7. $\sqrt{1461.7}$ 8. $\sqrt{0.0684}$

9. 4^2 10. 11^2 11. 1.6^2 12. 0.2^2

13. 0.7^2 14. 1.32^2 15. 4.061^2 16. 0.0684^2

17. $\sqrt{1.35 - 0.2}$ 18. $\sqrt{1.35} - \sqrt{0.2}$ 19. $\sqrt{6.1} \times \sqrt{2.3}$

20. $\sqrt{6.1 \times 2.3}$ 21. $\dfrac{8.36 - 2.07}{\sqrt{1.41 - 0.17}}$ 22. $\dfrac{11.72 - \sqrt{9.04}}{\sqrt{12} + 3.2}$

1

The Set of Real Numbers

NUMBERS AND SYMBOLS

We cannot begin to study mathematics without first thinking about numbers and the way they behave. The aim of this chapter is to introduce you to some of the properties of numbers. We also introduce the basic symbols used in mathematics and discuss some of their properties.

The treatment used is by no means exhaustive and we try to appeal to intuition wherever possible. We prove many things by using examples rather than using general proofs. The aim is to look at the development and use of numbers as an application of common sense rather than a set of mysterious processes.

For many of you, this chapter will be revision. Some readers will therefore be able to skim through it; others will need to take a little more time. However, no-one should underestimate the importance of this introduction as it is necessary for a full understanding of what follows in later chapters. It is a good idea to keep coming back to this chapter and *re-reading* it at regular intervals, especially if some of the ideas don't sink in the first time.

SETS AND SUBSETS

A *set* is *any* collection of objects such as a bunch of keys or bananas, a band of thieves or musicians, a gang, a class, a posse, a bouquet of flowers, a team of football players, a CD collection. Just about *anything* you can think of is contained in some set or other.

The individual items in a set are called the *elements* of the set.

Sometimes all the elements may be similar as with the individual keys in a bunch. Sometimes it may only be membership of the set itself that the elements have in common. A plumber's toolbox might contain a pipe-wrench, a saw, some nails, a piece of solder and a ham sandwich! The objects still form a set.

A *subset* is a set contained within some other set. For example, a mathematics class is a set. It will contain many subsets: The set of female students. The set of male students. The set of students (male and female) with blue eyes. The set of students who left their homework on the bus (again!). And so on.

We will confine the rest of our discussion to what this book is all about: *sets of numbers*.

THE NATURAL NUMBERS

The simplest numbers are the ones used for counting. Early civilisations all needed ways of keeping a check on numbers: how much grain was stored for the winter, how many soldiers in the army, how many more days left until winter sets in. Thus people would represent numbers by a notch on a stick, knots or beads on a cord or marks on a clay tablet. It was therefore natural to invent systems of marks and symbols to represent numbers and the relationships between them. Some of these systems have disappeared because they were too restricted or clumsy to advance beyond simple counting. Try multiplying XVIII by CXLIV using the Roman system and you will see what we mean! (The answer, by the way, is MMDXCII.)

The numbers used for counting, i.e. 1,2,3,4... etc. are called *natural numbers*. They are also called *integers* or *whole numbers*. We are used to counting in the denary or decimal scale which was invented by Hindu mathematicians and extended by the Arabs. Zero was invented later and is needed when *subtraction* causes no result.

The Number Line

We can represent numbers by using a line. The natural numbers are placed in a line with equal distances between them. Early civilisations would have written their number line like this; each natural number is a point on the line.

$$1 \qquad 2 \qquad 3 \qquad 4 \qquad 5 \quad \text{etc.}$$
$$\bullet \qquad \bullet \qquad \bullet \qquad \bullet \qquad \bullet$$

The first thing we notice is that the natural numbers have a particular *order*. Immediately we can begin to use statements like 'three is bigger than one' or 'seven is less than nine'. We can also invent shorthand symbols that say this for us:

> means 'is greater than'   and   < means 'is less than'

so: $3 > 1$ and $7 < 9$

Always remember that statements like those above are *English sentences* written in shorthand. They should therefore *make sense*!

$$3 > 1 \quad \text{means} \qquad \text{'three is greater than one'}$$

Note that the open end of the symbol faces the larger of the two numbers,

i.e. $\qquad\qquad 5 > 7 \qquad\qquad$ is meaningless!

Statements like this are called *inequalities.*

We can go further and write $1 < 2 < 3 < 4 < 5 < 6 < ...$ and so on.

Exercise 1.1

1.   Write the following inequalities as English sentences:

   (a) $3 < 4$ (b) $5 > 1$ (c) $11 < 13$

2.   Write the following English sentences as inequalities.

   (a)   Two is less than six.

   (b)   One is less than fifteen.

   (c)   Twenty is greater than nine.

THE BASIC LAWS OF NATURAL NUMBERS

Addition: A Binary Operator

Choose any two members from the set of natural numbers $1,2,3,4,5,6,7,8,9,10...$ etc. say, 2 and 7. We can make a third natural number by adding them together:

i.e. $\qquad\qquad\qquad\qquad 2 + 7 = 9$

'+' is known as a *binary operator*. If a binary operator gives a result which is a member of the same set, the operation is said to be *closed.* Thus a more sophisticated way of saying 'adding two whole numbers together gives another whole number' is to say that '*the natural numbers are closed under addition*'.

Note that we have introduced two symbols, the binary operator '+' which is an *instruction* that connects two quantities from the same set, and '=' which indicates a *result* and merely says 'is the same as'.

The *statement* '2 + 7 = 9' forms an *equation* which is an *English sentence* written in a shorthand form.

If two quantities are not the same, we use the symbol '$\neq$', which is the '=' sign crossed out. Thus, for example, $13 \neq 21$.

Commutative Law of Addition

If we choose the above numbers in a different order, the result remains the same,

so $7 + 2 = 9$ and also $2 + 7 = 9$

Hence $2 + 7 = 7 + 2$

This fact is true whatever natural numbers we choose. Changing the order of the numbers is called *commuting* and this *commutative law of addition* merely states that the order in which two natural numbers are added makes no difference to the result.

Associative Law of Addition. Use of Brackets

What happens if we want to add together *more* than two natural numbers, say 2, 3 and 4? Does the order in which we add them make a difference?

We can use *brackets* here to emphasise the order in which the operations are carried out. Brackets enable us to change the focus of our concentration and tell us *which operation to do first*. We will use them frequently throughout the text.

So, to add the three natural numbers together we can do

either $(2 + 3) + 4 = 5 + 4 = 9$ adding $2 + 3$ first
or $2 + (3 + 4) = 2 + 7 = 9$ adding $3 + 4$ first

and of course it makes no difference to the result!

Therefore $2 + 3 + 4 = (2 + 3) + 4 = 2 + (3 + 4) = 9$ and we can add in any order. This is known as the *associative law of addition*.

We can now combine the two laws to note that

$$2 + 3 + 4 = 2 + 4 + 3 \qquad \text{(commuting the second pair)}$$
$$= 4 + 2 + 3 \qquad \text{(commuting the new first pair)}$$
$$= 4 + 3 + 2 \qquad \text{etc.}$$

Hence when we are adding the natural numbers, we need not worry about the order in which addition is done and, in fact, the brackets are never necessary.

If we wish to extend this idea to four numbers we can nest brackets within brackets. In this case the calculations within the innermost brackets are done first.

So, to add four natural numbers 3, 5, 7 and 11

$$3 + 5 + 7 + 11 \quad = ((3 + 5) + 7) + 11$$
$$= (8 + 7) + 11$$
$$= 15 + 11$$
$$= 26$$

or $\qquad 3 + 5 + 7 + 11 \qquad = 3 + ((5 + 7) + 11)$
$$= 3 + (12 + 11)$$
$$= 3 + 23$$
$$= 26$$

or we can change the order and find many other ways of performing the calculation. They all give the same answer; so, again, we only need the brackets to help ourselves keep a check on the calculation.

Exercise 1.2

By calculating each side separately, confirm the following statements. Remember that any calculation within brackets is carried out first. Use a calculator if you wish.

1. $12 + (6 + 7) = (12 + 6) + 7$

2. $33 + (17 + 11) = (33 + 17) + 11$

3. $147 + (256 + 914) = (147 + 256) + 914$

Confirm the following statements by calculating each part separately:

4. $77 + (49 + 16) = (77 + 49) + 16 = 49 + (77 + 16)$

5. $18 + ((5 + 9) + 121) = ((18 + 5) + 121) + 9 = (5 + 9) + (18 + 21)$

6. $(((1 + 2) + 3) + 4) + 5 = 1 + (((2 + 3) + 4) + 5) = (1 + 2) + 3 + (4 + 5)$

Multiplication: Another Binary Operator

Multiplication is merely a shorthand way of doing addition. If we want to add a number to *itself* a few times, we can write

$$2 + 2 = 4 \qquad \text{or} \qquad 2 + 2 + 2 = 6$$

If we wish to do this more often it soon becomes clumsy:

for example, $2 + 2 + 2 + 2 + 2 + 2 + 2 + 2 + 2$ needs to be shortened!

We have chosen to add a set of twos together nine times and this is written as 9×2. We have therefore invented a new binary operation called *multiplication* by taking a pair of numbers from the set of natural numbers and combining them together in a certain order to produce a third natural number.

Thus $9 \times 2 = 18$ which is another natural number.

As with addition, natural numbers are *closed* under the binary operation of multiplication since when we multiply two natural numbers together, we produce another natural number.

Commutative Law of Multiplication

It is easy to see that

$$2 + 2 + 2 + 2 + 2 + 2 + 2 + 2 + 2 \text{ gives the same result as } 9 + 9$$

i.e. $9 \times 2 = 2 \times 9$

Similarly $3 + 3 + 3 + 3 \text{ gives the same result as } 4 + 4 + 4$

i.e. $4 \times 3 = 3 \times 4$

So, the order in which we multiply two natural numbers together does not matter and multiplication is commutative for the natural numbers.

Associative Law of Multiplication

Using brackets again to change precedence:

$$3 \times (4 \times 7) = 3 \times 28 = 84 \quad \text{(multiplying the second pair first)}$$

and $\quad (3 \times 4) \times 7 = 12 \times 7 = 84 \quad$ (multiplying the first pair first)

Hence $\qquad 3 \times 4 \times 7 = 3 \times (4 \times 7) = (3 \times 4) \times 7$

and multiplication is associative for the natural numbers. Again both laws can be combined and we can confidently say that the natural numbers can be multiplied in any order,

i.e. $\qquad 3 \times 4 \times 7 = 3 \times 7 \times 4 = 7 \times 3 \times 4 \quad$ etc.

Again, brackets are unnecessary.

Exercise 1.3

By calculating each part separately, use your calculators to verify the following statements:

1. $\quad 97 \times (14 \times 19) = (97 \times 14) \times 19 = (19 \times 97) \times 14$

2. $\quad (15 \times 17) \times (11 \times 23) = ((15 \times 17) \times 11) \times 23 = 15 \times (17 \times (11 \times 23))$

3. $\quad 2 \times ((3 \times 4) \times 5)) = ((2 \times 3) \times 4)) \times 5 = (2 \times 3) \times (4 \times 5)$

Factors, Multiples and Prime Numbers

As we see above, some natural numbers can be produced by *multiplying* smaller natural numbers together. These smaller numbers are called *factors,* for example:

$$12 = 3 \times 4 = 6 \times 2 = 12 \times 1$$

We have *factorised* the number 12 in three different ways. The number 12 is known as a *multiple* of 3 or 4 or 2 or 6, etc.

Some natural numbers have no factors, except themselves and one. They are called *prime numbers*,

e.g. $\qquad 13 = 13 \times 1 = 1 \times 13$

and there is no other way of producing 13 by multiplying natural numbers together.

Look again at the ways we have factorised 12 and keep on factorising:

$$3 \times 4 = 3 \times 2 \times 2 \qquad \text{and} \qquad 6 \times 2 = 2 \times 3 \times 2$$

We can keep on factorising until we can factorise no more. When we have done this, we have found all the factors of 12 which are *prime* numbers and the result is called a *prime factorisation*.

So $\qquad\qquad$ $12 = 2 \times 2 \times 3$ $\qquad\qquad$ is the prime factorisation of 12

$\qquad\qquad$ $35 = 5 \times 7$ $\qquad\qquad$ is the prime factorisation of 35

and $\qquad\qquad$ $120 = 12 \times 10$

$\qquad\qquad\qquad = 6 \times 2 \times 2 \times 5$

$\qquad\qquad\qquad = 2 \times 3 \times 2 \times 2 \times 5$ $\qquad$ is the prime factorisation of 120

All natural numbers which have 2 as a factor are called *even* numbers, the others are called *odd* numbers. The number 2 is the only even prime number.

Identity Element for Multiplication

Note the special property of the number '1'. It is the only number that can be used to multiply other numbers and leave them unchanged. For example:

$$13 = 13 \times 1 = 1 \times 13$$

$$37 = 37 \times 1 = 1 \times 37$$

$$672 = 672 \times 1 = 1 \times 672 \text{ etc.}$$

We can therefore multiply any natural number, *in any order*, by the number 1, without changing it. We call the number '1' the *identity element* for multiplication.

Squares, Cubes and Other Powers

Squaring a number simply means multiplying it by itself. If we square the natural numbers on our number line from 1 to 10, say, then we generate the natural numbers 1, 4, 9, 16, 25, 36, 49, 64, 81 and 100, called squares or *perfect squares*.

Cubing a number means multiplying it by itself and then by itself again. If we cube the first ten natural numbers, we generate the numbers 1, 8, 27, 64, 125, 216, 343, 512, 729 and 1000 known as *perfect cubes*.

If we write out the multiplications we are doing here, we have, for example:

$$6 \times 6 = 36 \qquad\qquad 7 \times 7 \times 7 = 343 \text{ and so on}$$

We can continue this process and multiply a number by itself as many times as we like. Just like continued addition, this can become a little tedious and clumsy:

for example, $4 \times 4 \times 4 \times 4 \times 4 \times 4 \times 4 \times 4 \times 4 \times 4 \times 4$ needs to be shortened!

Just as multiplication is a shorthand way of doing continued addition, we can invent a shorthand way of doing continued multiplication. The above is called the eleventh *power* of 4 and is equal to 4,194,304.

We write this eleventh power of 4 as 4^{11}. We call this '4 raised to the power 11' or simply '4 to the eleven' which means 4 multiplied eleven times. It does *not* mean 4×11, which is only 44. The power is also known as an ***index*** (plural: indices) or sometimes as an ***exponent***. The process of 'raising to powers' is called *exponentiation*.

Hence we also write $6 \times 6 = 6^2$ (six squared)

and $7 \times 7 \times 7 = 7^3$ (seven cubed)   etc.

          We will return to the idea of powers in a later chapter.

Example 1.1   Find a prime factorisation of 7875.

Solution   We need to factorise 7875 in terms of prime numbers. It is easy to do with a calculator by dividing 7875 by prime numbers of increasing size. Clearly 2 is not a factor since 7875 is odd.

Start with 3: $\dfrac{7875}{3} = 2625;$ and again $\dfrac{2625}{3} = 875$

But 825 is not divisible by 3 (you get 291.666666... if you try it!).

Now try 5: $\dfrac{875}{5} = 175;$ $\dfrac{175}{5} = 35$ and $\dfrac{35}{5} = 7$

and that's it, since 7 is itself a prime number.

So, we can write $7875 = 7 \times 5 \times 5 \times 5 \times 3 \times 3$

or, using the neater power notation: $7875 = 3^2 \times 5^3 \times 7$

Exercise 1.4

1. Write down the first 20 prime numbers. You will need to think of a systematic way of doing this.

2. Why is 2 the only even prime number?

3. Use a calculator to perform a prime factorisation of the following numbers. Write your answers in terms of powers of the prime factors, e.g. $12 = 2 \times 2 \times 3 = 2^2 \times 3$.

    (i)   360   (ii)   450   (iii)   6615

4. $4^5 = 4 \times 4 \times 4 \times 4 \times 4$ and $5^4 = 5 \times 5 \times 5 \times 5$. Use your calculator to confirm that these quantities are *not* the same and hence that 'raising to a power' is not commutative.

5. Repeat question 4 for the following:

    (i) 3^7 and 7^3 (ii) 8^9 and 9^8

Highest Common Factor (HCF)

The HCF of a set of integers is the largest integer which is a factor of each of them. For example, 12 and 24 have a common factor of 3 but this is not the highest common factor, which is 12 itself. How do we know when we have found the highest common factor?

> To find the HCF of a set of integers, do a prime factorisation of each integer and pick the *lowest* power of every *COMMON* factor.

Example 1.2   Find the HCF of 24, 28 and 40.

Solution $24 = 6 \times 4 = 2 \times 3 \times 2 \times 2 = 2^3 \times 3$

$28 = 7 \times 4 \qquad\qquad = 2^2 \times 7$

$40 = 4 \times 10 = 2 \times 2 \times 2 \times 5 = 2^3 \times 5$

Now 2 is common to all three numbers, so is 2^2, but 2^3 is not and neither are 3, 7 and 5.

Hence the HCF of 24, 28 and 40 is $2^2 = 4$ (note again that this is the *lowest* power of 2 because 2^3 will divide into 40 and 24 but not into 28).

Example 1.3   Find the HCF of 270, 450 and 1260.

Solution $270 = 5 \times 6 \times 9 = 2 \times 3^3 \times 5$

                         $450 = 10 \times 45 = 2 \times 5 \times 3 \times 3 \times 5 = 2 \times 3^2 \times 5^2$

                         $1260 = 4 \times 63 \times 5 = 2^2 \times 3^2 \times 5 \times 7$

Hence the HCF of these numbers is $2 \times 3^2 \times 5 = 90$. We have yet again used the *lowest* power of each factor.

Example 1.4   Find the HCF of 10, 11 and 21.

Solution $10 = 2 \times 5$ $11 = 11 \times 1$ $21 = 3 \times 7$

These numbers do not have a common factor; hence they cannot have an HCF. They are said to be *co-prime*. (Note that to be co-prime, they do not have to be prime!)

Exercise 1.5

Use your calculator (if necessary) to find the HCF of the following sets of integers. If the integers in any of the sets are co-prime, say so.

1.   8, 10, 12 and 16   2.   36, 42, 56 and 68   3.   37, 74, 148 and 296

4.   35, 140 and 420   5.   30, 120 and 840   6.   21, 44 and 65

Lowest Common Multiple (LCM)

The LCM of a set of integers is the smallest integer that they will all divide into.

For example, a common multiple of 3 and 6 is 18, but it is not the lowest. The number 6 is in fact the lowest. How do we know when we have found the lowest common multiple?

> To find the LCM of a set of integers, do a prime factorisation of each integer and pick the *highest* power of every factor (not just the common ones).

Example 1.5   Find the LCM of 10, 15, 45 and 75.

Solution $10 = 2 \times 5$ $15 = 3 \times 5$

$45 = 3^2 \times 5$ $75 = 3 \times 5^2$

Hence the lowest common multiple must have 2, 5^2 and 3^2 as its factors. These are the highest powers of every factor. Therefore the LCM of these numbers is $2 \times 3^2 \times 5^2 = 450$.

Exercise 1.6

Find the LCM of the following sets of integers:

1.   3, 4, 5 and 6   2.   7, 9, 10 and 12   3.   12, 15, 18 and 21

4.   24, 39, 56 and 78   5.   49, 98 and 121   6.   10, 33, 65 and 169

Combining Addition and Multiplication: the Distributive Law

What happens if we try to combine our two binary operations, addition and multiplication? Suppose we want to work out $7 \times (3 + 5)$. The brackets tell us to perform the addition first.

Hence $7 \times (3 + 5) = 7 \times 8 = 56$

However, we find that $(7 \times 3) + (7 \times 5) = 21 + 35 = 56$

This is the same answer; so if we wish, we can perform the multiplication first, provided we multiply *both* numbers in the brackets by the number outside.

This idea can be extended to any number of numbers within the brackets, for example:

$$3 \times (2 + 5 + 3 + 6) = 3 \times (16) = 48$$

and $(3 \times 2) + (3 \times 5) + (3 \times 3) + (3 \times 6) = 6 + 15 + 9 + 18 = 48$

This is known as the *distributive law* and is true for all the natural numbers. We say that multiplication is distributive over addition. Note that the distributive law is also true if the multiplication follows the bracket. (This is because of the commutative law!)

For example: $(3 + 4 + 7 + 5) \times 3 = (19) \times 3 = 57$

and $\qquad (3 \times 3) + (4 \times 3) + (7 \times 3) + (5 \times 3) = 9 + 12 + 21 + 15 = 57$

Therefore: $\qquad (3 + 4 + 7 + 5) \times 3 = (3 \times 3) + (4 \times 3) + (7 \times 3) + (5 \times 3)$

and so the number outside the brackets multiplies every number inside.

Multiplication Takes Precedence over Addition

Suppose we wish to calculate $3 \times 4 + 2$ and there are no brackets to tell us which operation to perform first. Then we could add first and calculate the answer to be

$$3 \times (4 + 2) = 3 \times (6) = 18$$

or we could multiply first and then we would end up with

$$3 \times 4 + 2 = (3 \times 4) + 2 = 12 + 2 = 14$$

These answers are not the same! We will therefore need a convention to help us when brackets are not provided. It has been agreed by mathematicians that multiplication takes precedence over addition; therefore:

> Unless brackets change the order of precedence,
> multiplications are always performed before additions.

Hence $\qquad 4 \times 5 + 7 = (4 \times 5) + 7 = 20 + 7 = 27 \qquad not \qquad 48$

and $\quad 4 \times 3 + 6 \times 7 + 5 \times 4 = (4 \times 3) + (6 \times 7) + (5 \times 4) = 12 + 42 + 20 = 74$

Exercise 1.7

By calculating both sides separately and using your calculator if necessary, confirm the distributive law in the following cases. Think carefully about what is happening in question 5. It will be important later!

1. $\quad 15 \times (7 + 9) = (15 \times 7) + (15 \times 9)$

2. $\quad 13 \times (3 + 7 + 11) = (13 \times 3) + (13 \times 7) + (13 \times 11)$

3. $\quad (5 + 17 + 16) \times 4 = (5 \times 4) + (17 \times 4) + (16 \times 4)$

4. $\quad (3 + 4) \times (5 + 6) = (3 \times 5) + (3 \times 6) + (4 \times 5) + (4 \times 6)$

Remembering that multiplication takes precedence, perform the following calculations:

5. $3 \times 4 + 7$ 6. $3 + 4 \times 7$

7. $7 \times 5 + 2 \times 6$ 8. $7 \times 5 + 12 \times 4 + 3$

9. $2 + 7 \times (8 + 5)$ 10. $11 + 12 \times 4 + 3 \times (5 + 3)$

11.   By calculating each side separately, verify the following statement and hence confirm that addition is not distributive over multiplication.

$$3 + (4 \times 5) \neq (3 + 4) \times (3 + 5)$$

Subtraction

Subtraction is the operation which performs the reverse or *inverse* process to addition. Thus choosing two natural numbers *in a particular order* we may write:

$$6 - 4 = 2$$

This is exactly equivalent to the statements

$$6 = 2 + 4 \quad \text{or} \quad 6 = 4 + 2$$

and we can think of the last two statements as being the result of having added 4 to both sides of the first equation:

i.e. $(6 - 4) + 4 = 2 + 4 = 6$

or $4 + (6 - 4) = 4 + 2 = 6$

Note that this might imply that subtraction is associative, but we must be *very* careful here, since we are *mixing* the binary operations '+' and '−'.

Consider the following and remember that anything placed in brackets is done first:

$$(7 - 4) - 2 = 3 - 2 = 1$$

and $7 - (4 - 2) = 7 - 2 = 5$

We see immediately that subtraction is *not* associative! *The order of subtraction is important!* It is perfectly legal to write $7 - 4 - 2$, without using brackets, but it means subtract *both the* 4 *and the* 2 from the 7. If you subtract the 2 from the 4 and then subtract the result from 7, you get the wrong answer!

So $\qquad\qquad\qquad\qquad 7 - 4 - 2 = 1$ *not* 5

However, we can show that multiplication is distributive over subtraction and combinations of addition and subtraction.

For example $\qquad\qquad\qquad 3 \times (9 - 4) = 3 \times (5) = 15$

and $\qquad\qquad\qquad\qquad (3 \times 9) - (3 \times 4) = 27 - 12 = 15$

which means that $\qquad\qquad\quad 3 \times (9 - 4) = (3 \times 9) - (3 \times 4)$

Similarly: $\qquad\qquad\qquad\quad 7 \times (8 + 2 - 3) = 7 \times (7) = 49$

and $\qquad\qquad (7 \times 8) + (7 \times 2) - (7 \times 3) = 56 + 14 - 21 = 49$

i.e. $\qquad\qquad\qquad 7 \times (8 + 2 - 3) = (7 \times 8) + (7 \times 2) - (7 \times 3)$

Finally, if no brackets are provided, we use the convention that multiplication takes precedence over subtraction and combinations of addition and subtraction.

For example, $4 \times 3 + 6 \times 7 - 5 \times 4 = (4 \times 3) + (6 \times 7) - (5 \times 4) = 12 + 42 - 20 = 34$.

Hence:

> Unless brackets change the order of precedence,
> multiplications are always performed before additions and subtractions.

Exercise 1.8

Evaluate the following:

1. $3 \times 4 - 7$ 　　　　 2. $\quad 7 \times 3 - 4$ 　　　　 3. $\quad 46 - 11 \times 4$

4. $(46 - 11) \times 4$ 5. $\quad 7 \times 5 - 12 + 4 \times 5$ 　 6. $\quad 7 \times (12 - 5) + 4 - 3$

Zero

Another problem with subtraction is that, if we try to subtract any natural number from itself, we end up with nothing! The answer *is not part of the set of natural numbers* and therefore we will have to invent a symbol meaning 'nothing' and extend the set of numbers,

so $\qquad\qquad\qquad\qquad 3 - 3 = 0 \qquad 5 - 5 = 0$ etc.

where the symbol '0' is called 'zero' or 'nought'.

The Identity Element for Addition

Note that the above statements are exactly equivalent to saying

$$0 + 3 = 3 \qquad \text{or} \qquad 3 + 0 = 3$$

and $\qquad\qquad\qquad\qquad 5 + 0 = 5 \qquad \text{or} \qquad 0 + 5 = 5$

This means that zero can be added to any number, in any order, without changing it. Thus, just as 1 is the identity element for multiplication, zero is known as the *identity element* for addition.

The Negative Integers

If we try to subtract a larger natural number from a smaller one, we again have no natural number that represents the solution! For example, suppose we try to calculate $4 - 6$. Young children would say 'that's impossible' and would not realise how wise the statement is! What they mean, of course, is that we cannot perform the calculation *using only the set of natural numbers*. The natural numbers are therefore *not* closed under subtraction.

Clearly, $\qquad\qquad\qquad 4 - 6$ is *not* the same as $6 - 4$

We write this as $\qquad\qquad\qquad 4 - 6 \neq 6 - 4$

so the operation of subtraction is *not* commutative.

So what *does* $4 - 6$ mean? Suppose we extend our number line to look like this:

$$-6 \quad -5 \quad -4 \quad -3 \quad -2 \quad -1 \quad 0 \quad 1 \quad 2 \quad 3 \quad 4 \quad 5 \quad 6$$

Now imagine the number line as stepping stones across a river, with zero in the middle of the stream. Adding means walking towards the right bank and subtracting means walking towards the left bank. At the end of the walk, give yourself the symbol '+' if you end up to the right of midstream and give yourself the symbol '−' if you end up to the left of midstream.

One way of picturing $6 - 4$ is to *start at zero*, walk six steps right and then four steps left. This is the same as starting from zero and moving two steps *right*.

$4 - 6$ can be pictured as *starting again at zero*, walking four steps right, followed by six steps left. We end up two steps to the *left* of midstream. Using this new idea, we have extended the set of natural numbers (the *positive integers*) to include zero and the *negative integers*.

Thus $$4 - 7 = -3$$

is a whole number to the left of zero on our extended number line.

Also $$7 - 4 = +3$$

is a whole number to the right of zero, as before.

Notice that in the equation $7 - 4 = -3$, the symbol '−' is being used in two different ways. On the left hand side (LHS) it is an *instruction* to subtract 7 from 4. On the right hand side (RHS) it serves to distinguish the number -3 from its positive counterpart 3 and tells us that the result of our calculation is not the same as 3, which can also be written as +3, as above. (Some authors actually write the numbers as $^{+}3$ and $^{-}3$.)

Because of the idea of positive and negative *directions*, numbers with a sign (+ or −) attached to them are called *directed numbers*.

Typical examples of when we need this idea of subtracting larger natural numbers from smaller ones are

(i)   When the temperature is 4 °C and drops overnight by 6 °C, we cannot say the temperature is 2 °C, we call it −2 °C. In this case the '−' sign means 'below zero'.

(ii)   You have £4 in the bank and decide to withdraw £6; it would hardly be fair to the bank to claim that you have £2 left! You now have −£2 in the bank. This time the '−' sign means 'overdrawn'.

(iii)   A kingfisher is flying 4 feet above a river and dives downwards into the stream. If the bird dives 6 feet altogether, it ends up 2 feet below the water, or we could say −2 feet above the water. Here, '−' means 'underwater'.

Exercise 1.9

1.   Think of three more examples of where we will need the idea of negative numbers.

2.   Is there an identity element for subtraction? Give a reason for your answer.

Division: Rational Numbers

Division is the inverse of multiplication. We reverse the multiplication of two numbers by dividing:

e.g. $8 \div 2 = 4$ is the same as saying $8 = 4 \times 2$ or $8 = 2 \times 4$

Division is obviously not commutative, since

$$2 \div 8 \neq 8 \div 2$$

$8 \div 2$ is fine! It gives the answer 4 which is another natural number, but we are going again to have to extend the set of numbers since we have to find a meaning for $2 \div 8$.

Think of placing extra small stones into the river (see below) between our main stepping stones. We can now take two large steps to get from 0 to 2 or eight small steps. The small stones have *divided* the distance between 0 and 2 into eight smaller steps. Each of these smaller steps represents one-eighth of the distance from 0 to 2 and so one small step is 'one-eighth of two' written as $2 \div 8$ or $\dfrac{2}{8}$ or $\dfrac{1}{8} \times 2$.)

$$\frac{2}{8} = \frac{1}{4}$$

The number $\dfrac{2}{8}$ is a *fraction* or *rational number*. All rational numbers can be written in the form $\dfrac{\text{integer}}{\text{integer}}$. The integer above the line is the *numerator* and that below the line is the *denominator*.

The line clearly shows that $\dfrac{2}{8}$ is the same as the number we get by dividing the

distance between 0 and 1 into four equal parts; so the fractions $\dfrac{2}{8}$ and $\dfrac{1}{4}$ are

equivalent. This is why we are allowed to do the following:

$$\frac{2}{8} = \frac{(2\times1)}{(2\times4)} = \frac{1}{4}$$

We have *factorised* here and *cancelled common factors*, to leave the fraction *in lowest terms*.

Note that all the natural numbers are rational numbers, since, for example:

$$\frac{8}{2} = \frac{(2\times4)}{(2\times1)} = \frac{4}{1} = 4$$

Hence the natural numbers form a subset of the rational numbers.

It is important to realise that cancelling common factors is equivalent to dividing the numerator and denominator by the same number. This process does not change the fraction.

Proper and Improper Fractions

A *proper* fraction is so named because it is a fraction of the number 1. We can find all proper fractions by placing stepping stones at appropriate distances between 0 and 1. If we perform a calculation such as dividing three cakes among four people, each person is given $\dfrac{3}{4}$ of a cake. The numerator is smaller than the denominator.

An *improper* fraction is bigger than one and is always the result of a division where the numerator is greater than the denominator. We naturally generate such fractions when we perform calculations like dividing four cakes among three people. This gives the fraction $4 \div 3$ or $\dfrac{4}{3}$ where the numerator is larger than the denominator.

Is division associative? A simple example should confirm that it is *not*:

e.g. $\qquad\qquad\qquad (8 \div 4) \div 2 = 2 \div 2 = 1$

and $\qquad\qquad\qquad 8 \div (4 \div 2) = 8 \div 2 = 4$

Reciprocals

Dividing by the number 4 gives the same result as multiplying by $\frac{1}{4}$. We will confirm this fact later. Hence, since 4 can be written as $\frac{4}{1}$, we may regard $\frac{1}{4}$ as being $\frac{4}{1}$ 'turned upside down' or inverted. Each number is known as the *reciprocal* of the other. We can extend the idea and say that, for example, $\frac{6}{5}$ is the reciprocal of $\frac{5}{6}$ and vice versa. We will also show later that multiplying by one of these numbers is the same as dividing by the other.

Decimal Fractions

These are not new numbers! Writing fractions in decimal form is merely an alternative way of presenting the same numbers. Thus $\frac{1}{2} = 0.5$ and $\frac{3}{4} = 0.75$, $\frac{5}{4} = 1.25$, etc. We will not go into how to combine decimals by addition, multiplication, etc. Your calculator will perform all its main functions and give answers in decimal form, unless you instruct it to use fractions or scientific notation etc.

The Distributive Law of Division

Division is distributive over addition and subtraction, providing we are very careful to understand how the law works. This law probably causes more problems than any other! It is very important to recognise that, unlike the distributive law involving addition and multiplication, we can only use this law *IN A CERTAIN ORDER*!

Consider $(6 + 9) \div 3 = 15 \div 3 = 5$

but $(6 \div 3) + (9 \div 3) = 2 + 3 = 5$

therefore $(6 + 9) \div 3 = (6 \div 3) + (9 \div 3)$

Similarly $(5 + 7) \div 4 = 12 \div 4 = 3$

and $(5 \div 4) + (7 \div 4) = 1.25 + 1.75 = 3$

giving $(5 + 7) \div 4 = (5 \div 4) + (7 \div 4)$

Rewriting this in a fractional form: $\dfrac{5+7}{4} = \dfrac{5}{4} + \dfrac{7}{4}$ and we see that a *SINGLE* denominator is allowed to divide into both (or all) numbers in the numerator which are added.

Similarly $\qquad\qquad (11 + 8 - 7) \div 4 = (12) \div 4 = 3$

and $\qquad\qquad (11 \div 4) + (8 \div 4) - (7 \div 4) = 2.75 + 2 - 1.75 = 3$

Thus $\qquad\qquad \dfrac{11+8-7}{4} = \dfrac{11}{4} + \dfrac{8}{4} - \dfrac{7}{4}$

and again a single denominator is allowed to divide each number in the numerator separately, providing the numbers in the numerator are connected by addition or subtraction.

What happens, however, when we try to perform the following division

$$3 \div (6 + 4) = 3 \div 10 = 0.3$$

in decimal notation?

Is this the same as $(3 \div 6) + (3 \div 4)$?

Well, $\qquad (3 \div 6) + (3 \div 4) = 0.5 \div 0.75 = 1.25$

in decimal notation.

They are not equal and thus $\qquad \dfrac{3}{6+4} \neq \dfrac{3}{6} + \dfrac{3}{4}$

A fraction with a denominator comprising more than a single number can *not* be split up into separate fractions. This fact is very important in algebra.

Exercise 1.10

1. Use your calculator to work out $(12 \div 4) \div 5$ and $12 \div (4 \div 5)$ and hence confirm that division is not associative.

2. By calculating both sides separately, show the following statements to be true and hence confirm that division is distributive over subtraction:

$$(i) \quad (11 - 8) \div 6 = (11 \div 6) - (8 \div 6)$$

$$(ii) \quad (3 - 5) \div 7 = (3 \div 7) - (5 \div 7)$$

3. By calculating both sides separately, confirm the truth of the following statements and hence verify that the distributive law of division only works in a certain order:

$$(i) \quad 6 \div (11 - 8) \neq (6 \div 11) - (6 \div 8)$$

$$(ii) \quad 7 \div (3 - 5) \neq (7 \div 3) - (7 \div 5)$$

4. Is there an identity element for division? Give a reason for your answer.

Division by Zero

The statement $4 \times 5 = 20$ is equivalent to the two statements

$$4 = 20 \div 5 \text{ and } 5 = 20 \div 4$$

We know that $7 \times 0 = 0$. Is this equivalent to the two statements

$$0 = 0 \div 7 \text{ and } 7 = 0 \div 0?$$

There is no problem with the first statement. If I have nothing to begin with and divide nothing between seven people, they each get nothing, so

$$7 \times 0 = 0 \text{ is equivalent to } 0 = 0 \div 7$$

Similarly, we could divide zero by any other number and still get the answer zero.
Hence: $$\frac{0}{\text{anything}} = 0$$

However, the second statement presents a problem. If $7 \times 0 = 0$ means that $7 = 0 \div 0$ then by similar reasoning $5 \times 0 = 0$ means that $5 = 0 \div 0$ and similarly for any number I can think of.

It is hardly reasonable to say that $\frac{0}{0} = 5$ or 7 or any other number I can think of.

Hence we must therefore conclude that division by zero has no meaning and so:

DIVISION BY ZERO IS NOT ALLOWED !

We will think about this again in a later chapter.

Order of Precedence

When we evaluate expressions containing a mixture of the four binary operations, $+$, $-$, $\times$ and $\div$, we have already decided that, in the absence of brackets, multiplication takes precedence over addition. Since division reverses multiplication and subtraction reverses addition, then it should be clear that both multiplications and divisions should be performed before additions and subtractions.

With expressions containing powers, the powers should be evaluated first, for example:

$$9 \times 4^2 = 9 \times 16 \text{ NOT } 36^2$$

This instruction means square the 4 *before* multiplying by 9.

Similarly, $7 + 4^3$ means $7 + 64$ not 11^3

Brackets will change this precedence and any calculation enclosed in brackets must be performed first.

Hence $(9 \times 4)^2$ means 36^2 and $(7 + 4)^3$ means 11^3

Remember that the order of precedence is

BRACKETS FIRST
POWERS
DIVISIONS AND MULTIPLICATIONS
ADDITIONS AND SUBTRACTIONS LAST

Example 1.6   Evaluate   (a) $9 \times 4 - 6 \div 3$ (b) $9 \times (4 - 6) \div 3$

(c) $6 \times 3^2 - 32 \div 2^3$

Solution   (a) $9 \times 4 - 6 \div 3 = (9 \times 4) - (6 \div 3) = 36 - 2 = 34$

(b) $9 \times (4 - 6) \div 3 = 9 \times -2 \div 3 = -18 \div 3 = -6$

(c) $6 \times 3^2 - 32 \div 2^3 = (6 \times 9) - (32 \div 8) = 54 - 4 = 50$

Note that in (b), when we had to choose between multiplication and division, the operations were performed from left to right.

Exercise 1.11

Perform the following calculations, using a calculator if necessary:

1. $3 + 7 \times 5$
2. $4 \times 5 - 11$
3. $6 \times 5 - 3 + 4 \times 5$
4. $6 \div 2 + 5$
5. $6 \times 5 - 11 \div 4 + 1$
6. $12 - 8 \div 4 + 7$
7. $4 \times (8 + 11)$
8. $3 + 7 \times (4 + 6)$
9. $17 - 3 \times (4 - 2)$
10. $1315 \div 5 + 3 \times (7 - 3)$
11. $31 - 5^2 + 9 \times 2^3$
12. $7 \times (11 - 5)^2 \div 9 + (11 - 5^2)$

13. Why is the following calculation impossible?

$$12 \times 7 - 3 \div (6 \times 7 - 42)$$

IRRATIONAL NUMBERS

Here is an update of the real number line, showing some of the types of numbers we have met so far.

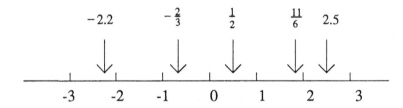

Does this mean that we have found all the numbers on the line? Have we placed enough stepping stones across the river to form a dam with no gaps in it? Unfortunately not. All rational numbers can be written as (integer/integer). Even the integers themselves can be written as $\frac{1}{1}, \frac{2}{1}, \frac{3}{1}$, etc., so the set of integers is a proper subset of the set of rational numbers.

However, there are certain numbers that cannot be written in this way. We can illustrate this by looking at square roots. Finding square roots is the opposite process to squaring. Since squaring is a multiplication process, then finding square roots must be a process of division.

Since, for example, $6^2 = 36$, we say that 6 is the square root of 36, written $\sqrt{36}$.

(By the way, did you know that the square root symbol $\sqrt{}$ is just an old-fashioned letter 'r'? It means '$\sqrt{}$ oot '!)

Remember that the first two natural numbers which are perfect squares are 1 and 4; therefore $\sqrt{1} = $ and $\sqrt{4} = 2$. It is hence pretty obvious that $\sqrt{2}$ cannot be a whole number! Is it, however, rational?

The answer is *no* but you will have to take our word for it at this stage. $\sqrt{2}$ is called an *irrational number* as it cannot be written as an exact fraction. Your calculator will give an approximate value of $\sqrt{2} = 1.4142$ to 4 decimal places, 1.414213 to 6 decimal places and 1.414213562 to 9 decimal places. No matter how accurate your calculator, you can never write the number down exactly. It is a non-repeating, non-terminating decimal. In other words the decimal places continue for ever and never form a repeating sequence. This is a feature of all irrational numbers.

Notice that $$\sqrt{2} \times \sqrt{2} = 2 \qquad \sqrt{3} \times \sqrt{3} = 3 \text{ etc.}$$

Therefore, to fill all the gaps in our number line, we will extend the set of numbers once more to include the irrational numbers. We now have a continuous set of numbers with no gaps in it and we have truly dammed the river with our stepping stones!

There are many important irrational numbers which we usually write using reserved letters of the alphabet (or Greek alphabet). One such number is π which is the number we always get when we divide the length of the circumference of a circle by the length of the diameter. An often used rational approximation to π is $\dfrac{22}{7}$, which is accurate to the first two decimal places. We also use $\pi = 3.14159$ for 5 decimal place accuracy.

Another irrational number we will meet later is known as 'e' and is the base of natural logarithms. It has the value 2.71828 to 5 decimal places.

Proof that these numbers are irrational is well beyond the scope of this book, so you will definitely have to take our word for it! In fact, it is generally difficult to establish whether or not a number is irrational but there are certain facts that it is useful to know. It can be shown that

Recurring and terminating decimals are rational numbers.

For example: $0.325 = \dfrac{325}{1000} = \dfrac{13}{40}$ is a terminating decimal

$0.333333333.... \dfrac{1}{3}$ is a recurring decimal

$0.231313131313131... = \dfrac{229}{990}$ is also a recurring decimal

(You can check the above with a calculator, using the fractions button.)

Non-recurring, non-terminating decimals are irrational.

If we try to find the value of π, we find that the decimal places *never* repeat. In other words, $\pi = 3.141592654...$ and we never get a repeating pattern of numbers to the right of the decimal point. (This is extremely difficult to prove.) Therefore π cannot be written as a rational number. However, we can always squeeze π between two rational numbers. The closer together we choose the rational numbers, the better is our approximation to π:

e.g. $3.1 < \pi < 3.2$

and $3.14 < \pi < 3.15$

and $3.141 < \pi < 3.142$ and so on

We can do this for any irrational number. Rational numbers such as $\sqrt{2}$ and $\sqrt{3}$ are known as surds. We will look at some of the properties of surds in a later chapter.

Exercise 1.12

Give brief explanations, in your own words, of the following words and phrases:

(a)   set,   (b)   natural number,   (c)   binary operator,

(d)   closed,   (e)   prime number   (f)   factor,

(g)   multiple,   (h)   identity element,

(i)   addition is commutative,   (j)   multiplication is associative,

(k)   multiplication is distributive over addition and subtraction.

SUMMARY

Many of the facts summarised below have only been shown to be true for the natural numbers. We therefore state them without proof. However, you will have an opportunity to verify some of them in the next chapter.

1. The real numbers consist of

    The natural numbers (positive integers) together with zero
    The negative integers
    Rational numbers   (positive and negative)
    Irrational numbers (positive and negative)

    Together with four binary operations $+$, $-$ and $\times$, $\div$.

2. The *complete* set of real numbers is closed under all these operations but subsets may not be closed. For example, a real number $+$ a real number $=$ a real number (always!).

    But the set of irrational numbers is not closed under multiplication, since, for example, $\sqrt{2} \times \sqrt{2} = 2$, which is rational!

    Similarly, the set of integers is not closed under division since, for example, $\dfrac{3}{4}$ is not an integer.

3. The real numbers are ordered. For example:

    $-3 < -2 < -\sqrt{2} < -1 < -0.5 < 0 < 0.5 < 1 < \sqrt{3} < 2 < e < 3 < \pi < 4$ and so on.

4. $+$ and $\times$ are commutative on any subset $-$ and $\div$ are not

    $+$ and $\times$ are associative on any subset $-$ and $\div$ are not

    This means that we can add and multiply numbers in any order but changing the order of subtractions and divisions will change the answer!

5. $\times$ is distributive over $+$ and $-$ (in both directions)

    So, if we want to multiply a list of numbers by a certain number, every member of the list is multiplied by that number.

    $\div$ is distributive over $+$ and $-$ (in one direction only)

So, if we want to divide a list of numbers by a certain number, every member of the list is divided by that number. However, if we want to divide a certain number by a list of numbers we do not divide separately by every member of the list. The list must be added together first and then a single division performed.

6. 0 is the identity element for addition. Adding zero *in any order* does not change the result.

For example: $0 + 3 = 3 + 0 = 3$

1 is the identity for multiplication. Multiplying by 1 *in any order* does not change the result

There is no identity element for either $-$ or $\div$.

Notice that $\quad (+3) - 0 = +3 \quad$ but $\quad 0 - (+3) \neq +3 \quad$ so 0 is not an identity for $-$

Also $\quad\quad\quad 3 \div 1 = 3 \quad\quad$ but $\quad 1 \div 3 \neq 3 \quad\quad$ so 1 is not an identity for $\div$

7. There is an order of precedence in which these operations must be done. Unless brackets change that order, powers are evaluated first and then multiplications and divisions are done before additions and subtractions.

2

Number Skills

WORKING WITH DIRECTED NUMBERS

Manipulating directed numbers is an important skill and there are various rules for combining directed numbers. The rules we deduce can be used with all real numbers but to illustrate the rules, we will use only integers. There will be an opportunity to use rational directed numbers in a later exercise.

Addition and Subtraction

Remember that we consider the number 3, or +3, as meaning 'start from zero and move right three places'; the number -3 is considered as meaning 'start from zero and move left three places'.

'Adding +3' has an obvious meaning. If we try to perform the calculation $5 + (+3)$, which could also be written $5 + +3$, we mean in words

'Starting from zero, move right five steps and then move right three more steps'.

Hence $\qquad 5 + +3 = 5 + 3 = 8$.

Illustration:

$5 + (+3) = 5 + 3 = 8$

'Adding -3' can be interpreted in a similar way.

$5 + (-3)$ which is the same as $5 + -3$ means 'Starting from zero, move right five steps and then move left three steps'.

Thus $5 + -3 = 5 - 3 = 2$

Illustration:

$$5 + (-3) = 5 - 3 = 2$$

To interpret subtraction it is best to approach the process from a slightly different viewpoint. We have already pointed out that the '−' sign has been used in two ways: as an instruction to subtract one number from another and as an indication of the result of subtracting a larger number from a smaller number. There is a third useful way to think of the '−' sign. It is the inverse or 'opposite' of addition.

The act of starting from zero and moving left three places has the opposite effect to starting from zero and moving right three places. In this sense the number -3 is the opposite of the number $+3$. We can reduce these wordy statements to simple shorthand:

so   'the opposite of $+3$ is -3' becomes $-(+3) = -3$ or $-+3 = -3$

and   'the opposite of -3 is $+3$' becomes $-(-3) = +3$ or $--3 = +3$

Hence we may easily interpret the meaning of

$5 - +3 =$ start from zero, move right 5 and then do the opposite of moving right 3

       $=$ start from zero, move right 5 and then move left 3

       $= 5 - 3$

Illustration:

$$5 - (+3) = 5 - 3 = 2$$

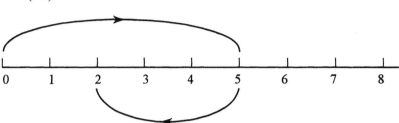

and also

$5 - -3$ = start from zero, move right 5 and then *do the opposite of moving left* 3
= start from zero, move right 5 and then *move right* 3
= 5 + 3

Illustration:

$$5 - (-3) = 5 + 3 = 8$$

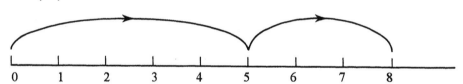

If we collect the four results together:

$$5 + +3 = 5 + 3$$
$$5 + -3 = 5 - 3$$
$$5 - +3 = 5 - 3$$
$$5 - -3 = 5 + 3$$

We may therefore deduce a rule:

> When directed numbers are added or subtracted, adjacent like signs give a positive result and adjacent unlike signs give a negative result.

Note that adding any number to its 'opposite' always gives zero.

Hence $\qquad\qquad\qquad\qquad\qquad$ $3 + (-3) = 0$ and $\quad -3 + (+3) = 0$

Example 2.1 $\qquad$ Evaluate the following, giving reasons for your answers.

(a) $\quad 7 + (-3)$ $\qquad$ (b) $\quad 11 + (-5)$ $\qquad$ (c) $\quad 14 - +6$ $\qquad$ (d) $\quad 9 - -7$

Solution $\qquad$ In each case, the reasoning begins 'starting from zero'.

(a) $\quad 7 + (-3) = 7 - 3 = 4$ $\qquad$ (move right 7 and then left 3)

(b) $\quad 11 + (-5) = 11 - 5 = 6$ $\qquad$ (move right 11 and then left 5)

(c) $\quad 14 - (+6) = 14 - 6 = 8$ $\qquad$ (move right 14 and the opposite of moving right 6)

(d) $\quad 9 - -7 = 9 + 7 = 16$ $\qquad$ (move right 9 and the opposite of moving left 7)

Example 2.2 $\qquad$ Repeat the previous example for

(a) $\quad -7 + (-3)$ $\qquad\qquad$ (b) $\quad -9 + (+4)$

(c) $\quad -5 - +6$ $\qquad\qquad$ (d) $\quad -10 - -4$

Solution

(a) $\quad -7 + (-3) = -7 - 3 = -10$ $\qquad$ (left 7 and then left another 3)

(b) $\quad -9 + (+4) = -9 + 4 = -5$ $\qquad$ (left 9 and then right 4)

(c) $\quad -5 - +6 = -5 - 6 = -11$ $\qquad$ (left 5 and the opposite of right 6)

(d) $\quad -10 - -4 = -10 + 4 = -6$ $\qquad$ (left 10 and the opposite of left 4)

In practice, it is best when performing more involved calculations to gather together positive and negative numbers separately, as illustrated below. All the following should be checked using your calculator.

Example 2.3

Evaluate the following:

(a) $17 + -6 - 4 + 12$

(b) $28 - -4 + 12 - 7$

(c) $-12 - 4 + -3 - 7$

(d) $32 + -16 - -41 + 18$

Solution

(a) $17 + -6 - 4 + 12 = 17 - 6 - 4 + 12 = (17+12) - (6 + 4) = 29 - 10 = 19$

(b) $28 - -4 + 12 - 7 = 28 + 4 + 12 - 7 = (28 + 4 + 12) -7 = 44 - 7 = 37$

(c) $-12 - 4 + -3 - 7 = -12 - 4 - 3 - 7 = -(12 + 4 + 3 + 7) = -26$

(d) $32 + -16 - -41 + 18 = 32 - 16 + 41 + 18 = (32 + 41 + 18) - 16$
$= 91 - 16 = 75$

Exercise 2.1

1. Evaluate the following. Give reasons for your answers and illustrate the calculations on the number line.

   (a) $17 + 6$ (b) $14 - 8$ (c) $-4 + 7$ (d) $7 + -7$
   (e) $-7 - 5$ (f) $5 + -9$ (g) $-2 + -7$ (h) $12 - -6$

2. Evaluate the following by collecting together positive and negative numbers.

   (a) $3 - -2 - 1$ (b) $-2 - -4 + 3$ (c) $23 + -5 + 3$

   (d) $2 - -8 + -4 - 7$ (e) $-17 + 30 - 22 + -11$

   (f) $19 - -8 + -6 - 3$ (g) $-10 - 12 + 11 - -4 + 5$

Multiplication and Division

When multiplying directed numbers, we can use the fact that multiplication is a shorthand way of doing addition.

So, remembering that $4 \times 5 = 5 + 5 + 5 + 5 = 20$

what do we mean by $4 \times (-5)$?

$$4 \times (-5) = (-5) + (-5) + (-5) + (-5) = -20$$

What about $(-4) \times 5$?

$$(-4) \times 5 = 5 \times (-4) \quad \text{because we can multiply in any order}$$
$$= (-4) + (-4) + (-4) + (-4) + (-4)$$
$$= -20$$

Finally $\quad (-4) \times (-5)$

This one is trickier to justify but considering the last three results, it would be rather neat and tidy if the answer were $+20$. This is in fact the case and we can justify it as follows:

Since $\quad\quad 4 \times (-5) \quad = -20$
then $\quad\quad -4 \times (-5) \quad = -(-20)$
$\quad\quad\quad\quad\quad\quad\quad\quad = +20$
Hence $\quad\quad (-4) \times (-5) = \quad 20$

By similar reasoning, we can show that the same results hold for division and so we may summarise as follows.

When numbers are involved in multiplications or divisions:

> If the signs are the same, the result is positive.
> If the signs are different, the result is negative.

Example 2.4 Evaluate the following:

(a) $\quad (-4) \times (-7)$ (b) $\quad +3 \times -11$ (c) $\quad -7 \times 5$

(d) $\quad 9 \div (-3)$ (e) $\quad (-7) \div (-21)$ (f) $\quad \dfrac{-15}{35}$

Solution

(a) $\quad (-4) \times (-7) = +(4 \times 7) = +28 = 28$

(b) $\quad +3 \times -11 = -(3 \times 11) = -33$

(c) $\quad -7 \times 5 = -(7 \times 5) = -35$

(d) $\quad 9 \div (-3) = -\left(\dfrac{9}{3}\right) = -3$

(e) $\quad (-7) \div (-21) = +\left(\dfrac{7}{21}\right) = +\dfrac{1}{3} = \dfrac{1}{3}$

(f) $\quad \dfrac{-15}{35} = -\dfrac{15}{35} = -\dfrac{3}{7}$

We can extend the idea to more than two numbers.

Example 2.5 Evaluate:

(a) $\quad (-3) \times (-7) \times (+4)$ (b) $\quad \dfrac{(-12) \times (-5)}{(-10)}$ (c) $\quad \dfrac{(-9)}{3 \times (-12)}$

Solution

(a)
$$(-3) \times (-7) \times (+4) = +(3 \times 7) \times (+4)$$
$$= (+21) \times (+4) = +(21 \times 4)$$
$$= 84$$

(b)
$$\dfrac{(-12) \times (-5)}{(-10)} = \dfrac{+(12 \times 5)}{-10} = \dfrac{60}{-10} = -\dfrac{60}{10} = -6$$

(c)
$$\dfrac{(-9)}{3 \times (-12)} = \dfrac{-9}{-(3 \times 12)} = +\dfrac{9}{36} = \dfrac{1}{4}$$

Hence all we have to do is count the number of '−' signs. If there are an even number, the result is positive, otherwise the result is negative.

Exercise 2.2

Evaluate the following:

1.	$7 \times (-5)$	2.	-5×7	3.	7×5
4.	$(-5) \times (-7)$	5.	$(-3) \times (-4) \times (-5)$	6.	$(-1)^2$
7.	$(-1)^3$	8.	$(-3)^2$	9.	$(-4)^3$
10.	$4 \times (-3) \times (5) \times (-6)$	11.	$16 \div (-4)$	12.	$-16 \div 4$
13.	$1 \div -1$	14.	$-1 \div -1$	15.	$-2 \div -4$
16.	$\dfrac{(-6) \times (-8)}{-4}$	17.	$\dfrac{-20}{(-5) \times (-7)}$	18.	$\dfrac{4 \times -6 \times -8}{-3 \times -2 \times -4}$

WORKING WITH FRACTIONS

Addition and Subtraction of Fractions

> When fractions are added and subtracted, we find a common denominator and then add or subtract the numerators.

Why should this be so?

We have already touched upon the fact that fractions with the same denominator can be added immediately. Thus

$$\frac{2}{9} + \frac{5}{9} = \frac{7}{9}$$

The reason for this should be obvious:

Just as   2 apples + 5 apples = 7 apples

then   2 ninths + 5 ninths = 7 ninths

Similarly,

just as   5 apples − 2 apples = 3 apples

then   5 ninths − 2 ninths = 3 ninths

i.e. $$\frac{5}{9} - \frac{2}{9} = \frac{3}{9}$$ (which cancels down to $\frac{1}{3}$)

So, fractions with a *common denominator* can be added or subtracted by *adding or subtracting their numerators*.

Suppose, however, that the denominators are different. For example, how do we calculate $\frac{3}{5} + \frac{1}{10}$? We need to use the idea of *equivalent fractions*. If we can make the denominators the same, we can add the fractions. In the example above, we cancelled $\frac{3}{9}$ down to become $\frac{1}{3}$ by the following process:

$$\frac{3}{9} = \frac{1 \times 3}{3 \times 3} = \frac{1}{3}$$

remembering that we can divide the numerator and denominator by the same number without changing the fraction.

We could now reverse the process, starting with $\frac{1}{3}$ and multiplying the numerator and denominator by 3 to get back to $\frac{3}{9}$. In other words:

$$\frac{1}{3} = \frac{1 \times 3}{3 \times 3} = \frac{3}{9}$$

This means we can multiply the numerator and denominator by the same number without changing the fraction. We have actually only multiplied the fraction by 1.

We can do this with any fraction, so $\frac{3}{5} = \frac{3 \times 2}{5 \times 2} = \frac{6}{10}$

and therefore $$\frac{3}{5} + \frac{1}{10} = \frac{6}{10} + \frac{1}{10} = \frac{7}{10}$$

All we actually did here was find a number which 5 and 10 both divide into. This can be any *common multiple* but the best one to choose is the LCM. We then convert the fractions so that they have the LCM as a common denominator. The calculation is normally arranged as in the following examples.

Example 2.6   Evaluate $\dfrac{2}{3} + \dfrac{5}{6}$

Solution   The LCM of the denominators is 6 since 3 is a factor of 6:

$$\frac{2}{3} + \frac{5}{6} = \frac{(2 \times 2)}{(3 \times 2)} + \frac{5}{6} \qquad \text{this step is usually left out}$$

$$= \frac{(2 \times 2) + 5}{6} \qquad \text{making 6 a common denominator}$$

$$= \frac{4 + 5}{6} = \frac{9}{6} = \frac{3}{2}$$

Example 2.7   Evaluate $\dfrac{2}{5} + \dfrac{3}{7}$

Solution   The LCM of the denominators is $35 = 5 \times 7$

$$\frac{2}{5} + \frac{3}{7} = \frac{(2 \times 7)}{35} + \frac{(3 \times 5)}{35} \qquad \text{making 35 a common denominator}$$

$$= \frac{14 + 15}{35} = \frac{29}{35}$$

We can use the same method for subtraction and combinations of addition and subtraction.

Example 2.8   Evaluate $\dfrac{3}{7} + \dfrac{1}{4} - \dfrac{1}{6}$

Solution   The LCM of the denominators is $84 = 7 \times 2^2 \times 3$

$$\frac{3}{7} + \frac{1}{4} - \frac{1}{6} = \frac{(3 \times 12) + (1 \times 21) - (1 \times 14)}{84}$$

$$= \frac{36 + 21 - 14}{43} = \frac{43}{84}$$

Expressing Mixed Numbers as Improper Fractions

A mixed number is a number such as 5½ or 3¾. Numbers like this are extremely clumsy to deal with and we urge you to avoid them if possible. The two examples given are better written in decimal form, as 5.5 and 3.75. Alternatively they can be written as improper fractions, because

$$5\tfrac{1}{2} = 5 + \tfrac{1}{2} = \tfrac{5}{1} + \tfrac{1}{2} = \tfrac{10+1}{2} = \tfrac{11}{2} \qquad \text{and} \qquad 3\tfrac{3}{4} = 3 + \tfrac{3}{4} = \tfrac{3}{1} + \tfrac{3}{4} = \tfrac{12+3}{4} = \tfrac{15}{4}$$

Exercise 2.3

Perform the following calculations *without* using a calculator. Leave your answers as proper or improper fractions in their lowest terms.

1. $\dfrac{3}{4} + \dfrac{5}{6}$

2. $\dfrac{3}{4} + \dfrac{5}{7}$

3. $\dfrac{5}{9} - \dfrac{5}{12}$

4. $\dfrac{3}{13} - \dfrac{5}{26}$

5. $\dfrac{1}{5} - \dfrac{1}{4}$

6. $\dfrac{3}{8} + \dfrac{5}{11} - \dfrac{3}{44}$

7. $\dfrac{7}{16} - \dfrac{5}{6} - \dfrac{11}{12}$

8. $3 + \dfrac{1}{4} - \dfrac{2}{3}$ (write 3 as $\dfrac{3}{1}$)

Write the following mixed numbers as improper fractions:

9. $7\tfrac{1}{7}$

10. $9\tfrac{3}{16}$

11. $13\tfrac{1}{6}$

12. $51\tfrac{17}{35}$

Multiplication of Fractions

> When two fractions are multiplied together, their numerators are multiplied together and their denominators are multiplied together.

Why should this be so?

Since multiplication is a shortand way of doing addition

$$3 \times \frac{1}{4} = \frac{1}{4} + \frac{1}{4} + \frac{1}{4} = \frac{1+1+1}{4} = \frac{3}{4}$$

We have simply multiplied the *numerator* by 3.

What about $\frac{3}{4} \times 3$? This is the same as $3 \times \frac{3}{4}$ so again using the idea of continued addition

$$3 \times \frac{3}{4} = \frac{3}{4} + \frac{3}{4} + \frac{3}{4} = \frac{9}{4} \text{ which is } \frac{3 \times 3}{4}$$

So, if a fraction and an integer are multiplied together, the integer multiplies *only the numerator* of the fraction.

Suppose we want to multiply two fractions together. Firstly, let both numerators be 1. How can we work out $\frac{1}{2} \times \frac{1}{5}$?

In words, this means 'a half of a fifth'. Suppose we rewrite $\frac{1}{5}$ as its equivalent fraction $\frac{2}{10}$ which is the same as $\frac{1}{10} + \frac{1}{10}$. Then half of this quantity is simply $\frac{1}{10}$ and clearly we may write

$$\text{'a half of a fifth'} = \frac{1}{2} \times \frac{1}{5} = \frac{1}{10}$$

So, if the numerators are both 1, we just multiply the denominators together!

It is now a simple step to show how to multiply more general fractions together. Study the following argument carefully. It is a good example of applying our basic multiplication laws and some of the rules learned above.

$$\frac{5}{6} \times \frac{7}{8} = \left(5 \times \frac{1}{6}\right) \times \left(7 \times \frac{1}{8}\right)$$

$$= 5 \times \frac{1}{6} \times 7 \times \frac{1}{8} \qquad \text{removing brackets}$$

$$= 5 \times 7 \times \frac{1}{6} \times \frac{1}{8} \qquad \text{we can multiply in any order}$$

$$= 35 \times \frac{1}{48} \qquad \text{the denominators can be multiplied}$$

$$= \frac{35}{48} \qquad \text{the integer multiplies numerator only}$$

This means that $\frac{5}{6} \times \frac{7}{8} = \frac{5 \times 7}{6 \times 8} = \frac{35}{48}$ and we have thus multiplied both numerators and both denominators.

Example 2.9 Evaluate:

(a) $\frac{2}{3} \times \frac{4}{5} \times \frac{7}{9}$

(b) $\frac{5}{6} \times \frac{2}{15} \times \frac{7}{11}$

(c) $\left(\frac{2}{3}\right)^2$

(d) $\frac{7}{9} \times \left(-\frac{4}{5}\right)$

Solution

(a)

$$\frac{2}{3} \times \frac{4}{5} \times \frac{7}{9} = \frac{2 \times 4 \times 7}{3 \times 5 \times 9} = \frac{56}{135}$$

(b) This time we cancel before multiplying.

$$\frac{5}{6} \times \frac{2}{15} \times \frac{7}{11} = \frac{5^1}{6_3} \times \frac{2^1}{15_3} \times \frac{7}{11} = \frac{1 \times 1 \times 7}{3 \times 3 \times 11} = \frac{7}{99}$$

(c)

$$\left(\frac{2}{3}\right)^2 = \frac{2}{3} \times \frac{2}{3} = \frac{2 \times 2}{3 \times 3} = \frac{4}{9}$$

So, to square a fraction, square the numerator and the denominator. The idea can be extended to cubing a fraction and to higher powers.

(d)

$$\frac{7}{9} \times \left(-\frac{4}{5}\right) = -\left(\frac{7}{9} \times \frac{4}{5}\right) = \frac{28}{45}$$

using the rules of signs developed earlier.

Exercise 2.4

Evaluate the following products *without* using a calculator. You should then *confirm* your answers using a calculator in as efficient a way as possible.

1. $\dfrac{3}{5} \times \dfrac{7}{11}$ 2. $\dfrac{5}{6} \times \left(-\dfrac{4}{9}\right)$ 3. $-3 \times \dfrac{11}{12}$

4. $-\dfrac{6}{7} \times -\dfrac{21}{18}$ 5. $\left(\dfrac{5}{6}\right)^2$ 6. $\left(\dfrac{4}{5}\right)^3$

7. $\left(-\dfrac{2}{3}\right)^2$ 8. $-\left(\dfrac{2}{3}\right)^2$ 9. $\left(-\dfrac{3}{5}\right)^3$

10. $-\dfrac{7}{2} \times -4 \times -\dfrac{3}{14}$ 11. $(7\frac{1}{2}) \times (-1\frac{1}{4}) \times (2\frac{1}{3})$

In question 11, convert to improper fractions before multiplying.

Dividing by Fractions

> When we divide by a fraction, we invert
> the fraction and multiply by its reciprocal.

Why should this be so?

Consider $2 \div \dfrac{1}{5}$ which can also be written as $\dfrac{2}{\left(\frac{1}{5}\right)}$. In this form, it is extremely clumsy but remember that we can multiply numerator and denominator by the same number without changing the fraction. Let us multiply them both by 5.

Then $\qquad\qquad \dfrac{2}{\left(\frac{1}{5}\right)} = \dfrac{2 \times 5}{\left(\frac{1}{5}\right) \times 5} = \dfrac{10}{1} = 10$

so, we have proved that $\dfrac{2}{\left(\frac{1}{5}\right)} = 2 \times 5 = 10$. We have inverted the fraction and multiplied by its reciprocal.

The essence of the above process is finding a number to multiply numerator and denominator by so that we get an equivalent fraction with denominator 1.

Now consider:

$$\frac{5}{6} \div \frac{7}{8} = \frac{\frac{5}{6}}{\frac{7}{8}} = \frac{\frac{5}{6} \times \frac{8}{7}}{\frac{7}{8} \times \frac{8}{7}} = \frac{\frac{5}{6} \times \frac{8}{7}}{1} = \frac{5}{6} \times \frac{8}{7}$$

Thus dividing by a fraction is the same as multiplying by its reciprocal.

Example 2.10   Perform the following divisions, without using a calculator:

$$\text{(a)} \quad \frac{3}{5} \div \frac{5}{9} \qquad \text{(b)} \quad -\frac{22}{7} \bigg/ -\frac{11}{21} \qquad \text{(c)} \quad 1\tfrac{1}{3} \div 3\tfrac{1}{2}$$

Solution

(a)   In this problem, you must resist the temptation to cancel before inverting:

$$\frac{3}{5} \div \frac{5}{9} = \frac{3}{5} \times \frac{9}{5} = \frac{27}{25}$$

(b)   This is just an alternative way of writing division:

$$-\frac{22}{7} \bigg/ -\frac{11}{21} = +\frac{22}{7} \times \frac{21}{11} = \frac{2}{1} \times \frac{3}{1} = 6$$

(c)   Convert to improper fractions first:

$$1\tfrac{1}{3} \div 3\tfrac{1}{2} = \frac{4}{3} \div \frac{7}{2} = \frac{4}{3} \times \frac{2}{7} = \frac{8}{21}$$

Exercise 2.5

Perform the following divisions without using a calculator. Then confirm your answers by using your calculator in as efficient a way as possible.

1. $\dfrac{5}{7} \div \dfrac{15}{28}$ 2. $-\dfrac{35}{51} \div \dfrac{7}{102}$ 3. $-\dfrac{15}{64} \div -\dfrac{25}{256}$

4. $7\tfrac{1}{4} \div 4\tfrac{1}{5}$ 5. $\left(\dfrac{2}{3} \div \dfrac{4}{5}\right) \div \dfrac{6}{7}$ 6. $\dfrac{2}{3} \div \left(\dfrac{4}{5} \div \dfrac{6}{7}\right)$

7. $\dfrac{19}{20} \bigg/ \dfrac{57}{45}$ 8. $\dfrac{13}{15} \bigg/ \dfrac{39}{95} \bigg/ \dfrac{6}{7}$ (evaluate from left to right)

Mixed Operations with Fractions

There are too many different ways of using fractions to go into all the possible combinations. Generally, we still need to follow the precedence rule. Unless brackets change the order of precedence, multiplications and divisions are done before additions and subtractions.

Example 2.11   Simplify:

(a) $\dfrac{1}{4} \div \left(\dfrac{1}{5} \div \dfrac{1}{6} \right)$

(b) $\dfrac{\left(\dfrac{1}{2} + \dfrac{2}{3} \right) \times \left(\dfrac{3}{7} - \dfrac{2}{5} \right)}{2\frac{1}{2} \times \dfrac{4}{5} + \dfrac{1}{3}}$

Solution

(a) $\quad \dfrac{1}{4} \div \left(\dfrac{1}{5} \div \dfrac{1}{6} \right) = \dfrac{1}{4} \div \left(\dfrac{1}{5} \times 6 \right) = \dfrac{1}{4} \div \dfrac{6}{5} = \dfrac{1}{4} \times \dfrac{5}{6} = \dfrac{5}{24}$

(b)   This example is so complicated that it is best to break it down into smaller stages. This is a useful technique in mathematics. Make a large problem into several smaller, easier problems.

Numerator $= \left(\dfrac{1}{2} + \dfrac{2}{3} \right) \times \left(\dfrac{3}{7} - \dfrac{2}{5} \right) = \left(\dfrac{3+4}{6} \right) \times \left(\dfrac{15-14}{35} \right)$

$$= \dfrac{7}{6} \times \dfrac{1}{35} = \dfrac{1}{30}$$

Denominator $= 2\frac{1}{2} \times \dfrac{4}{5} + \dfrac{1}{3} = \left(\dfrac{5}{2} \times \dfrac{4}{5} \right) + \dfrac{1}{3} = 2 + \dfrac{1}{3} = \dfrac{7}{3}$

Therefore: $\quad \dfrac{\left(\dfrac{1}{2} + \dfrac{2}{3} \right) \times \left(\dfrac{3}{7} - \dfrac{2}{5} \right)}{2\frac{1}{2} \times \dfrac{4}{5} + \dfrac{1}{3}} = \dfrac{1}{30} \div \dfrac{7}{3} = \dfrac{1}{30} \times \dfrac{3}{7} = \dfrac{1}{70}$

You should now confirm our answers using your calculator. Example 2.11(b) could be done by calculating the denominator first. This result can then be put in memory. Then calculate the denominator and divide by 'memory return'. Alternatively, think how you might perform the calculation by using the bracket buttons on your calculator.

Exercise 2.6

Simplify the following. Try to perform the calculations without a calculator and then find an efficient way of using your calculator to verify your answers.

1. $$\dfrac{2\frac{1}{4}+\dfrac{2}{3}}{3\frac{1}{2}}-\dfrac{3}{16}$$

2. $$\dfrac{3}{16}\div\left(\dfrac{1}{8}\times\dfrac{2}{5}\right)$$

3. $$\dfrac{5}{3}\div\left(\dfrac{1}{2}-\dfrac{4}{5}\times2\right)$$

4. $$\dfrac{3\frac{1}{4}\div\left(\dfrac{3}{4}+\dfrac{5}{6}\right)}{\dfrac{5}{3}\div\left(\dfrac{2}{3}\div\dfrac{7}{8}\right)}$$

5. $$\dfrac{\left(\dfrac{5}{9}+\dfrac{7}{16}\right)\times\left(3\times\dfrac{4}{11}\right)}{\left(\dfrac{2}{3}-\dfrac{1}{6}\right)\div\dfrac{3}{16}}$$

In the following exercise you will have the opportunity to verify some of the statements made in the summary at the end of Chapter 1. It will be good revision!

Exercise 2.7

1. Complete the following statement:

$$-3-(-4)=\ldots$$

   Are the negative integers closed under subtraction?

2. Can you think of any two negative integers which give a positive answer when *added together*? Hence can you state whether or not the negative integers are closed under addition?

3. Are the negative real numbers closed under addition?

4. Complete the following statement:

$$(-3)\times(-4)=\ldots$$

   Are the negative integers closed under multiplication?

5. Complete the following statement:

$$-\dfrac{3}{4}\Big/-\dfrac{5}{6}=\ldots$$

   Are the negative rational numbers closed under division?

6.   Complete the following statement

$$\sqrt{5} \times \sqrt{5} = ...$$

What does this tell us about the irrational numbers?

7.   Show why it would have been wise to include a pair of brackets in the following (incomplete) statement.

$$3/4/5 = ...$$

8.   By calculating both sides separately, confirm the commutative law of addition in the following cases. Try *not* to use a calculator.

(a) $\dfrac{5}{6} + \dfrac{7}{8} = \dfrac{7}{8} + \dfrac{5}{6}$ (b) $(-5) + (-8) = (-8) + (-5)$

(c) $3\frac{1}{2} + \left(-\dfrac{1}{5}\right) = \left(-\dfrac{1}{5}\right) + 3\frac{1}{2}$

9.   By calculating both sides separately, confirm the associative law of addition in the following cases. Try *not* to use a calculator.

(a) $3 + (-4 + 5) = (3 + -4) + 5$ (b) $\dfrac{7}{11} + \left(\dfrac{3}{5} + \dfrac{7}{2}\right) = \left(\dfrac{7}{11} + \dfrac{3}{5}\right) + \dfrac{7}{2}$

10.   Similarly, confirm that multiplication is distributive over addition in the following cases:

(a) $\dfrac{7}{9} \times \left(3\frac{1}{2} - \dfrac{9}{10}\right) = \left(\dfrac{7}{9} \times 3\frac{1}{2}\right) - \left(\dfrac{7}{9} \times \dfrac{9}{10}\right)$

(b) $\left(\dfrac{14}{11} + \dfrac{3}{2}\right) \times \dfrac{7}{3} = \left(\dfrac{14}{11} \times \dfrac{7}{3}\right) + \left(\dfrac{3}{2} \times \dfrac{7}{3}\right)$

11.   Verify the following statements and hence confirm that division is distributive over addition 'in one direction' only.

$$\left(\dfrac{7}{10} + \dfrac{3}{4}\right) \Big/ \dfrac{5}{6} = \dfrac{7}{10} \Big/ \dfrac{5}{6} + \dfrac{3}{4} \Big/ \dfrac{5}{6} \qquad \text{but} \qquad \dfrac{5}{6} \Big/ \left(\dfrac{7}{10} + \dfrac{3}{4}\right) \neq \dfrac{5}{6} \Big/ \dfrac{7}{10} + \dfrac{5}{6} \Big/ \dfrac{3}{4}$$

3

Algebra: a Basic Toolkit

WHY LEARN ALGEBRA?

Algebra is the underlying language on which mathematics is built. In order to communicate ideas mathematically, it is essential that we learn the skills required to use this language. When doing algebra, we must never forget that we are dealing with *numbers* and in Chapter 1 we saw that numbers obey certain basic laws. A knowledge of these laws should now enable you to understand why we need algebra.

These days, working with algebra is not just a case of pencil and paper. There are sophisticated new tools to help us, in the form of well-written computer programs such as DERIVE®. It is also possible to buy hand-held algebraic manipulation calculators but these are expensive at present. In a few years they should come down in price.

Twenty five years ago, electronic calculators were not available to the general public and many calculations had to be performed using tedious and dreary methods involving conversion of all the numbers to logarithms and looking them up in books of tables. Often this made mathematics seem totally baffling (and boring!), since the method of calculating results obscured the simplicity of the basic ideas.

Calculators have freed us from this form of drudgery and have enabled us to spend more of our valuable time in actually getting on with it! Modern aids to algebraic manipulation will also have their place in the near future and will allow us to perform certain repetitive tasks quickly and easily. The beauty of programs such as DERIVE is that they enable teachers to explain quickly how and why algebra works. The danger of algebraic manipulating machines is that they become black boxes. All understanding of numbers becomes unimportant and we are reduced to the mathematical equivalent of monkeys with typewriters.

So, 'do we still need to learn algebra?', I hear you ask. The answer is a very firm *yes*!

There are many reasons for studying algebra, amongst which are:

- Algebra enables us to see more clearly how the basic laws work and enables us to manipulate numbers to discover deeper relationships and more complex laws.

- Algebra enables us to reduce long-winded statements about numbers to brief, simple statements that clarify the relationships between those numbers.

- Algebra enables us to communicate with other people in other countries, who speak different languages. The language of algebra, with minor differences, is the same all over the world.

- Algebra enables us to write down *in one single statement* (known as a formula) how to calculate literally millions of different but related problems.

The purpose of algebra is *simplification*. Good mathematicians who have the skill to manipulate very complex algebra very rarely need actually to do so! They know how to avoid trouble by keeping things simple. An inexperienced user of algebra will often end up with a tangled mess of garbage which is difficult to unravel, even though no real mistakes may have been made. Knowledge of what algebra can and cannot do will help you to keep things simple.

THE FOUR RULES

Algebra is a way of using numbers where letters of the alphabet take the place of the numbers themselves. The advantage of this is that we can build *formulae* to help us solve many problems of similar type all in one go. Once we have used letters to show the relationships between numbers, we can substitute real numbers and generate real answers.

There are only four basic ways of combining numbers: addition, subtraction, multiplication and division. Hence we already know that the numbers 2 and 5 can be combined in eight possible ways, i.e.

SUMS	$2 + 5 = 7$	$5 + 2 = 7$
DIFFERENCES	$2 - 5 = -3$	$5 - 2 = 3$
PRODUCTS	$2 \times 5 = 10$	$5 \times 2 = 10$
QUOTIENTS	$2 \div 5 = 0.4$	$5 \div 2 = 2.5$

Note that these eight ways generate just six 'new' numbers since we can add or multiply in any order but the order of subtracting or dividing is important.

In algebra we would say 'Suppose we have two numbers x and y'. We can combine them in eight ways as follows, again giving six 'new' numbers:

$x + y$	$y + x$	these give the same number
$x - y$	$y - x$	
$x \times y$	$y \times x$	these give the same number
$x \div y$	$y \div x$	

Immediately we can see a problem with the multiplication symbol: it looks like a letter x so the third item in our list is confusing. We will therefore agree, here and now, *not to use the multiplication sign*. Hence in future:

> If a sign does not appear between two quantities, they are multiplied together.

Example 3.1   What do we mean by the following?

(a) xy (b) yx (c) xyz

(d) $3x$ (e) $x(y + z)$ (f) x^2

Solution

(a) xy means 'x times y'

(b) yx means 'y times x'

(c) xyz means 'x times y times z'

(d) $3x$ means '3 times x'

(e) $x(y + z)$ means 'x times everything in the brackets'

(f) x^2 means 'x times x'

Notice that xy is exactly the same as yx, since the commutative law tells us that we can multiply in any order. It is common practice, however, to write products with the letters in alphabetical order.

We also rarely use the division symbol and

$$\text{'}x \text{ divided by } y\text{' is usually written as } \frac{x}{y} \text{ or as } x/y.$$

Life becomes more complicated when we need to know how to handle combinations of the four rules. This was discussed in Chapter 1. If you are still unsure of how to combine the four binary operations, you need to read Chapter 1 again.

Another important aspect of learning algebra is knowing the meaning of certain key words. Often people are confused because they are not familiar with the words we use to describe particular combinations of numbers. With this in mind we will begin with a few important words that we will meet often in the rest of the book. You may well know the meaning of some or even all of the following. In which case, it will be good revision!

Constants and Variables

We will meet many quantities that will not change. They are called *constants*. There are two kinds of constants, numerical constants such as 3, 17, 2.95 and ½ and also algebraic constants which are represented by letters. The convention we use is that, unless told otherwise, letters from the beginning of the alphabet are usually regarded as constants. Thus a, b, c, d, . . . up to about m are usually constant. There are also special constants such as π which has the value 3.14159 to 5 decimal places.

Quantities that *can* change their values are called *variables*. Not surprisingly, we choose letters from the end of the alphabet to represent variable quantities. So, x, y and z will never be regarded as constant. We *never* use the letter o or O since it might be confused with zero.

Example 3.2   What do we normally mean by: (a) xy (b) ax?

Solution

If we meet xy, this usually means 'the variable x times the variable y' but ax means 'the constant a times the variable x'.

Terms

In algebra, a **term** is any quantity that is added to or subtracted from another quantity.

For example, $xy + a - pqr + 7$ contains four terms, namely xy, a, pqr and 7. The last term is a constant numerical term and will always have the value 7, no matter what values the other quantities take on.

We refer to the above as 'the term in xy', 'the term in a', 'the term in pqr' and 'the constant term'.

Similarly, $x + 5(xy - 3) - 45a + 17(p - 7q)$ is composed of four terms, namely x, $5(xy -3)$, $45a$ and $17(p -7q)$. They are terms because they obey our basic definition, i.e. they are all quantities that are added to or subtracted from other quantities. The fact that some of them are not very simple makes no difference.

A little thought will tell you that the term $(xy -3)$ is itself made up of the terms xy and 3. Similarly for $(p -7q)$.

Expressions

We have seen examples of expressions in the work above. An **expression** is any collection of terms.

Hence $x + y$ is an expression

                                   $xy + a -pqr + 7$ is an expression

                       so is $x + 5(xy - 3) - 45a + 17(p - 7q)$

                   and so is $5 (xy -3)$

Coefficients

A *coefficient* is a quantity (usually numerical) which multiplies another quantity. Consider the expression

$$3x + 4y + z -6pq + 9$$

           3 is the coefficient of x 4 is the coefficient of y

           -6 is the coefficient of pq $-6p$ is the coefficient of q

           $-6q$ is the coefficient of p

           9 is known as the constant coefficient in this expression.

           The coefficient of z is *one* (*not zero*)   (See below)

N.B.1 Numerical coefficients are *always* placed in front of a term. For example, suppose we wanted to multiply x by $2y$. We would *not* write this as $x2y$. Since numbers can be multiplied in any order (commutative law), it is sensible always to write this as $2xy$.

N.B.2 If a number is multiplied by the number 1, the number remains unchanged. Hence if we want to write 1 times x we simply write x, not $1x$, which is clumsy.

Thus $x + y$ means 'one x and one y' and in the expression $x + y$, the coefficients of x and y are both 1.

Indices

In Chapter 1 we defined multiplication as a shorthand way of doing addition.

For example, suppose we want to add the number a to itself seven times. It is much less tedious to write $7a$ than $a+a+a+a+a+a+a$.

When we wished to *multiply* a number by itself several times, we used a similar shorthand and introduced the idea of powers.

Just as 4×4 is written as 4^2 and $4 \times 4 \times 4$ is written as 4^3, we can write

$$a \times a = a^2 \quad \text{and} \quad a \times a \times a = a^3$$

We can continue this idea:

Thus $a \times a \times a \times a \times a \times a \times a$ is written a^7 (read 'a to the seven')

It is clearly not the same as seven times a.

For example, 7×5 means $5 + 5 + 5 + 5 + 5 + 5 + 5 = 35$

but 5^7 means $5 \times 5 \times 5 \times 5 \times 5 \times 5 \times 5 = 78,125$

a^7 is said to be in *exponential* form. a is known as the *base*; 7 is called the *power* or *index* (plural: *indices*) or sometimes the *exponent*.

Example 3.3   Write in exponential form: (i) 8   (ii) 125   (iii) 243

Solution

(i) $8 = 2.2.2 = 2^3$ base 2   power 3

(ii) $125 = 5.5.5 = 5^3$ base 5   power 3
(iii) $243 = 3.3.3.3 = 3^4$ base 3   power 4

All numbers can be written in such a form but we will find that the power is not always a whole number. As we already know, the powers 2 and 3 are given special names so a^2 and a^3 are read as 'a squared' and 'a cubed'. We will return to indices in the next chapter and again in a later chapter.

Brackets

Brackets are often used as a convenient way of grouping terms together so that we can concentrate on them. We have already used them several times above and in Chapter 1. Often we will want to multiply the whole of an expression by the same number and having it in brackets reminds us that *all* of its terms must be multiplied. Suppose we want to multiply the expression $x + 4y - 7$ by 5.

We would write this as $5(x + 4y - 7)$

Since we are not going to use the $\times$ symbol, we will sometimes use brackets around single numerical quantities when multiplication is involved. Thus '3 times 4' will be written 3(4) or sometimes (3)(4). We can also use '.' to represent multiplication. So 3.4 can also mean '3 times 4'.

Clearly, some confusion can arise here since, in decimal notation, 3.4 means '3 point 4'. However, we always make it clear in the text when we are using '.' to represent multiplication.

Remember also that brackets can be used to change the order of precedence and that with numerical expressions, calculations within brackets are carried out first.

Equations

An *equation* is simply a statement that two quantities are equal.

Hence $x = 3$ is an equation

               $x^2 + 7x - 6 = 0$ is an equation

               $x + y = 3x + 4$ is an equation

               $3x + 1 = 9y - 5$ is an equation

Novices often confuse expressions with equations but *an expression never contains an equals sign!*

Formulae

A formula is an equation connecting two or more variable quantities. The plural of formula is formulae but it is acceptable to use the word formulas if you prefer.

A formula shows the relationship between the variables and gives us a rule for calculating one of them from a combination of the others.

Formulae can be useful in providing us with a shorthand way of writing complicated statements.

For example, look at this wordy statement:

'To calculate the average speed of a car we need to find the distance travelled by the car in a certain time period and divide it by the actual time taken.'

This can be simplified to

$$\text{average speed of car} = \frac{\text{distance travelled}}{\text{time taken}}$$

or, better still:

Let s = average speed d = distance travelled t = time taken

Then: $s = \dfrac{d}{t}$

Here is an even more wordy statement:

'The time taken (in seconds) for a pendulum to move from one end of its swing to the other and back again is six point two eight times the square root of its length (in metres) divided by ten.'

This time is known as the time period.
Let the time period be T and the length of the pendulum be l.

Then l divided by 10 is written as $\dfrac{l}{10}$. The square root of this quantity is $\sqrt{\dfrac{l}{10}}$ and when we multiply this by 6.28, we get

$$T = 6.28\sqrt{\frac{l}{10}} \text{ seconds}$$

Exercise 3.1

1.   From the following expressions, write down the required quantity:

   (a) $3x + 4y - 5z$ write down the term in y and the coefficient of z

   (b) $3x + 5xy + 7xyz$ write down the term in xy and the coefficient of z

   (c) $xyz + pqr - 15$ write down the constant coefficient

   (d) $4x^3 - 7x^2 + 12x - 4$ write down the coefficient of x^2

   (e) $3x + 5xy - 10ab$ write down the term that would always be constant

2.   By using suitable letters of the alphabet, build formulae from the following statements. (Remember not to use o or O.) You may not understand what some of the quantities mean. This does not matter. Start all your answers with 'Let ...'.

   (a)   'The profit made is the selling price minus the cost price.'

   (b)   'The error in a calculation is the calculated value minus the true value.'

   (c)   'To find the percentage error, divide the actual error by the true value and then multiply by one hundred.'

   (d)   'The area of a circle is 3.142 times the radius squared.'

   (e)   'The acceleration of an object moving in a circle is given by the velocity squared divided by the radius.'

   (f)   'To find the focal length of a lens, find the sum of object distance and image distance and divide this into the product of object distance and image distance.'

   (g)   'The total cost of an article is the sum of a certain fixed cost and five times a certain variable cost.'

   (h)   'Final velocity is the initial velocity plus the acceleration times the time taken.'

   (i)   'To convert temperatures from Celsius to Fahrenheit, multiply the number of degrees Celsius by 1.8 and then add 32.'

(j)   'To convert temperatures from Fahrenheit to Celsius, subtract 32 from the number of degrees Fahrenheit, then multiply the whole of this quantity by 5 and finally divide the whole lot by 9.'

SIMPLE SUBSTITUTION

Substituting in Expressions

Suppose we know the numerical values of x and y. Then we can substitute these values into any expression containing x and y, as the following example shows. When substituting known numerical values into expressions, we can simply use the rules learned in Chapter 1.

Example 3.4   If $x = 5$ and $y = 2$, find:

(a) $x + y$ (b) $x - y$ (c) $4x$ (d) xy

(e) x^3 (f) $6x^2y$ (g) $4x + 3y$ (h) $\dfrac{3x}{y+4}$

Solution   When $x = 5$ and $y = 2$, then:

(a) $x + y = 5 + 2 = 7$ (b) $x - y = 5 - 2 = 3$

(c) $4x = 4.(5) = 20$ (d) $xy = (5).(2) = 10$

(e) $x^3 = (5)^3 = 125$ (f) $6x^2y = 6.(5)^2.2 = 6.25.2 = 300$

(g) $4x + 3y = 4.(5) + 3.(2) = 20 + 6 = 26$ (h) $\dfrac{3x}{y+4} = \dfrac{3.(5)}{2+4} = \dfrac{15}{6} = \dfrac{5}{2} = 2.5$

Example 3.5   Repeat Example 3.4 for $x = -5$ and $y = 3$.

Solution   When $x = -5$ and $y = 3$, then:

(a) $x + y = -5 + 3 = -2$ (b) $x - y = -5 - 3 = -8$

(c) $4x = 4.(-5) = -20$ (d) $xy = (-5).(3) = -15$

(e) $x^3 = (-5)^3 = -125$ (f) $6x^2y = 6.(-5)^2.3 = 6.25.3 = 450$

(g) $4x + 3y = 4.(-5) + 3.(3) = -20 + 9 = -11$ (h) $\dfrac{3x}{y+4} = \dfrac{3.(-5)}{3+4} = \dfrac{-15}{7} = -\dfrac{15}{7}$

Exercise 3.2

1. Given that $x = -4$, $y = 7$ and $z = -3$, find the numerical values of the following expressions.

   (a) $x + y$ (b) $y + z$ (c) $z + x$ (d) $x - y$

   (e) $y - z$ (f) $z - y$ (g) $x + y + z$ (h) $2x + 3y$

   (i) $3x - 4z$ (j) $x^2 + y^2 + z^2$ (k) $2xy$ (l) $x(y + z)$

   (m) $\dfrac{x+y}{z}$ (n) $\dfrac{y}{x+z}$ (o) $\dfrac{xy}{x+y}$ (p) $\dfrac{3xy - 4yz}{5xz}$

2. Repeat question 1 for $x = \dfrac{1}{2}$, $y = \dfrac{1}{3}$, $z = \dfrac{1}{4}$.

Substituting in Formulae

Since a formula gives us a rule for calculating a variable from a combination of others, we often need to use the formula to work out the numerical value of our variable in particular cases.

If a formula contains three variables, we will need to know numerical values of *two* of them in order to calculate the third. If it contains four variables, we need to know three of them and if it contains n variables, we need to know $(n - 1)$ of them!

For example, suppose we have an expression

$$E = 4x - 3y$$

Then x and y can take any values we like. Changing the numerical values of x and y will also change the calculated value of E.

So, if $x = 2$ and $y = 1$: $E = 4(2) - 3(1) = 8 - 3 = 5$

and if $x = \dfrac{1}{2}$ and $y = \dfrac{3}{5}$ $E = 4\left(\dfrac{1}{2}\right) - 3\left(\dfrac{3}{5}\right) = 2 - \dfrac{9}{5} = \dfrac{10-9}{5} = \dfrac{1}{5}$

and if $x = 3$ and $y = 4$ $\hspace{3cm}$ $E = 4(3) - 3(4) = 12 - 12 = 0$

We often meet units of measurement when dealing with formulae. We must be aware of these units when calculating our answers. The next example shows how the formula $s = \dfrac{d}{t}$ developed above enables us to calculate *any* speed, in *any* units.

Example 3.6 $\hspace{2cm}$ Calculate the average speed when:

$\hspace{2.5cm}$ (a) $\hspace{1cm}$ a car travels 200 miles in 4 hours

$\hspace{2.5cm}$ (b) $\hspace{1cm}$ a car travels 256 kilometres in 4 hours

$\hspace{2.5cm}$ (c) $\hspace{1cm}$ a cyclist cycles 100 metres in 5 seconds

$\hspace{2.5cm}$ (d) $\hspace{1cm}$ a jet plane flies so that $d = 24$ miles, $t = 2$ minutes

Solution

All these problems involve different units. Do *not* substitute units into the formula. You need to be aware of the units used and give the correct units in your answer.

(a) $\hspace{0.5cm}$ If $d = 200$ miles and $t = 4$ hours, then

$$s = \frac{200}{4} = 50 \text{ miles per hour (mph)}$$

(b) $\hspace{0.5cm}$ $d = 256$ km and $t = 4$ hours, then

$$s = \frac{256}{4} = 64 \text{ km per hour (km/h)}$$

(c) $\hspace{0.5cm}$ $d = 100$ m and $t = 5$ seconds, then

$$s = \frac{100}{5} = 20\text{m per second (m/s)}$$

(d) $\hspace{0.5cm}$ $d = 24$ miles and $t = 2$ minutes

$$s = \frac{24}{2} = 12 \text{ miles per minute}$$

Example 3.7   If a stone is thrown vertically downwards with a speed of 4 m/s from the top of a high building, the distance travelled (s) in time t is given by

$$s = 5t^2 + 4t$$

where s is measured in metres and t is measured in seconds. Find the distance travelled when $t = 1, 2$ and 3 seconds. How far does the stone travel in the third second?

Solution   When $t = 1$ $s = 5(1)^2 + 4(1) = 5 + 4 = 9$ metres

                 When $t = 2$ $s = 5(2)^2 + 4(2) = 5(4) + 8 = 28$ metres

                 When $t = 3$ $s = 5(3)^2 + 4(3) = 5(9) + 12 = 57$ metres

distance travelled in the third second

     = distance travelled in first 3 seconds – distance travelled in first 2 seconds

     = 57 – 28 = 29 metres

Example 3.8   In the simple pendulum formula developed above, find the time period when the length is   (a)   5 metres,   (b)   50 cm.

Solution   The formula is $T = 6.28\sqrt{\dfrac{l}{10}}$ but we were told that, to get an answer in seconds, we need the length to be in metres, so we must be careful to use 0.5 metres in (b).

(a)   When $l = 5$ m,

$$T = 6.28\sqrt{\frac{5}{10}} = 6.28\sqrt{0.5}$$

$$= 6.28(0.7071)$$

$$= 4.44 \; seconds$$

(b)   When $l = 0.5$ m,

$$T = 6.28\sqrt{\frac{0.5}{10}} = 6.28\sqrt{0.05}$$

$$= 6.28(0.2236)$$

$$= 1.40 \; seconds$$

In both cases we have used a calculator and rounded to 2 decimal places.

Example 3.9 u, v and f are connected by the formula $f = \dfrac{uv}{(u+v)}$. Find the value of f when   (a) $u = 15$ and $v = 25$,   (b) $u = \dfrac{1}{3}$ and $v = \dfrac{1}{2}$.

Solution   This time, there are no units to worry about therefore:

(a)   when $u = 15$ and $v = 25$

$$f = \frac{(15)(25)}{(15+25)} = \frac{375}{40} = 9.375$$

(b)   when $u = \dfrac{1}{3}$ and $v = \dfrac{1}{2}$

$$f = \frac{\left(\frac{1}{3}\right)\left(\frac{1}{2}\right)}{\left(\frac{1}{3}+\frac{1}{2}\right)} = \frac{\frac{1}{6}}{\left(\frac{2+3}{6}\right)} = \frac{1}{6} \div \frac{5}{6} = \frac{1}{6} \times \frac{6}{5} = \frac{1}{5}$$

Exercise 3.3

By substituting the values given, find the unknown quantities.

1.   If $C = 2\pi r$,   find C when $\pi = 3.14$ and $r = 6$.

2.   If $I = \dfrac{PTR}{100}$,   find I when $P = 500$, $T = 3$ and $R = 8$.

3.   If $A = \pi r^2$,   find A when $\pi = 3.14$ and $r = 10$.

4. If $p = \dfrac{RT}{V}$, find p when $R = 40$, $T = 300$ and $V = 60$.

5. If $A = 2\pi R(R + H)$, find A when $\pi = 3.14$, $R = 7$ and $H = 2$.

6. If $C = \dfrac{nV}{R + nr}$, find C when $V = 3$, $n = 7$, $R = 7.5$ and $r = 1.5$.

7. If $x = \dfrac{4y}{\sqrt{z}}$, find x when $y = 20$ and $z = 16$.

8. If $v^2 = 2\left(\dfrac{1}{x} - \dfrac{1}{y}\right)$, first find v^2 and hence find v when $x = 2$ and $y = 3$.

9. Repeat question 8 if $x = \dfrac{1}{2}$ and $y = \dfrac{2}{3}$.

10. A car accelerates from initial velocity u (m/s) with acceleration a (m/s^2) for a fixed distance s (m). The final velocity v (m/s) can be calculated from the formula

$$v^2 = u^2 + 2as$$

If $u = 5$ m/s, $a = 2$ m/s^2 and $s = 300$ m, find v^2 and hence find v.

11. You do not need to understand the terms used in this question. Units are given in brackets.

The flux density B (tesla) between the poles of a powerful magnet can be measured using a small search coil attached to a tangent galvanometer.

B can be calculated by dividing the product of R and Q by the product of N and A, where R (ohm) is the total resistance of the coil and galvanometer, Q (coulomb) is the charge that passes round the circuit, N is the number of turns on the coil and A is the area of the coil in square metres.

Use the information given to build a formula connecting the five quantities mentioned and hence find B when $R = 0.2$ ohm, $Q = 0.000003$ coulomb, $N = 10$ turns and $A = 0.00004$ m^2.

ADDITION AND SUBTRACTION OF ALGEBRAIC TERMS

Like and Unlike Terms

$3x$ and $5x$ are like terms because they are both multiples of the number x. $3x$ and $5y$ are unlike terms because they are multiples of different numbers, x and y. Like terms can be added and subtracted in order to simplify expressions. This is known as collecting like terms. For example, because multiplication is just a shorthand way of doing addition,

$$3x + 5x = (x + x + x) + (x + x + x + x + x)$$
$$= x + x + x + x + x + x + x + x$$
$$= 8x$$

Hence $\qquad 3x + 5x = 8x$

However, $3x + 5y$ cannot be simplified. The expression must be kept as two separate terms.

Example 3.10   Simplify by collecting like terms:

(a) $3x + 4y - 5z + 7x - 2y + 6z$ (b) $3x - 4xy + 12y - 2x + 7y - 5xy$

(c) $3xy + 4yx$

Solution

(a) $\qquad 3x + 4y - 5z + 7x - 2y + 6z$

$\qquad = (3x + 7x) + (4y - 2y) + (-5z + 6z)$

$\qquad = 10x + 2y + z$

(b) $\qquad 3x - 4xy + 12y - 2x + 7y - 5xy$

$\qquad = (3x - 2x) + (-4xy - 5xy) + (12y + 7y)$

$\qquad = x - 9xy + 19y$

(c)   Since we can multiply in any order, xy and yx represent the same number and so $3xy$ and $4yx$ are like terms!

Hence $3xy + 4yx = 7xy$ (or $7yx$)

Notice that even though the middle term contains x and y, these three terms are unlike terms and cannot be further simplified.

Notice also that we have made use of the rules of signs covered in the previous chapter. This is studied in more detail below.

N.B.   Often confusion arises in peoples' minds since we use the equals sign in different ways. When we collected like terms above, we used the '=' sign to link together the *logical steps* used to simplify the expression on the first line. Each line in the argument gives a different expression which is slightly simpler than the one on the line above it.

Thus there are no equations involved in this argument. An equation will always have a quantity on both sides of it, *even if that quantity is zero!*

Exercise 3.4

Simplify by collecting like terms:

1. $7x + 12x$ 2. $6x - 4x$ 3. $7x - 4x$

4. $-2x - 7x$ 5. $-11x + 4x$ 6. $-3x + 6x$

7. $8y - 6y + 4y$ 8. $5n + 11n - 6n$ 9. $3x^2 + 4x^2 - 5x^2$

10. $5pq + 7pq - 3pq$ 11. $3x - 2y + 4z - 7x + 3y - 2z$

12. $3x + 4x^2 - 7x^3 + 5x - 3x^2 + 9x^3$ 13. $xy + 2yx + 6yx - 3xy$

14. $xy^2 + 3x^2y + 7xy^2 - 2x^2y$ 15. $ab^3 - 4a^2b^2 + 7a^2b$

MULTIPLICATION AND DIVISION OF ALGEBRAIC QUANTITIES

We will often have to multiply quantities together where some of them are positive and some are negative. Similarly, we will need to divide such quantities. All this has been covered in Chapter 1, where we showed that when numbers are multiplied or divided, if the signs are the same, the result is positive, and if the signs are different, the result is negative. A few examples will illustrate how to perform these calculations.

Example 3.11   Perform the following multiplications:

(a) $(+a)(+b)$ (b) $(p)(-q)$ (c) $(5x)(7y)$

(d) $(3a)(-4b)$ (e) $(-16m)(2n)$ (f) $(-ab)(-xy)$

Solution   In the following, we have made use of the fact that the order of multiplication can be changed. Also, a '.' has been used for multiplication.

(a) $(+a)(+b) = +(ab) = ab$

(b) $(p)(-q) = -(pq) = -pq$

(c) $(5x)(7y) = 5.x.7.y = 5.7.x.y = 35xy$

(d) $(3a)(-4b) = -(3a)(4b) = -(3.4.a.b) = -12ab$

(e) $(-16m)(2n) = -(16m)(2n) = -(16.2.m.n) = -32mn$

(f) $(-ab)(-xy) = +(ab)(xy) = abxy$

In each case, we have simply worked out the sign first, then multiplied coefficients together and finally multiplied the 'letters' together. Thus for example:

$$(-12ab)(3xy) = -36abxy \text{ can be done 'all in one go'}$$

When multiplying terms containing the same symbol, we will need to use powers such as x^2 and x^3.

Example 3.12   Perform the following multiplications:

(a) $(-x)(x^2)$ (b) $(5a^2)(4a^2b)$

(c) $(3p^2q^2)(7pq^3)$ (d) $7x\left(\dfrac{p}{q}\right)$

Solution

(a) $(-x)(x^2) = -(x)(x^2) = -(x.x.x) = -x^3$

(b) $(5a^2)(4a^2b) = (5.a.a.4.a.a.b) = 20a^4b$

(c) $(-3p^2q^2)(-7pq^3) = +(3p.p.q.q.7.p.q.q.q) = 21p^2q^5$

(d)   In a problem like this, it is best to treat the $7x$ as if it were an integer and write it with a denominator of 1. We can then perform the multiplication by multiplying numerators and denominators together:

i.e. $$7x\left(\frac{p}{q}\right) = \frac{7x}{1} \cdot \frac{p}{q} = \frac{7xp}{q}$$

Hence the $7x$ term only multiplies the *numerator* of the fraction.

Division obeys the same rules of signs. This time, if the same quantity appears in the numerator and denominator, it may be cancelled.

Example 3.13   Perform the following divisions:

(a) $\dfrac{+a}{+b}$ (b) $\dfrac{+5a}{-3b}$ (c) $\dfrac{-7x}{4y}$ (d) $\dfrac{-11m}{-8n}$

(e) $\dfrac{xy}{x}$ (f) $\dfrac{xy}{xyz}$ (g) $\dfrac{4a^2b}{8b^2a}$ (h) $\dfrac{35x^2y^2z^2}{7xyz^2}$

Solution

(a) $\dfrac{+a}{+b} = +\dfrac{a}{b} = \dfrac{a}{b}$ (b) $\dfrac{+5a}{-3b} = -\dfrac{5a}{3b}$

(c) $\dfrac{-7x}{4y} = -\dfrac{7x}{4y}$ (d) $\dfrac{-11m}{-8n} = +\dfrac{11m}{8n} = \dfrac{11m}{8n}$

(e) $\dfrac{xy}{x} = \dfrac{x^1y}{x_1} = \dfrac{y}{1} = y$ (f) $\dfrac{xy}{xyz} = \dfrac{^1xy^1}{_1xy_1z} = \dfrac{1}{z}$

(g) $\dfrac{4a^2b}{8b^2a} = \dfrac{^14.a.a.b}{_28.b.b.a} = \dfrac{a}{2b}$ (h) $\dfrac{35x^2y^2z^2}{7xyz^2} = \dfrac{^535.x.x.y.y.z.z}{_17.x.y.z.z} = \dfrac{5xy}{1} = 5xy$

Exercise 3.5

Simplify the following multiplications and divisions. Remember that an odd number of '−' signs gives a negative answer.

1. $(3x)(4y)$ 2. $(-5p)(-6q)$ 3. $(ab)(-c)$

4. $\left(\dfrac{x}{4}\right)16y$ 5. $7x(-3y)(-2z)$ 6. $15a\left(\dfrac{b}{c}\right)$

7. $7a \div (-5b)$ 8. $8xy/4xz$ 9. $-3pqr(-7rqp)$

10. $11mn^2(2m^2n)$ 11. $\dfrac{(5x^2yz^3)(3xy^2z)}{10xyz}$

12. $(-8pqr \div 4p)/7qr^2$

4

Algebra: More Tools

THE THREE RULES OF INDICES

In the previous chapter, we multiplied powers of the same number together by writing the products out in full. This soon becomes tedious and fortunately we can simplify such calculations by developing three rules of indices.

First Rule of Indices

Suppose we wish to multiply together a^4 and a^3. We can write the product out in full:

i.e.
$$a^4.a^3 = (a.a.a.a).(a.a.a)$$
$$= a.a.a.a.a.a.a$$
$$= a^7$$

Therefore $a^4.a^3 = a^{(4+3)} = a^7$ and this gives us an amazingly quick way of multiplying when we have powers of the same number.

Rule 1.   When multiplying powers of a number together, we add the indices.

If $a = 5$, $5^3.5^4 = 5^{(3+4)} = 5^7$ which can now be found from your calculator. This is a shorthand way of writing $125 \times 625 = 78,125$.

Exercise 4.1

Verify the first rule in the following cases, by writing the LHS out in full and simplifying.

1. $2^3.2^7 = 2^{10}$ 2. $a^4 a^6 = a^{10}$

3. $x^7 x^5 = x^{12}$ 4. $a^3 a^4 a^5 = a^{12}$

Second Rule of Indices

Similarly, if we want to divide powers of the same number, one example should be enough to show what is happening:

$$\frac{a^6}{a^2} = \frac{a^1.a^1.a.a.a.a}{a_1.a_1} \quad \text{we can cancel two of the } as$$

$$= \frac{a^4}{1} = a$$

Hence
$$\frac{a^6}{a^2} = a^{(6-2)} = a^4$$
giving:

> Rule 2. When dividing powers of a number, subtract indices.

Substituting $a = 5$ above:

$$\frac{5^6}{5^2} = 5^4 \text{ is a shorthand way of writing } \frac{15,625}{25} = 625$$

N.B. The number a itself can be written as a^1 since, for example, $\frac{a^3}{a^2} = a^{3-2} = a^1$ and, of course, a^3 divided by a^2 is simply a.

Hence

$$\boxed{a^1 = a}$$

Simplification of Products and Quotients

We can now use the two rules in order to simplify terms.

Example 4.1 Simplify the following. Note that we have used a dot '.' to denote multiplication:

(a) $a^5.a^3$ (b) $a^2.a.a^7$ (c) $\frac{5a^7.a^9}{a^4}$

(d) $a^2b.ab^2$ (e) $\frac{10a^4b^3.a^2b^3}{5ab}$

Solution We have put many steps in that you can leave out once you are confident with the rules. Note in (d) and (e) that although it is possible to add and subtract powers of a and b separately, we can go no further. It is not possible to combine powers of a with powers of b.

(a) $a^5.a^3 = a^{5+3} = a^8$ using the first rule of indices

(b) $a^2.a.a^7 = a^{2+1+7} = a^{10}$ remembering that a can be written as a^1

(c) $\dfrac{5a^7.a^9}{a^4} = \dfrac{5a^{16}}{a^4}$ using rule 1

$= 5a^{16-4}$ using rule 2

$= 5a^{12}$

(d) $a^2b.ab^2 = a^2.a.b.b^2$ since we can multiply in any order

$= a^{2+1}b^{1+2}$

$= a^3b^3$

(e) $\dfrac{10a^4b^3.a^2b^3}{5ab} = \dfrac{10a^4.a^2.b^3.b^3}{5ab} = \dfrac{10}{5} \cdot \dfrac{a^6b^6}{ab} = 2a^5b^5$

Exercise 4.2

1. Verify the second rule in the following cases by writing out the LHS in full and cancelling:

(a) $\dfrac{2^9}{2^7} = 2^2$ (b) $\dfrac{3^{12}}{3^7} = 3^5$ (c) $\dfrac{a^9}{a^3} = a^6$ (d) $\dfrac{x^{11}}{x^4} = x^7$

2. Simplify:

(a) $a^6.a^2$ (b) $a^2.a^2.a$ (c) $a^5.a.a^3$

(d) $ab^4.a^2b^2$ (e) $2a^2.4a^2$ (f) $5a^2b.3ab^2$

(g) $6x.6x$ (h) $(6x)^2$ (i) $(6x)^3$

3.   Simplify:

(a) $\dfrac{a^7}{a^3}$ (b) $\dfrac{a^9}{a^3}$ (c) $\dfrac{a^{10}}{a^5}$

(d) $\dfrac{a^2b^3.a^2b^5}{b^4}$ (e) $\dfrac{a^3.a^4.a^8}{a^5.a^7}$ (f) $\dfrac{3x^2.5y^3.7z^4}{x^3.6y^2.11z}$

(g) $\dfrac{3abc}{15abc}$ (h) $\dfrac{4xyz}{12x^2y^2z^2}$ (i) $\dfrac{3x}{5y}.\dfrac{15xy}{18y^2z}.7xyz$

The Third Rule of Indices

There is a third rule of indices, needed when we have to take a 'power of a power'.

What does $(a^2)^3$ mean?

$$(a^2)^3 = (a^2).(a^2).(a^2) = (a.a).(a.a).(a.a) = a.a.a.a.a.a = a^6$$

Note that $(a^3)^2$ gives the same answer

since $(a^3)^2 = (a.a.a).(a.a.a) = a.a.a.a.a.a = a^6$

so we have multiplied the powers together.

> **Rule 3.**   When taking powers of powers, we multiply the indices together..

Exercise 4.3

Verify the third rule in the following cases, by writing out the LHS in full and simplifying:

(a) $(a^4)^3 = a^{12}$ (b) $(x^3)^4 = x^{12}$ (c) $(a^5)^3 = a^{15}$ (d) $(x^3)^5 = x^{15}$

Power of a Product

Be aware of the difference between expressions like $5a^2$ and $(5a)^2$. In the first one, the index is attached to the a only. Remember that if no sign appears between two numbers, then they are multiplied together.

$5a^2$ means $5.a^2 = 5.a.a$ and the a is squared first, before being multiplied by 5

However with $(5a)^2$, we have deliberately bracketed $5a$ to show that the index is attached to *everything* in the brackets. The 5 and the a are multiplied together first and then the whole lot is squared.

$(5a)^2$ means $(5a).(5a) = 5.a.5.a$

$$= 5.5.a.a \qquad \text{(changing the order of multiplication)}$$

In other words $(5a)^2 = 25a^2$

This is an example of how to deal with finding the power of a product. It can be extended to any number of terms within the product, as the following examples show:

Example 4.2   Multiply out the following:

(a) $(3x)^3$ (b) $(-abc)^3$ (c) $(p^2q^3)^4$

Solution

(a) $(3x)^3 = (3x).(3x).(3x) = 3.x.3.x.3.x = 3.3.3.x.x.x = 27x^3$

(b) $(-abc)^3 = (-abc).(-abc).(-abc) = -(abc.abc.abc) = -a^3b^3c^3$

(c) $(p^2q^3)^4 = (p^2q^3).(p^2q^3).(p^2q^3).(p^2q^3) = p^8q^{12}$

Hence in future you may leave out all the intermediate steps and use the following rule:

> If a product is raised to a certain power,
> *each part* of the product is raised to that power.

Power of a Quotient

We may perform similar manipulations with quotients. For example:

$$\left(\frac{x}{y}\right)^3 = \left(\frac{x}{y}\right).\left(\frac{x}{y}\right).\left(\frac{x}{y}\right)$$

$$= \frac{x.x.x}{y.y.y} = \frac{x^3}{y^3}$$

This can again be extended to more complicated examples.

Example 4.3   Perform the following multiplications:

(a) $\quad \left(\frac{3x}{4y^2}\right)^2$ (b) $\quad \left(-\frac{5p^2q^2}{ab}\right)^3$

Solution

(a) $\quad \left(\frac{3x}{4y^2}\right)^2 = \left(\frac{3x}{4y^2}\right)\left(\frac{3x}{4y^2}\right) = \frac{3x.3x}{4y^2.4y^2} = \frac{9x^2}{16y^4}$

(b) $\quad \left(-\frac{5p^2q^2}{ab}\right)^3 = \left(-\frac{5p^2q^2}{ab}\right)\left(-\frac{5p^2q^2}{ab}\right)\left(-\frac{5p^2q^2}{ab}\right)$

$$= -\left(\frac{5p^2q^2.5p^2q^2.5p^2q^2}{ab.ab.ab}\right)$$

$$= -\frac{125p^6q^6}{a^3b^3}$$

Again, in future, you may leave out the intermediate steps and use the following:

> If a quotient is raised to a certain power, both the
> numerator and denominator are raised to that power.

Exercise 4.4

1.   Simplify the following products:

    (a) $(3x^3)^2$ (b) $(xy^2)^3$ (c) $(xy^2z^3)^4$

    (d) $(4abc)^4$ (e) $(2a^2b^3c^4)^5$ (f) $((x^2)^3)^4$

2.   Simplify the following:

    (a) $\left(\dfrac{a^2}{b^3}\right)^4$ (b) $\left(\dfrac{3a^2}{4b^3}\right)^3$

    (c) $\left(\dfrac{1}{ab^2c}\right)^3$ (d) $\left(\dfrac{2xy}{abc}\right)^3 \cdot \left(\dfrac{3abc}{2xy}\right)^2$

Zero Index

It is useful at this stage to ask the question: 'What happens if we divide a number in power form by itself ?'

For example: divide a^3 by a^3

Clearly $\dfrac{a^3}{a^3} = 1$ (since anything divided by itself is one)

But using rule 2 of indices gives $\dfrac{a^3}{a^3} = a^{3-3} = a^0$ which leads us to the conclusion that we can extend our original ideas about indices to include zero index as long as we define:

$$\boxed{a^0 = 1}$$ 'anything to the power zero is one'

Hence, for example,

    $3^0 = 1$ $17^0 = 1$ $1^0 = 1$ $(-6)^0 = 1$ $52,376,492^0 = 1$ etc.

We will return to this idea in Chapter 11.

Things we *cannot* do with indices

It is very easy to go wrong when using indices and yet, if you learn and understand the rules, they are not difficult to manipulate. There are certain things we can *not* do with indices. Take note of the following:

N.B.1 As in the last example, index rules only work for powers of the same number.

Remember that, for example, a^3b^4 means $a.a.a.b.b.b.b$ so there is no way we can combine powers of a with powers of b.

It can also be seen that if we use numerical bases, we cannot combine the indices for $2^5.3^4$.

N.B.2 Another common misconception is in thinking that expressions like $a^2 + a^3$ can be simplified.

$a^2 + a^3$ means $(a.a) + (a.a.a)$ and we can see immediately that a^2 and a^3 must be treated as *unlike terms*.

Thus, just as $a + b$ cannot be simplified any further, neither can $a^2 + a^3$.

N.B.3 Finally, be very careful with $(a + b)^2$. We will deal with such expressions in the next chapter. It is not equal to $a^2 + b^2$ as a simple numerical example will show:

e.g. $(3 + 4)^2 = 7^2 = 49$

But $3^2 + 4^2 = 9 + 16 = 25$

Hence $(3 + 4)^2 \neq 3^2 + 4^2$

and in general $(a + b)^2 \neq a^2 + b^2$

Exercise 4.5

Simplify, if possible. If not possible, say so!

1. $(2a)^3$ 2. $(3ab)^2$ 3. $(ab)^3.(a^2b)^2$

4. $\dfrac{a^3b}{a^2b^2}$ 5. $a + b^2$ 6. $ab + b$

7. $\dfrac{(a+b)}{a}$ 8. $\dfrac{a^2}{(a^3+b)}$ 9. $3a^2.(3a)^2$

10. $3a^2 + (3a)^2$

SURDS

This section can safely be omitted until studying trigonometry.

A surd is an irrational number such as $\sqrt{2}$ or $\sqrt{3}$. Such numbers crop up regularly in many branches of mathematics, particularly trigonometry. The cube roots, fourth roots, etc., of many numbers are also surds but we will restrict our attention to the manipulation of square root surds only and can now use some of the techniques learned earlier.

The word 'surd', by the way, comes from the Latin word *surdus* meaning 'deaf'. One can only presume that, since they are irrational, they are 'deaf to reason'.

Since we know that these numbers are irrational, the only way to write them *exactly* is to leave our answers in a form involving the root sign. For example, if $x = \sqrt{2}$ is one solution to a quadratic equation (see later), it is often best to leave the answer in this form. As soon as we write $x = 1.4...$ or $x = 1.41...$ or $x = 1.4142...$, we are making an approximation.

Addition and Subtraction of Terms Containing Surds

Just as $3x + 4x = 7x$ and $9x - 4x = 5x$, we may add and subtract like terms containing surds.

Hence $3\sqrt{2} + 4\sqrt{2} = 7\sqrt{2}$ and $9\sqrt{2} - 4\sqrt{2} = 5\sqrt{2}$

However, just as we cannot simplify $3x + 4y$, we cannot simplify $3\sqrt{2} + 4\sqrt{3}$.

Example 4.4   Simplify $3\sqrt{2} + 4\sqrt{3} + 7 + 2\sqrt{2} - 3\sqrt{3} + 1$.

Solution   We merely collect together like terms:

$$3\sqrt{2} + 4\sqrt{3} + 7 + 2\sqrt{2} - 3\sqrt{3} + 1$$
$$= (3\sqrt{2} + 2\sqrt{2}) + (4\sqrt{3} - 3\sqrt{3}) + (7 + 1)$$
$$= 5\sqrt{2} + \sqrt{3} + 8$$

This cannot be further simplified since the three terms are unlike.

Exercise 4.6

1. Simplify:

    (a) $3\sqrt{2} + 5\sqrt{2} - \sqrt{3} + 3\sqrt{3} - 6$ (b) $1 + 2\sqrt{2} - 3\sqrt{3} + \sqrt{2} - 5\sqrt{3}$

    (c) $3\sqrt{7} + 2\sqrt{7} - (7\sqrt{2} + 5\sqrt{7} - 7\sqrt{5})$

    (d) $6\sqrt{6} - 3\sqrt{3} - 2\sqrt{2} - (4\sqrt{6} + \sqrt{2} - 3\sqrt{3})$

2. Find the sum of:

    (a) $2\sqrt{2} + 3\sqrt{3} - 5\sqrt{5},\ 5\sqrt{2} - 3\sqrt{3} + 2\sqrt{5} + 4$ and $3 - \sqrt{2} - \sqrt{3} - \sqrt{5}$

    (b) $4 + 5\sqrt{2} - 6\sqrt{3},\ 4\sqrt{5} - 6\sqrt{2} + 5\sqrt{3},\ 2\sqrt{5} + \sqrt{2} + 5$ and $\sqrt{5} - 1 - \sqrt{3}$

3. In each case below, subtract the first expression from the second:

    (a) $2\sqrt{3} - 5\sqrt{2} + \sqrt{6}$ from $5\sqrt{3} + 4\sqrt{2} - 6$

    (b) $10\sqrt{6} - 3\sqrt{5} - 4\sqrt{10}$ from $5\sqrt{6} + 6\sqrt{5} - 10$

Multiplication and Division of Surds

We know that $\sqrt{2} \times \sqrt{2} = 2$ and $\sqrt{3} \times \sqrt{3} = 3$ because of the definition of a square root. Similarly for any positive number a:

$$\boxed{\sqrt{a} \times \sqrt{a} = a}$$ which is the same thing as $$\boxed{\frac{a}{\sqrt{a}} = \sqrt{a}}$$

These facts can be used to simplify calculations involving surds.

Example 4.5   Simplify the following:

(a) $5\sqrt{3} \times 2\sqrt{3}$ (b) $(\sqrt{2})^5$ (c) $\dfrac{15}{\sqrt{5}}$ (d) $\dfrac{42}{3\sqrt{7}}$

Solution

(a) $5\sqrt{3} \times 2\sqrt{3} = (5 \times 2) \times (\sqrt{3} \times \sqrt{3}) = 10 \times 3 = 30$

(b) $(\sqrt{2})^5 = (\sqrt{2} \times \sqrt{2}) \times (\sqrt{2} \times \sqrt{2}) \times \sqrt{2} = 2 \times 2 \times \sqrt{2} = 4\sqrt{2}$

(c) $\quad \dfrac{15}{\sqrt{5}} = \dfrac{3 \times 5}{\sqrt{5}} = 3 \times \dfrac{5}{\sqrt{5}} = 3\sqrt{5}$

(d) $\quad \dfrac{42}{3\sqrt{7}} = \dfrac{6 \times 7}{3\sqrt{7}} = \dfrac{6}{3} \times \dfrac{7}{\sqrt{7}} = 2 \times \sqrt{7} = 2\sqrt{7}$

Example 4.6   Simplify   (a) $(2\sqrt{3})^2$ (b) $(3\sqrt{5})^3$

Solution   (a) $(2\sqrt{3})^2 = 2\sqrt{3} \times 2\sqrt{3} = 2 \times 2 \times \sqrt{3} \times \sqrt{3} = 4 \times 3 = 12$

(b) $(3\sqrt{5})^3 = 3\sqrt{5} \times 3\sqrt{5} \times 3\sqrt{5} = 3 \times 3 \times 3 \times \sqrt{5} \times \sqrt{5} \times \sqrt{5}$

$$= 27 \times 5\sqrt{5} = 135\sqrt{5}$$

Exercise 4.7

Simplify:

1.	$2\sqrt{5} \times \sqrt{5}$	2.	$3\sqrt{3} \times \sqrt{3}$	3.	$6\sqrt{x} \times \sqrt{x}$
4.	$2\sqrt{y} \times 3\sqrt{y}$	5.	$(\sqrt{2})^4$	6.	$(\sqrt{3})^5$
7.	$(3\sqrt{2})^2$	8.	$(7\sqrt{5})^2$	9.	$(2\sqrt{7})^3$
10.	$\dfrac{15}{\sqrt{3}}$	11.	$\dfrac{20}{\sqrt{5}}$	12.	$\dfrac{20}{2\sqrt{10}}$
13.	$\dfrac{30}{5\sqrt{3}}$	14.	$\dfrac{40x}{8\sqrt{x}}$	15.	$\dfrac{60x}{15\sqrt{x}}$

Further multiplication and division properties are needed when we wish to deal with different surds.

For any two positive numbers a and b:

$$\boxed{\sqrt{a} \times \sqrt{b} = \sqrt{ab}} \qquad \text{and} \qquad \boxed{\dfrac{\sqrt{a}}{\sqrt{b}} = \sqrt{\dfrac{a}{b}}}$$

For example: $\sqrt{2} \times \sqrt{3} = 1.4142 \times 1.7321 = 2.4495 = \sqrt{6}$ working to 4 d.p.

Example 4.7   Simplify the following:

(a) $3\sqrt{5} \times 4\sqrt{7}$ (b) $2\sqrt{5} \times 4\sqrt{6} \times 3\sqrt{7}$

(c) $\dfrac{\sqrt{21}}{\sqrt{7}}$ (d) $\dfrac{8\sqrt{10}}{2\sqrt{5}}$

Solution   (a) $3\sqrt{5} \times 4\sqrt{7} = 3 \times 4 \times \sqrt{5} \times \sqrt{7} = 12\sqrt{35}$

(b) $2\sqrt{5} \times 4\sqrt{6} \times 3\sqrt{7} = 2 \times 4 \times 3 \times \sqrt{5} \times \sqrt{6} \times \sqrt{7} = 24\sqrt{210}$

(c) $\dfrac{\sqrt{21}}{\sqrt{7}} = \sqrt{\dfrac{21}{7}} = \sqrt{3}$

(d) $\dfrac{8\sqrt{10}}{2\sqrt{5}} = \dfrac{8 \times \sqrt{10}}{2 \times \sqrt{5}} = \dfrac{8}{2} \times \dfrac{\sqrt{10}}{\sqrt{5}} = 4 \times \sqrt{\dfrac{10}{5}} = 4\sqrt{2}$

Exercise 4.8

Simplify

1. $2\sqrt{3} \times 3\sqrt{2}$ 2. $5\sqrt{2} \times 3\sqrt{3}$ 3. $\sqrt{2} \times 2\sqrt{3} \times 3\sqrt{5}$

4. $\dfrac{8\sqrt{15}}{2\sqrt{5}}$ 5. $\dfrac{10\sqrt{6}}{5\sqrt{3}}$ 6. $\dfrac{7}{\sqrt{7}}$

7. $\dfrac{10}{\sqrt{2}}$ 8. $\dfrac{36\sqrt{xy}}{12\sqrt{x}}$ 9. $\dfrac{40\sqrt{xyz}}{5\sqrt{x}}$

Reducing Surds to Their Simplest Form

Often we take the square root of a number which has factors. If one of these factors is a perfect square, we may use the above multiplication property (backwards!).

So $\sqrt{27} = \sqrt{9 \times 3} = \sqrt{9} \times \sqrt{3} = 3\sqrt{3}$

also $\sqrt{32} = \sqrt{16 \times 2} = \sqrt{16} \times \sqrt{2} = 4\sqrt{2}$

and $\sqrt{20} = \sqrt{4 \times 5} = \sqrt{4} \times \sqrt{5} = 2\sqrt{5}$

Example 4.8   Simplify:

     (a) $3\sqrt{6} \times 2\sqrt{10}$ (b) $3\sqrt{8} + 4\sqrt{18} - \sqrt{50}$ (c) $5\sqrt{27} + \sqrt{15}\sqrt{5}$

Solution

(a) $3\sqrt{6} \times 2\sqrt{10}$ $= 3 \times 2 \times \sqrt{6} \times \sqrt{10} = 6\sqrt{60}$

$$= 6\sqrt{4 \times 15} = 6 \times \sqrt{4} \times \sqrt{15}$$

$$= 6 \times 2 \times \sqrt{15}$$

$$= 12\sqrt{15}$$

(b) $3\sqrt{8} + 4\sqrt{18} - \sqrt{50}$ $= 3\sqrt{4 \times 2} + 4\sqrt{9 \times 2} - \sqrt{25 \times 2}$

$$= 3 \times 2\sqrt{2} + 4 \times 3\sqrt{2} - 5\sqrt{2}$$

$$= 6\sqrt{2} + 12\sqrt{2} - 5\sqrt{2}$$

$$= 13\sqrt{2}$$

(c) $5\sqrt{27} + \sqrt{15}\sqrt{5}$ $= 5\sqrt{9 \times 3} + \sqrt{15 \times 5}$

$$= 5 \times 3\sqrt{3} + \sqrt{3 \times 5 \times 5} = 15\sqrt{3} + 5\sqrt{3} = 20\sqrt{3}$$

Exercise 4.9
Reduce the following to their simplest form:

1. $\sqrt{50}$ 2. $\sqrt{40}$ 3. $\sqrt{300}$ 4. $3\sqrt{27}$

5. $\dfrac{2\sqrt{343}}{7}$ 6. $\sqrt{8x^3}$ 7. $2\sqrt{2} \times \sqrt{6}$ 8. $4\sqrt{3} \times \sqrt{15}$

9. $3\sqrt{5} \times 2\sqrt{10}$ 10. $2\sqrt{2} \times 5\sqrt{6}$ 11. $\sqrt{2} \times \sqrt{6} \times \sqrt{15}$

12. $\sqrt{5} \times \sqrt{35} \times \sqrt{14}$ 13. $\sqrt{24} + \sqrt{54} - \sqrt{96}$

14. $2\sqrt{18} - 3\sqrt{8} + 2\sqrt{50}$ 15. $3\sqrt{18} - 4\sqrt{2} - \sqrt{50} + 8\sqrt{98}$

Rationalising Denominators

Imagine how we might try to find the value of $\dfrac{1}{\sqrt{2}}$ if we did not have calculators.

Since we cannot write $\sqrt{2}$ down exactly, then we will have to make a decision on how many decimal places to use. We might therefore choose any of the following divisions:

$$\frac{1}{\sqrt{2}} = \frac{1}{1.4\ldots} \quad \text{or} \quad \frac{1}{1.41\ldots} \quad \text{or} \quad \frac{1}{1.414\ldots} \quad \text{or} \quad \frac{1}{1.4142\ldots} \quad \text{etc.}$$

The more accurate we make the denominator, the harder we make the problem!

In addition, none of these answers is accurate. Even using a calculator will not give an exact answer. If, for example, we used a 5 d.p. approximation to $\sqrt{2}$ in the denominator, we certainly could not quote our calculated answer to more than 5 d.p.

However, we have a rather neat trick at our disposal. Remember that 1 is the identity element for multiplication (Chapter 1). Thus we can multiply $\dfrac{1}{\sqrt{2}}$ by 1 without changing its value. If we are clever enough to write 1 as $\dfrac{\sqrt{2}}{\sqrt{2}}$, we can do the following:

$$\frac{1}{\sqrt{2}} = \frac{1}{\sqrt{2}} \times \frac{\sqrt{2}}{\sqrt{2}} = \frac{1 \times \sqrt{2}}{\sqrt{2} \times \sqrt{2}} = \frac{\sqrt{2}}{2}$$

This is much easier to deal with. The denominator is now rational and we can now be as accurate as we like. Whatever value we choose for $\sqrt{2}$, we only have to divide it by 2, which can even be done without a calculator!

For example, $\dfrac{\sqrt{2}}{2} = \dfrac{1.4}{2} \quad \text{or} \quad \dfrac{1.41}{2} \quad \text{or} \quad \dfrac{1.414}{2} \quad \text{or} \quad \dfrac{1.4142}{2} \quad \text{etc.}$

approximately, giving successive approximations:

$$\frac{1}{\sqrt{2}} = \frac{\sqrt{2}}{2} = 0.7 \quad \text{or} \quad 0.705 \quad \text{or} \quad 0.707 \quad \text{or} \quad 0.7071 \quad \text{etc.}$$

So:

> To rationalise a denominator containing a single surd, multiply numerator and denominator by that surd.

By similar reasoning:

$$\frac{1}{\sqrt{3}} = \frac{\sqrt{3}}{3}$$

and in general:

$$\frac{1}{\sqrt{a}} = \frac{\sqrt{a}}{a}$$

Example 4.9   Rationalise the denominators and simplify:

(a) $\dfrac{10}{\sqrt{2}}$ (b) $\dfrac{11}{\sqrt{11}}$ (c) $\dfrac{5}{2\sqrt{3}}$ (d) $\dfrac{3\sqrt{2}}{5\sqrt{5}}$

Solution

(a) $\dfrac{10}{\sqrt{2}} = \dfrac{10 \times \sqrt{2}}{\sqrt{2} \times \sqrt{2}} = \dfrac{10\sqrt{2}}{2} = 5\sqrt{2}$

(b) $\dfrac{11}{\sqrt{11}} = \dfrac{11 \times \sqrt{11}}{\sqrt{11} \times \sqrt{11}} = \dfrac{11\sqrt{11}}{11} = 11$

(c) $\dfrac{5}{2\sqrt{3}} = \dfrac{5 \times \sqrt{3}}{2\sqrt{3} \times \sqrt{3}} = \dfrac{5\sqrt{3}}{2 \times 3} = \dfrac{5\sqrt{3}}{6}$

(d) $\dfrac{3\sqrt{2}}{5\sqrt{5}} = \dfrac{3\sqrt{2} \times \sqrt{5}}{5\sqrt{5} \times \sqrt{5}} = \dfrac{3\sqrt{10}}{5 \times 5} = \dfrac{3\sqrt{10}}{25}$

Exercise 4.10

Rationalise the denominators and simplify:

1. $\dfrac{3}{\sqrt{2}}$ 2. $\dfrac{1}{\sqrt{8}}$ 3. $\dfrac{5}{\sqrt{3}}$ 4. $\dfrac{2}{\sqrt{32}}$

5. $\dfrac{9\sqrt{2}}{\sqrt{3}}$ 6. $\sqrt{\dfrac{2}{5}}$ 7. $\dfrac{1}{\sqrt{192}}$ 8. $\dfrac{2\sqrt{2}}{3\sqrt{5}}$

5

Products

We saw in Chapter 1 that multiplication is distributive over addition. In other words,

$$x(y + z) = xy + xz$$

and $$a(b + c + d) = ab + ac + ad$$

No matter how many terms appear in the expression within the brackets, any number multiplying that expression must multiply *each and every term.*

In the above examples we pre-multiplied by x and by a. The law also works for post-multiplication,

i.e. $$(y + z)x = yx + zx$$

and $$(b + c + d)a = ba + ca + da$$

When dealing with such products, we need to take into account the rules of signs and the rules of indices, as the following worked examples show. We have used a dot to indicate multiplication.

Example 5.1   Multiply out the following:

(a) $4(3y - 5z)$ (b) $xy(xy + x^2 - 3y^2)$

(c) $(3a^2b - 5ab)4a$ (d) $-5m(n - 2m + 3p)$

Solution

(a) $4(3y - 5z) = 4.3y - 4.5z = 12y - 20z$

(b) $xy(xy + x^2 - 3y^2) = xy.xy + xy.x^2 + xy.(-3y^2) = x^2y^2 + x^3y - 3xy^3$

(c) $(3a^2b - 5ab)4a = 3a^2b.4a - 5ab.4a = 12a^3b - 20a^2b$

(d) $-5m(n - 2m + 3p) = -5m.n + (-5m).(-2m) + (-5m).(3p)$

$$= -5mn + 10m^2 - 15mp$$

Note that we have shown intermediate steps in each calculation. With practice, these can be left out. Note also that, if a negative term outside the brackets multiplies the expression in the brackets, all the signs are changed.

Example 5.2   Simplify $-(x - y)$.

Solution   This one has been known to cause problems!

It simply means that the bracketed expression is multiplied by *minus one*. We could also read it as 'the opposite' of $(x - y)$.

In either case, $-(x + y) = -x + y$ which is the same as $y - x$.

Hence $-(x - y) = y - x$

And we again notice that a negative term multiplying the expression in the brackets changes all the signs.

Example 5.3   Simplify $5 - (4y + 3z)$.

Solution   This is another expression that can cause problems. Many people interpret it as meaning 'minus 5' times the expression in the brackets, but this cannot be true since the minus sign follows the 5.

If we break the problem down, $-(4y + 3z) = -4y - 3z$

and hence $5 - (4y + 3z) = 5 - 4y - 3z$

Often we meet several such products in an expression. Once the products have been performed, we can collect like terms.

Example 5.4   Simplify:

(a) $3(x + y + z) - 2(x - 4y + 3z)$

(b) $3a^2(ab + b^2c) - 5a^2b^2(ab + c)$

Solution

(a) $3(x + y + z) - 2(x - 4y + 3z)$ $= 3x + 3y + 3z - 2x + 8y - 6z$

$$= x + 11y - 3z$$

(b) $3a^2(ab + b^2c) - 5a^2b^2(ab + c) = 3a^3b + 3a^2b^2c - 5a^3b^3 - 5a^2b^2c$

$$= 3a^3b - 2a^2b^2c - 5a^3b^3$$

Exercise 5.1

1.   Perform the following multiplications:

     (a) $3(x + 2)$ (b) $2(a + b)$ (c) $3(4x + 7y)$

     (d) $5(3m - 4n)$ (e) $7(x - 3y)$ (f) $-(x + y)$

     (g) $-(2x - 3y)$ (h) $-(3x - 4)$ (i) $-2(3f + 2g)$

     (j) $4x(3y + 7)$ (k) $5xy(a + 2b - 3c)$ (l) $3x^2y(2xy^2 - 3y)$

     (m) $3a(1 + a + a^2)$ (n) $-3x^2y(4ab^2 - 7bcd)$

2.   Simplify the following:

     (a) $x^2 + 3(x^2 + 7y)$ (b) $3p - 2(2q - 6p)$

     (c) $7a^3 + 2(4 - a^3)$ (d) $5x^2 - 3x(x + x^2)$

     (e) $a(a - b) + 2b(b - a)$ (f) $a(a - b) + a(b - a)$

     (g) $4 - (x + y - z)$ (h) $5(x + 2) + 6 - (x + 1)$

3.   Simplify the following:

     (a) $3(x + 2) + 4(x + 1)$ (b) $5(2x + 4) - 3(x + 2)$

     (c) $6 (x - 2) - 2(x - 5)$ (d) $3(x + 2y) - 4(2x - 7y)$

     (e) $4(x - y + 2z) + 3(2x - 5y + 3z)$

     (f) $2x(x + 5) + x(x - 2) + 4x(x + 1)$

     (g) $5x(x - 7) + x(2x - 1) - 7x(x + 3)$

     (h) $3x(2x^2 + 5x - 4) - x^2(x - 1) + x(x^2 + 3x)$

The Product of Two Binomial Expressions

An expression containing two terms, such as $(a + b)$, is known as a binomial expression or simply a binomial. This particular type of binomial is also known as a linear expression for reasons that will become clear in a later chapter. How can we perform the multiplication $(a + b)(c + d)$?

We can show how this is done by using the distributive law three times.

When performing the multiplication, the single number $(c + d)$ is post – multiplying both terms in the first brackets; so

$$(a + b)(c + d) = a.(c + d) + b.(c + d)$$

But we know that

$$a.(c + d) = ac + ad \qquad \text{and} \qquad b.(c + d) = bc + bd$$

Combining these facts,

$$(a + b)(c + d) = ac + ad + bc + bd$$

We do not need to go through all this every time we perform such a multiplication, as long as we can spot what has actually been done.

If we rewrite the LHS as

$$(a + b)(c + d)$$

each line indicates a product that has been performed and, therefore, each term in the first brackets multiplies each term in the second. There are four multiplications performed and four terms in the resulting expression.

We must again be careful to use the rules of signs.

Example 5.5   Confirm the above rule for the product $(3 + 4)(5 + 6)$.

Solution $(3 + 4)(5 + 6) = 3.5 + 3.6 + 4.5 + 4.6$

$LHS = (3 + 4)(5 + 6) = (7).(11) = 77$

$RHS = 3.5 + 3.6 + 4.5 + 4.6 = 15 + 18 + 20 + 24 = 77$

Hence the rule is confirmed.

Example 5.6   Perform the following multiplications:

(a) $(a + b)(c - d)$ (b) $(a - b)(c + d)$ (c) $(a - b)(c - d)$

Solution

(a)

$$(a + b)(c - d)$$

$$= a.c + a.(- d) + b.c + b.(- d)$$
$$= ac - ad + bc - bd$$

(b)

$$(a - b)(c + d)$$

$$= a.c + a.d + (- b).c + (- b).d$$
$$= ac + ad - bc - bd$$

(c)

$$(a - b)(c - d)$$

$$= a.c + a.(- d) + (- b).c + (- b).(- d)$$
$$= ac - ad - bc + bd$$

We have explored all possibilities above. Do not learn these facts 'parrot fashion'. You should understand *why* the results are true.

Example 5.7   Perform the following multiplications:

(a) $(a + 1)(b + 2)$ (b) $(2x + 5)(3y - 7)$

(c) $(2m + 3n)(5p - 6q)$ (d) $3(x - 5y)(a + 2b)$

(e) $-(2a + 3b)(3c - 2d)$

Solution

(a) $(a + 1)(b + 2) = ab + 2a + b + 2$ (Remember not to write $a2$ or $1b$)

(b) $(2x + 5)(3y - 7) = 6xy - 14x + 15y - 35$

(c) $(2m + 3n)(5p - 6q) = 10mp - 12mq + 15np - 18nq$

(d) $3(x - 5y)(a + 2b) = 3(xa + 2xb - 5ya - 10yb)$

$$= 3xa + 6xb - 15ya - 30yb$$

(e) $-(2a + 3b)(3c - 2d) = -(6ac - 4ad + 9bc - 6bd)$

$$= -6ac + 4ad - 9bc + 6bd$$

Exercise 5.2

1.   Verify the following by calculating each side separately:

(a) $(3 + 5)(4 - 3) = 3.4 + 3.(-3) + 5.4 + 5.(-3)$

(b) $(10 - 15)(9 + 11) = 10.9 + 10.11 + (-15).9 + (-15).11$

(c) $(20 - 11)(25 - 10) = 20.25 + 20.(-10) + (-11).25 + (-11).(-10)$

2.   Perform the following multiplications:

(a) $(x + 4)(y - 3)$ (b) $(f - 7)(g - 6)$

(c) $(2x - 3)(y + 6)$ (d) $(4z - 5)(5y + 7)$

(e) $(m + 2n)(2p - q)$ (f) $(a^2 + b^2)(x - y)$

(g) $5(2x + 3y)(a + b)$ (h) $-2(p + q)(b - a)$

(i) $(ab + 4c)(2bc - 3a)$ (j) $(a - 3)(a + 6)$

(k) $(x - 5)(x - 4)$ (l) $(a + b)(a + 2b)$

Products Leading to Quadratic Expressions

Very often, we will meet products where multiples of the same quantity or quantities occur in both brackets. This was so in the last three parts of question 2 in the exercise above. In these cases, two of the terms generated will always be like terms and our final answer will contain just three terms. For example:

$$(x + 2)(x + 3) = x.x + x.3 + 2.x + 2.3$$
$$= x^2 + 3x + 2x + 6$$
$$= x^2 + 5x + 6 \text{ since } 3x \text{ and } 2x \text{ are like terms}$$

This type of expression is very important and is known as a *quadratic expression* in x or simply a *quadratic* in x, from the Latin word, *quadratus*, meaning a square. Note that there are three terms: a term in x^2, a term in x and a constant term.

Similarly: $(2x + 3y)(3x + 4y) = 2x.3x + 2x.4y + 3y.3x + 3y.4y$
$$= 6x^2 + 8xy + 9yx + 12y^2$$

Again, the two middle terms are like terms, thus:

$$(2x + 3y)(3x + 4y) = 6x^2 + 17xy + 12y^2$$

Although more complicated, this is still a quadratic expression. It may be regarded as either a quadratic in x or in y,

i.e. $6x^2 \quad + \quad (17y)x + \quad 12y^2$

      term in x^2 term in x 'constant term'

or $12y^2 \quad + \quad (17x)y + \quad 6x^2$

      term in y^2 term in y 'constant term'

Exercise 5.3

1.   Form quadratic expressions from the following products:

(a) $(x + 1)(x + 2)$ (b) $(x + 2)(x + 3)$

(c) $(x + 3)(x + 4)$ (d) $(2x + 1)(3x + 2)$

(e) $(5x + 4)(2x - 7)$ (f) $(x - 1)(x + 2)$

(g) $(x - 1)(x - 2)$ (h) $(3x - 1)(2x - 5)$

(i) $(2x - 4)(3x - 2)$ (j) $(6x - 7)(2x + 3)$

2. Repeat question 1 for the following products:

(a) $(x - 2y)(x + 3y)$ (b) $(3p + q)(2p - 3q)$

(c) $(2y - 3z)(3y - 2z)$ (d) $(3x + 4y)(2x - 3y)$

(e) $(ab + cd)(2ab - 3cd)$ (f) $(x^2 + 3)(x^2 + 4)$

(g) $(x^2 + 2y^2)(3x^2 - 2y^2)$ (h) $(a + b)(a + b)$

(i) $(a - b)(a - b)$ (j) $(a - b)(a + b)$

Three Important Products

The two most basic expressions containing more than one term are the sum of two terms $(a + b)$ and the difference of those terms $(a - b)$. The last three parts of question 2 above illustrate what happens when we multiply combinations of these sums and differences. We will look at them in detail.

The square of $(a + b)$

$$(a + b)(a + b) = a.a + a.b + b.a + b.b$$
$$= a^2 + 2ab + b^2 \qquad \text{since } ab \text{ and } ba \text{ are like terms}$$

But $(a + b)(a + b)$ can be written as $(a + b)^2$ so we have the following identity :

$$\boxed{(a+b)^2 = a^2 + 2ab + b^2}$$

The square of $(a - b)$

This is done by similar reasoning:

$$(a-b)^2 = (a-b)(a-b)$$
$$= a.a + a.(-b) + (-b).a + (-b).(-b)$$
$$= a^2 - ab - ba + b^2$$
$$= a^2 - 2ab + b^2 \qquad \text{since } ab \text{ and } ba \text{ are like terms}$$

Therefore:
$$\boxed{(a-b)^2 = a^2 - 2ab + b^2}$$

The product $(a + b)(a - b)$

Finally,

$$(a+b)(a-b) = a.a + a.(-b) + b.a + b.(-b)$$

$$= a^2 - ab + ba - b^2$$

$$= a^2 - ab + ab - b^2$$

$$= a^2 - b^2 \qquad \text{since } ab \text{ cancels}$$

giving the result:

$$\boxed{(a+b)(a-b) = a^2 - b^2}$$

Note that we would get the same result if the order of multiplication were reversed.

The above products crop up regularly and are worth memorising. There is no need to multiply them out 'from scratch'. If you recognise them, the answer can simply be written down, even in more complicated cases, as shown below.

Example 5.8   Find the following products:

    (a) $(2x + 3y)^2$ (b) $(3p - 5)^2$ (c) $(3m + 4n)(3m - 4n)$

Solution

(a)   This is of the form $(a + b)^2$:

$$(2x + 3y)^2 = (2x)^2 + 2.(2x).(3y) + (3y)^2$$
$$= 4x^2 + 12xy + 9y^2$$

(b)   This one is like $(a - b)^2$:

$$(3p - 5)^2 = (3p)^2 - 2.(3p).(5) + 5^2$$
$$= 9p^2 - 30p + 25$$

(c)   Finally, this is similar to $(a + b)(a - b)$:

$$(3m+4n)(3m-4n) = (3m)^2 - (4n)^2$$
$$= 9m^2 - 16n^2$$

Exercise 5.4

Find the following products:

1. $(x + 1)^2$ 2. $(x - 5)^2$ 3. $(2y + 3)^2$

4. $(3z - 7)^2$ 5. $(-a - b)^2$ 6. $(m + 3n)^2$

7. $(x^2 + y^2)^2$ 8. $(ab - cd)^2$ 9. $(2x + 3)(2x - 3)$

10. $(2x - 1)(2x + 1)$ 11. $(5a + 2b)(5a - 2b)$ 12. $(3x - 6)(3x + 6)$

THE BINOMIAL SERIES

The motivation for this section is twofold. We will need these results when looking at the law of natural growth and for the binomial distribution in the statistics chapters. Students may therefore safely leave out this work until studying those chapters.

We are going to try to find an easy method for expanding, in powers of x, expressions such as $(1+3x)^{10}$. We are not going to give a formal proof of the general formula we will develop here. We will give an intuitive argument based on known facts. Should you wish to see more formal proofs, they can be found in many advanced textbooks.

We already know that

$$(1 + x)^0 = 1 \qquad \text{(since anything to the power of zero equals 1)}$$

and $(1+x)^1 = 1+x$

and $(1+x)^2 = 1+2x+x^2$

It is now easy to extend this. Multiply both sides of the previous line by $(1 + x)$ and collect like terms, as follows:

$$(1+x)^3 = (1+x)(1+2x+x^2)$$
$$= 1+2x+x^2+x+2x^2+x^3$$
$$= 1+3x+3x^2+x^3$$

Similarly, using this result

$$(1+x)^4 = (1+x)(1+3x+3x^2+x^3)$$
$$= 1+3x+3x^2+x^3+x+3x^2+3x^3+3x^4$$
$$= 1+4x+6x^2+4x^3+x^4$$

Pascal's Triangle

The above method is already becoming tedious! It will therefore be useful to focus our attention on the coefficients in these expansions. This is best done by rearranging the coefficients in a triangular pattern:

Expression   Coefficients

$(1+x)^0$ 1

$(1+x)^1$ 1   1

$(1+x)^2$ 1   2   1

$(1+x)^3$ 1   3   3   1

$(1+x)^4$ 1   4   6   4   1

This pattern of numbers is known as *Pascal's triangle*. Note that each coefficient is the sum of the two coefficients immediately above it.

This *suggests* coefficients in the expansion of $(1+x)^5$ to be 1 5 10 10 5 1.

Exercise 5.5

1. What would you expect the coefficients to be in the expansions of $(1+x)^6$ and $(1+x)^7$? How many terms would you expect to generate in each of these cases?

2. Write down Pascal's triangle as far as the coefficients for $(1+x)^8$.
    Use this result to deduce the coefficients when the power is 9.
    Use this new result to deduce the coefficients when the power is 10.

Finding a General Formula

Pascal's triangle is very useful in working out the coefficients when the power is small (up to about 8). However, suppose you wanted the coefficients in the expansion of $(1+x)^{18}$. It would be very time consuming to have to keep calculating each line of the triangle. Also, we cannot use this method if the power is negative or fractional.

We therefore need a *general formula* which we can use whatever the value of the power. This again depends on recognising patterns.

We can check that

$$(1+x)^3 = 1 + \frac{3}{1}x + \frac{3.2}{1.2}x^2 + \frac{3.2.1}{1.2.3}x^3$$

$$(1+x)^4 = 1 + \frac{4}{1}x + \frac{4.3}{1.2}x^2 + \frac{4.3.2}{1.2.3}x^3 + \frac{4.3.2.1}{1.2.3.4}x^4$$

$$(1+x)^5 = 1 + \frac{5}{1}x + \frac{5.4}{1.2}x^2 + \frac{5.4.3}{1.2.3}x^3 + \frac{5.4.3.2}{1.2.3.4}x^4 + \frac{5.4.3.2.1}{1.2.3.4.5}x^5$$

All these cancel down to the expressions already found and the pattern *suggests* that, for power n:

$$(1+x)^n = 1 + \frac{nx}{1} + \frac{n(n-1)}{1.2}x^2 + \frac{n(n-1)(n-2)}{1.2.3}x^3 + \frac{n(n-1)(n-2)(n-3)}{1.2.3.4}x^4 + \Lambda$$

N.B.1   The factors of the numerator decrease in steps of 1: the factors of the denominator increase in steps of 1.

N.B.2   In each term, the coefficient has the same number of factors in the numerator and denominator as the power of x, e.g.

coefficient of x^4 is $\dfrac{n(n-1)(n-2)(n-3)}{1.2.3.4}$ numerator has four factors
denominator has four factors

N.B.3   We abbreviate   1.2   to 2! and read 'factorial 2'

1.2.3   to 3! and read 'factorial 3'

1.2.3.4 to 4! and read 'factorial 4' etc.

Thus we may rewrite the general result as:

$$(1+x)^n = 1 + \frac{nx}{1!} + \frac{n(n-1)}{2!}x^2 + \frac{n(n-1)(n-2)}{3!}x^3 + \frac{n(n-1)(n-2)(n-3)}{4!}x^4 + \Lambda$$

This formula was discovered by Sir Isaac Newton and is called *the binomial series*.

In the above form, this formula can also be used for indices other than positive integers. However, for the applications in this book, we will only consider positive integer values of n.

If n is a positive integer:

- the series ends
- it has $n + 1$ terms
- it is true for *any* value of x

Example 5.9   Write down the first four terms of the expansion of $(1 + x)^{10}$.

Solution   Here $n = 10$

$$(1+x)^{10} = 1 + \frac{10}{1}x + \frac{10.9}{1.2}x^2 + \frac{10.9.8}{1.2.3}x^3 + \cdots$$

(N.B. Don't cancel at this stage. Leave the coefficients clear for checking.)

Hence $$(1+x)^{10} = 1 + \frac{10}{1}x + \frac{\overset{5}{10}.9}{1.2}x^2 + \frac{\overset{3}{10}.9.\overset{4}{8}}{1.2.3}x^3 + \cdots$$

$$= 1 + 10x + 45x^2 + 120x^3 + \cdots$$

Example 5.10   Give the first four terms of $(1 - 2x)^5$.

Solution   Here x in the standard formula is replaced by $(-2x)$ and $n = 5$

$$(1-2x)^5 = \left[1+(-2x)\right]^5$$

$$= 1 + \frac{5}{1}(-2x) + \frac{5.4}{1.2}(-2x)^2 + \frac{5.4.3}{1.2.3}(-2x)^3 + \cdots$$

$$= 1 - 10x + \frac{5.4.2.2}{1.2}x^2 - \frac{5.4.3.2.2.2}{1.2.3}x^3 + \cdots$$

$$= 1 - 10x + 40x^2 - 80x^3 + \cdots$$

Exercise 5.6

Use the binomial expansion to give the first four terms of the following:

1. $(1+x)^6$ 2. $(1-x)^6$ 3. $(1+x)^8$

4. $(1-x)^8$ 5. $(1+2x)^4$ 6. $\left(1-\dfrac{3}{2}x\right)^4$

7. $(3+x)^5 = 3^5\left(1+\dfrac{x}{3}\right)^5$

8. Write down (without simplifying) the following:

    (i)   the fourth term of $(1+2x)^7$;
    (ii) the fifth term of $(1-x)^{11}$.

Extending the Idea

We have seen how the binomial expansion works for expressions like $(1+x)^n$. However, suppose there are two numbers in the brackets, neither of which is one. In question 7 of Exercise 5.6, you met $(3+x)^5$ and were given a hint on how to deal with it.

$$(3+x)^5 = 3^5\left(1+\tfrac{x}{3}\right)^5$$

$$= 3^5\left[1+5\left(\tfrac{x}{3}\right)+10\left(\tfrac{x}{3}\right)^2+10\left(\tfrac{x}{3}\right)^3+5\left(\tfrac{x}{3}\right)^4+\left(\tfrac{x}{3}\right)^5\right]$$

$$= 3^5+5.3^4.x^1+10.3^3.x^2+10.3^2.x^3+5.3^1.x^4+x^5$$

and we can take this one step further and write

$$(3+x)^5 = 3^5.x^0+5.3^4.x^1+10.3^3.x^2+10.3^2.x^3+5.3^1.x^4+3^0.x^5$$

where we have used a dot to indicate multiplication. There are several things to notice:

- The coefficients are the same as for $(1+x)^5$.

- The powers of 3 reduce from 5 to 0 and the powers of x increase from 0 to 5.

- The sum of the powers is always 5. e.g. $3^1 x^4 : 1 + 4 = 5$.

Hence we should be able to take a good guess at how to expand expressions such as $(a + b)^n$.

The first few powers of $(a + b)$ are

Expression	Expansions
$(a+b)^0$	$1a$
$(a+b)^1$	$1a + 1b$
$(a+b)^2$	$1a^2 + 2ab + 1b^2$
$(a+b)^3$	$1a^3 + 3a^2b^1 + 3a^1b^2 + 1b^3$
$(a+b)^4$	$1a^4 + 4a^3b^1 + 6a^2b^2 + 4a^1b^3 + 1b^4$

Note that:

- The coefficients are just the same as with $(1 + x)^n$. In other words, they can be chosen from Pascal's triangle.

- Powers of a reduce from n to 0.

- Powers of b increase from 0 to n.

- The sum of the powers is always n.

We will meet this expansion again when we study the binomial distribution. In all cases n will be a positive integer and will be small enough for us to use Pascal's triangle without further elaboration.

Example 5.11   Use the above remarks to generate expansions of

     (a) $(a + b)^7$ (b) $(2 + x)^4$ (c) $(3 + x)^6$

Solution

(a) $(a+b)^7 = 1.a^7 b^0 + 7a^6 b^1 + 21a^5 b^2 + 35a^4 b^3 + 35a^3 b^4 + 21a^2 b^5 + 7ab^6 + 1.a^0 b^7$

(b) $(2+x)^4 = 1(2)^4 x^0 + 4(2)^3 x^1 + 6(2)^2 x^2 + 4(2)^1 x^3 + 1(2)^0 x^4$

             $= 16 + 32x + 24x^2 + 8x^3 + x^4$

(c) $(3+x)^6 = 1(3)^6 x^0 + 6(3)^5 x^1 + 15(3)^4 x^2 + 20(3)^3 x^3 + 15(3)^2 x^4 + 6(3)^1 x^5 + 1(3)^0 x^6$

$$= 729 + 1458x + 1215x^2 + 540x^3 + 135x^4 + 18x^5 + x^6$$

For n a positive integer, we will therefore use:

$$(a+b)^n = a^n + na^{n-1}b + \frac{n(n+1)}{2!}a^{n-2}b^2 + \frac{n(n+1)(n+2)}{3!}a^{n-3}b^3 + \Lambda + b^n$$

Exercise 5.7

1.   Expand   (a) $(x-2)^5$ (b) $(5-x)^4$

2.   Write down the first three terms in the expansion of $(3-2x)^8$.

3.   Write down the first four terms in the expansion of $(2+x)^{10}$.

4.   Find the coefficient of x^6 in the expansion of

         (a) $(2-x)^7$ (b) $(3+2x)^{11}$

5.   Find the coefficient of x^4 in the expansion of $(2x+5)^6$.

6.   Write down and simplify

         (a)   the third term of $(x+5)^{10}$

         (b)   the fourth term of $(2-x)^8$

         (c)   the sixth term of $(2x-0.5)^9$

6

Factors

HCF of Algebraic Terms

We will spend this chapter reversing the processes learned in the last one. Factorising means breaking a term or expression down into components that are *multiplied* together. Like most reversing operations in mathematics, it is a little more difficult than the basic operation itself.

Just as the number 70 can be factorised into $2 \times 5 \times 7$, we can factorise simple algebraic terms such as $3x^2yz$.

The term $3x^2yz$ has 5 factors since $3x^2yz = 3.x.x.y.z$ where we have used a dot to represent multiplication. It is often necessary to find common factors of two or more algebraic terms and, just as with integers, we are usually interested in the highest common factor (HCF).

Remember that to find the HCF of two integers, we do a prime factorisation and then choose the *lowest* power of each common factor. We use a similar process to find the HCF of algebraic terms.

Example 6.1   Find the HCF of $3ab^2c$ and $6a^2b^3c^2$.

Solution

(i)   The HCF of 3 and 6 is 3 since 3 divides into both terms, 6 does not.

(ii)   Note that a divides exactly into both terms, a^2 does not

b^2 divides exactly into both terms, b^3 does not

c divides exactly into both terms, c^2 does not

Hence we need the *lowest* power of each common algebraic factor and therefore the HCF is $3ab^2c$.

So, to find the HCF of two (or more) algebraic terms:

(i) Find the HCF of the numerical coefficients.

(ii) Find the lowest power of each common algebraic factor.

(iii) Multiply together.

Example 6.2 Find the HCF of $21ay^2z^3$ and $35by^3z$.

Solution The HCF of the coefficients is 7. Neither a nor b divides into both terms, y^2 divides both and z divides both. Hence:

$$HCF = 7y^2z$$

Exercise 6.1

Find the HCF of the following sets of algebraic terms:

1. x^3y^2, x^2y^3 and x^2y
2. $x^2y^3z^3$, x^3y^3 and xy^2z^2
3. $4xy^2$, $8xyz$ and $12x^2yz^2$
4. $3xy, 4y$ and $7xy^2$
5. xy^2z^2, $x^2y^3z^3$ and $x^2y^4z^4$
6. a^2b^3, $a^3b^2c^2$ and ab^2c^3
7. $3a^2bc^3$, $6a^3b^2c^2$ and $18a^3c^4$
8. $5a^2b$, $15b^2c^2$ and $25a^3c^4$

Removing the HCF from Algebraic Expressions

We normally need to find common factors for each term in a complete expression. When such a factor is found, it can be removed or 'taken out'. We cannot just throw it away, however, as the following shows.

Consider the following product:

$$abc(a + 2b + 3c) = a^2bc + 2ab^2c + 3abc^2$$

Suppose that we had begun with the RHS. To reverse the above process, we look for the HCF of each term, which in this case is abc.

Therefore $\qquad a^2bc + 2ab^2c + 3abc^2 = abc(a + 2b + 3c)$

The RHS is the factorised form of the LHS. It is a product of abc and $(a + 2b + 3c)$ and has three simple factors a, b and c, together with the fourth factor $(a + 2b + 3c)$. This last expression is called a trinomial, because it consists of three numbers.

In the above, when the HCF abc is 'taken out', the parts of each term left are found by dividing each term by the HCF. Thus, the terms left in the brackets are

$$\frac{a^2bc}{abc} = a \qquad \frac{2ab^2c}{abc} = 2b \quad \text{and} \quad \frac{3abc^2}{abc} = 3c$$

Example 6.3   Factorise $3pq^3r + 12p^2q^2rs$.

Solution   The HCF of the two terms is $3pq^2r$; hence the terms left when this is 'taken out' are

$$\frac{3pq^3r}{3pq^2r} = q \qquad \text{and} \qquad \frac{12p^2q^2rs}{3pq^2r} = 4ps$$

Hence $3pq^3r + 12p^2q^2rs = 3pq^2r(q + 4ps)$

Example 6.4   Factorise $5xy + 10x^2y^2 - 15x^3y^3$.

Solution   The HCF is $5xy$ and hence the three terms left are

$$\frac{5xy}{5xy} = 1 \qquad \frac{10x^2y^2}{5xy} = 2xy \quad \text{and} \quad -\frac{15x^3y^3}{5xy} = -3x^2y^2$$

Therefore the expression factorises to $5xy(1 + 2xy - 3x^2y^2)$.

A common mistake above is not realising that when the factor $5xy$ is removed from the first term, there is another factor of 1. This must therefore be left in the brackets. Notice also how we handled the minus sign.

Occasionally, we might want to factorise the HCF out to the back of the brackets. This is perfectly legal because numbers can be multiplied in any order. Hence we can write, for example:

$$xy + 3y^2 = y(x + 3y) \qquad \text{or} \qquad xy + 3y^2 = (x + 3y)y$$

Exercise 6.2

Factorise the following:

1. $2x + 4$ 2. $10x + 25$ 3. $8x + 12y$

4. $4x - 4y$ 5. $ab - 3ac$ 6. $6x - 6$

7. $x - 3x^2$ 8. $6x - 3y^2$ 9. $5x - 10xy$

10. $4y - 16y^2$ 11. $8a^3 + 14a^2$ 12. $a^2b - 5a^3b^5$

13. $py^2 + py$ 14. $2\pi r^2 + 2\pi rh$ 15. $5x - 10y + 20z$

16. $px + qx + rx$ 17. $8x^2y^4 - 16x^3y^3 + 24x^2y^5$

18. $9x^2y + 6xy^2 - 3xy$ 19. $5ab + 10a^2b^2 - 15a^3b^3$

20. $-pqr - 2pq^2r - 3pqr^2$ 21. $-11xyz - 22x^2yz^3 - 66xy^3z^2$

Factorisation in Pairs

Consider the expression $x(a + b) + y(a + b)$.

The expression consists of two terms. Each term has $(a + b)$ as a factor and this may be 'taken out' as we did above. When this is done, the terms left behind are

$$\frac{x(a+b)}{(a+b)} = x \quad \text{and} \quad \frac{y(a+b)}{(a+b)} = y$$

and hence the above expression factorises to $(a + b)(x + y)$.

This looks familiar. We have merely reversed the process of multiplying out binomial expressions. We can go further and use this idea to factorise some expressions comprising four terms.

Now look at this expression:

$$ac + ad + bc + bd$$

There is no common factor for all four terms. Sometimes, but not always, expressions like this can be factorised in pairs. We might then be able to reduce it to a product of two binomial expressions.

The first two terms have a common factor a and the second pair have a common factor b.

Hence $\qquad ac + ad + bc + bd = a(c + d) + b(c + d)$

We now notice that $(c + d)$ is a common factor of both terms on the RHS and hence we may write

$$ac + ad + bc + bd = (c + d)(a + b) \text{ which is the same as } (a + b)(c + d)$$

It is important to realise that this does not always work, but if four terms do not have a common factor, it is always worth trying.

Example 6.5 $\qquad$ Factorise the following expressions, if possible:

(a) $\quad ab - 4a + 3b - 12$ $\qquad$ (b) $\quad xp - 2py + 3qx - 6qy$

(c) $\quad 2mc - 10md - nc + 5nd$ $\qquad$ (d) $\quad ax + ay + ba + bx$

Solution

(a) $\quad ab - 4a + 3b - 12 = a(b - 4) + 3(b - 4) = (b - 4)(a + 3)$

(b) $\quad xp - 2py + 3qx - 6qy = p(x - 2y) + 3q(x - 2y) = (x - 2y)(p + 3q)$

(c) $\quad 2mc - 10md - nc + 5nd = 2m(c - 5d) - n(c - 5d) = (2m - n)(c - 5d)$

Note that to ensure that we had a common factor at the intermediate stage, we had to use '$-n$' as a common factor for the second pair.

(d) $\quad ax + ay + ba + bx = a(x + y) + b(a + x)$

Note that we do not have a common factor at this stage. Even if you rearrange the terms and try different pairs, it is not possible to factorise this expression.

Hence $ax + by + ba + bx$ will not factorise.

Exercise 6.3

1. $\quad$ Factorise:

$\qquad$ (a) $\quad x(y + 6) + 2(y + 6)$ $\qquad$ (b) $\quad x(y + 1) - 3(y + 1)$

$\qquad$ (c) $\quad x(y - 5) + (y - 5)$ $\qquad$ (d) $\quad x(y - 2) - (y - 2)$

$\qquad$ (e) $\quad x(y + 5) + 3(5 + y)$ $\qquad$ (f) $\quad x(y - 5) + 3(5 - y)$

$\qquad$ (g) $\quad (x + 2)y + 3(x + 2)$ $\qquad$ (h) $\quad (x - 1)y - 5(x - 1)$

2.   By first factorising in pairs, reduce the following to forms similar to those in question 1. Hence fully factorise.

(a) $ax + by + bx + ay$ (b) $xp + yp - xq - yq$

(c) $a^2x^2 + axz + axz + z^2$ (d) $2rx - 4sx + ry - 2sy$

(e) $4ax + 6ay - 4bx - 6by$ (f) $p^2q^2 - xyq - p^2q + xy$

(g) $xy + y^2 - x - y$ (h) $wz + z^2 - 2z - 2w$

(i) $bx - 3ax + 2by - 6ay$ (j) $x^2 - 4xy - xy + 4y^2$

(k) $4a^2 + 4b^2 + ab^2 + a^3$ (l) $ax^2 + bx + ax + b$

Factorising Quadratic Expressions

We saw in Chapter 5 that when two linear binomial expressions containing the same quantity are multiplied together, we end up with a quadratic expression containing three distinct unlike terms. For example:

$$(x+1)(x+2) = x^2 + 3x + 2$$
$$(2x+1)(3x-4) = 6x^2 - 5x - 4$$

Many (but not all) such expressions can be factorised back into a product of two nice, easy, linear factors.

There are two distinct types of quadratic expression: those with coefficient of x^2 equal to one and those with coefficient of x^2 not equal to one. In either case, if we write the quadratic as $ax^2 + bx + c$, it can be shown (later!) that a quick and easy way to tell whether or not a quadratic will factorise into nice, simple, linear factors is:

The quadratic $ax^2 - bx + c$ will have simple linear factors if $\sqrt{b^2 - 4ac}$ is an integer.

Example 6.6   Will $x^2 - 3x - 28$ have simple linear factors?

Solution   In this expression, $a = 1, b = -3$ and $c = -28$

Hence $b^2 = (-3)^2 = 9$ and $4ac = 4(1)(-28) = -112$

So $9 - (-112) = 121 = 11^2$ and $\sqrt{b^2 - 4ac} = 11$

Therefore $x^2 - 3x - 28$ will factorise.

Example 6.7   Can we factorise $3x^2 + 7x + 2$?

Solution   This time, $a = 3, b = 7, c = 2$

Therefore, $b^2 = 7^2 = 49$ and $4ac = 4(3)(2) = 24$

$b^2 - 4ac = 49 - 24 = 25$ and $\sqrt{b^2 - 4ac} = 5$

So $3x^2 + 7x + 2$ also factorises into linear factors.

Example 6.8   What about $3x^2 + 7x + 3$?

Solution   We have $a = 3, b = 7, c = 3$. So

$$b^2 = 49 \qquad\qquad 4ac = 4(3)(3) = 36$$

Hence $b^2 - 4ac = 13$, and $\sqrt{b^2 - 4ac}$ is not an integer; so this quadratic will not give simple linear factors.

With experience, you should be able to recognise which quadratics 'obviously' give simple factors. If in doubt, you can easily use the above rule to check with a calculator before trying to factorise and this will save you valuable time.

Exercise 6.4

By testing $\sqrt{b^2 - 4ac}$ state whether or not the following quadratic expressions will factorise into a product of two simple linear factors:

1.	$x^2 + 2x - 15$	2.	$x^2 - x - 20$	3.	$x^2 - 4x - 2$
4.	$x^2 + 2x - 2$	5.	$x^2 + 6x + 7$	6.	$x^2 + 4x + 4$
7.	$x^2 - 10x + 15$	8.	$4x^2 + 4x - 3$	9.	$3x^2 - 2x + 4$

Let us now look at how actually to perform the factorisation.

Quadratics with coefficient of x^2 equal to one

The easiest way to see what happens when we multiply two linear expressions is to give some examples and try to spot a pattern:

$(x + 1)(x + 2) = x^2 + 3x + 2$ $1 + 2 = 3$ and $1 \times 2 = 2$

$(x + 4)(x + 3) = x^2 + 7x + 12$ $4 + 3 = 7$ and $4 \times 3 = 12$

$(x - 2)(x + 5) = x^2 + 3x - 10$ $-2 + 5 = 3$ and $-2 \times 5 = -10$

$(x - 3)(x - 7) = x^2 - 10x + 21$ $-3 + (-7) = -10$ and $-3 \times -7 = 21$

In each case, the *algebraic sum* (i.e. the sum including the signs) of the two numerical terms in the brackets gives the coefficient of x in the quadratic and their product gives the constant term. To reverse the process we therefore simply need to:

> Find the factors of the constant term whose algebraic sum is the coefficient of x.

Example 6.9   Factorise $x^2 + 11x + 18$.

Solution   We need the factors of 18 whose sum is 11. Possible pairs of factors are

(1×18) or (2×9)or (3×6)

or (-1×-18) or (-2×-9) or (-3×-6)

But only $2 + 9 = 11$

Hence the numerical parts of the two linear factors are 2 and 9, and so

$$x^2 + 11x + 18 = (x + 2)(x + 9)$$

You can now easily check this (and you should always do so!) by multiplying out.

Example 6.10   Factorise $x^2 - 11x + 18$.

Solution   We must take the signs of the coefficients into account.

This time we need the factors of 18 whose sum is *minus* 11. With a little thought we can see that the two factors needed are -2 and -9, since

$$(-2)(-9) = 18 \text{ and } -2 + (-9) = -11$$

and so $x^2 - 11x + 18 = (x - 2)(x - 9)$

Example 6.11   Factorise $x^2 + 3x - 10$.

Solution   The possible factors of -10 are

$$(-10 \times 1) \quad \text{or} \quad (-5 \times 2) \quad \text{or} \quad (10 \times -1) \quad \text{or} \quad (5 \times -2)$$

but 5 and -2 are the required pair, since

$$(5)(-2) = -10 \qquad \text{and} \qquad 5 + (-2) = 3$$

Hence $x^2 + 3x - 10 = (x - 2)(x + 5)$

Example 6.12   Factorise $3x^2 - 6x - 24$.

Solution   Watch out for common factors in the coefficients.

The coefficient of x^2 is clearly not one; but the coefficients have a common factor:

$$3x^2 - 6x - 24 = 3[x^2 - 2x - 8] \qquad \text{since 3 is a common factor}$$

$$= 3(x - 4)(x + 2)$$

Example 6.13   Factorise $x^2 - 17xy + 60y^2$.

Solution   This can be thought of as a quadratic in either x or y. If it factorises, the brackets will contain both x and y. In this case we need the factors of $60y^2$ whose sum is $-17y$, i.e.

$$x^2 - 17xy + 60y^2 = (x - 15y)(x - 2y) \qquad \text{(factors of } 60y^2 \text{ are } -15y \text{ and } -2y)$$

Exercise 6.5

Factorise the following quadratic expressions. If you think an answer is obvious, just write down the two linear factors and check by multiplying out. Otherwise use the above technique of finding the factors of the constant term whose algebraic sum is the coefficient of x.

1. $x^2 + 5x + 6$ 2. $x^2 + 4x + 3$ 3. $x^2 + 6x + 8$

4. $x^2 + 10x + 24$ 5. $x^2 + 15x + 54$ 6. $x^2 + 20x + 51$

7. $x^2 - 3x + 2$ 8. $x^2 - 10x + 16$ 9. $x^2 - 4x + 3$

10. $x^2 - 12x + 32$ 11. $x^2 + x - 20$ 12. $x^2 - 3x - 40$

13. $x^2 - 7x + 10$ 14. $x^2 - 17x - 60$ 15. $x^2 + 20x + 36$

Factorise the following expressions:

16. $y^2 + 5y + 4$ 17. $3a^2 - 6a - 9$ 18. $a^2 + 3a - 4$

19. $p^2 - 6p + 8$ 20. $2p^2 - 12p + 18$ 21. $z^2 + 2z - 8$

22. $z^2 - z - 12$ 23. $5x^2 - 5xy - 30y^2$ 24. $x^2 + 6xy - 7y^2$

25. $x^2 - 10xy + 9y^2$ 26. $x^2 - 7xy + 10y^2$ 27. $p^2 + 5pq - 14q^2$

28. $p^2 - 2pq - 15q^2$ 29. $h^2 - 8hk + 15k^2$ 30. $h^2 + 9hk + 14k^2$

Quadratics with coefficient of x^2 not equal to one

This time it is not as easy to spot the relationship between the coefficients.

Consider the product $(2x + 3)(5x + 4) = 10x^2 + 23x + 12$

How can we reverse this process? We begin by rewriting the above product in full:

$$(2x + 3)(5x + 4) = 10x^2 + 8x + 15x + 12$$

Notice that $8 \times 15 = 120$ and $10 \times 12 = 120$

This gives us a clue of how to proceed. We need the factors of 10×12 whose sum is 23.

We can set the problem out as follows:

$10x^2 + 23x + 12$ factors of 120 whose sum is 23 are 8 and 15

$= 10x^2 + 8x + 15x + 12$ splitting the term in x into $8x + 15x$

Since there is no common factor for all four terms, we now factorise in pairs:

Hence $10x^2 + 8x + 15x + 12$ four terms, factorise in pairs
$= 2x(5x + 4) + 3(5x + 4)$ both terms have factor $(5x + 4)$
$= (5x + 4)(2x + 3)$

Let us see if this works with another example.

Example 6.14   Factorise $10x^2 + 29x + 21$.

Solution   Multiply the coefficient of x^2 by the constant coefficient. We must find the factors of 210 whose sum is 29.

To do this, it is easiest to factorise 10 and 21 into prime factors, so

$$10 \times 21 = 5 \times 2 \times 3 \times 7$$

We now try adding various combinations of these factors:

$$10 + 21 = 31 \qquad 35 + 6 = 41 \qquad 15 + 14 = 29 \quad \text{etc.}$$

No need to go further; we have found the correct pair! We now split the middle term into $15x + 14x$,

i.e. $\qquad 10x^2 + 29x + 21$

$\qquad\qquad = 10x^2 + 15x + 14x + 21 \qquad$ splitting the middle term

$\qquad\qquad = 5x(2x + 3) + 7(2x + 3) \qquad$ factorising in pairs

$\qquad\qquad = (2x + 3)(5x + 7) \qquad$ since $(2x + 3)$ is a common factor

Example 6.15 $\qquad$ Factorise $6x^2 - 7x - 5$.

Solution $\qquad$ We will now need to choose factors of *minus* 30 whose algebraic sum is *minus* 7. Thus we need one positive and one negative factor and since this sum is negative, the negative factor will have to be the numerically bigger.

$$6 \times (-5) = 3 \times 2 \times (-5) = 3 \times (-10) \qquad \text{and} \quad 3 + (-10) = -7$$

so split $-7x$ into $-10x + 3x$:

$$6x^2 - 7x - 5 = 6x^2 - 10x + 3x - 5$$
$$= 2x(3x - 5) + (3x - 5)$$
$$= 2x(3x - 5) + 1(3x - 5)$$
$$= (3x - 5)(2x + 1)$$

Hence if the coefficient of x^2 is not one, a simple procedure will always be as follows:

1. $\quad$ Multiply together the coefficient of x^2 and the constant coefficient.
2. $\quad$ Find the factors of this number whose algebraic sum is the coefficient of x.
3. $\quad$ Split the term in x into two appropriate terms.
4. $\quad$ These four terms now factorise in pairs.

Example 6.16   Factorise $30x^2 - 5x - 10$.

Solution   Again watch out for a common numerical factor:

$$30x^2 - 5x - 10 = 5\,[6x^2 - x - 2] \qquad \text{(required factors of } -12 \text{ are } -4 \text{ and } 3)$$
$$= 5[6x^2 - 4x + 3x - 2]$$
$$= 5[2x(3x - 2) + 1(3x - 2)]$$
$$= 5(3x - 2)(2x + 1)$$

Example 6.17   Factorise $4 - 4x - 3x^2$.

Solution   Don't worry if the coefficient of x^2 is negative. If so, it is often better to proceed as follows. With practice, the first stage is not really necessary.

$$4 - 4x - 3x^2 \;\; = -[3x^2 + 4x - 4] \qquad\qquad \text{factorising out } -1$$
$$= -[3x^2 - 2x + 6x - 4] \qquad \text{required factors of } -12 \text{ are } -2 \text{ and } 6$$
$$= -[x(3x - 2) + 2(3x - 2)]$$
$$= -(3x - 2)(x + 2)$$
$$= (2 - 3x)(x + 2) \qquad\qquad \text{tidy up by multiplying } (3x - 2) \text{ by } -1$$

Exercise 6.6

Factorise.

1.	$3x^2 + 10x + 3$	2.	$6x^2 + 5x + 1$	3.	$3x^2 + 22x + 35$
4.	$6x^2 + 19x + 10$	5.	$6x^2 + 7x - 3$	6.	$5x^2 + 33x - 14$
7.	$20x^2 + x - 1$	8.	$6x^2 + 11x + 3$	9.	$7x - 6 - 2x^2$
10.	$13x - 14 - 3x^2$	11.	$15 - 4x - 4x^2$	12.	$7x - 6x^2 + 20$
13.	$6x^2 + 7xy + 2y^2$	14.	$6x^2 - xy - 12y^2$	15.	$36x^2 + 19x - 6$
16.	$6x^2 - 11xy - 10y^2$	17.	$9x^2 + 9xy - 4y^2$	18.	$1 + xy - 20x^2y^2$
19.	$15x^2y^2 - 2xy - 8$	20.	$4x^2 - 20xy + 25y^2$	21.	$2p^2 - 7pq - 9q^2$
22.	$3b^2 - 13bc - 16c^2$	23.	$10x - x^2 - 21$	24.	$84 - 5x - x^2$

SPECIAL CASES

If the Constant Term Is Not Present

In this case, x is *always* a common factor. We can always factorise out the x and can often find a common factor of the coefficients.

Example 6.18   Factorise:

(a) $x^2 + x$ (b) $x^2 + 5x$ (c) $3x^2 + 7x$ (d) $3x^2 - 6x$

Solution

(a) $x^2 + x = x(x + 1)$

(b) $x^2 + 5x = x(x + 5)$

(c) $3x^2 + 7x = x(3x + 7)$

(d) $3x^2 - 6x = 3x(x - 2)$

If the Term in x Is Missing

There are several possibilities:

$$x^2 + 7 \qquad \text{(cannot be factorised further)}$$
$$2x^2 + 6 = 2(x^2 + 3) \qquad \text{(common factor)}$$
$$35x^2 - 28 = 7(5x^2 - 4) \qquad \text{(common factor)}$$
$$x^2 + 9 \qquad \text{(cannot be factorised)}$$

In addition, there is one extra special case which we will treat separately.

The Difference of Two Squares

We saw earlier in this chapter that if we multiply $(a + b)$ by $(a - b)$ we get the result, $(a+b)(a-b) = a^2 - b^2$. If this equation is reversed, it may be rewritten as:

$$\boxed{a^2 - b^2 = (a+b)(a-b)}$$

Therefore any expression which is the difference of two squares can be factorised into a product of a sum and a difference. Because of the squared terms, this has applications for factorising certain quadratic expressions, as shown below.

Example 6.19 Factorise:

(a) $x^2 - 9$ (b) $3x^2 - 12$ (c) $x^2 - 2$

Solution

(a) $x^2 - 9 = x^2 - 3^2$ (difference between x^2 and $3^{2)}$

$\quad\quad\quad = (x + 3)(x - 3)$

(b) $3x^2 - 12 = 3(x^2 - 4)$ (3 is a common factor)

$\quad\quad\quad\quad = 3(x - 2)(x + 2)$ (difference of two squares)

(c) $x^2 - 2 = (x - \sqrt{2})(x + \sqrt{2})$ (difference of two squares)

Note that the *sum* of two squares has no real factors. There is no real binomial product that leads to such a sum of squares.

Perfect Squares

By similar reasoning to the above, we can reverse the results for $(a + b)^2$ and $(a - b)^2$ to produce

$$a^2 + 2ab + b^2 = (a + b)^2$$

and

$$a^2 - 2ab + b^2 = (a - b)^2$$

If you have learned how to recognise *perfect squares*, you would be wise to look out for them when trying to factorise quadratic expressions. For example,

$\quad\quad x^2 + 4x + 4$ is a perfect square since it may be written as $x^2 + 2.(2).x + (2)^2$

and so $\quad\quad\quad\quad x^2 + 4x + 4 = (x + 2)^2$

Similarly

$$x^2 + 6x + 9 = x^2 + 2(3)x + (3)^2 \quad\quad = (x + 3)^2$$

$$x^2 - 10x + 25 = x^2 - 2(5)x + (5)^2 \quad\quad = (x - 5)^2$$

$$4x^2 + 28x + 49 = (2x)^2 + 2(2x)(7) + (7)^2 = (2x + 7)^2 \quad etc.$$

Exercise 6.7

1.   Factorise the following differences of two squares:

(a) $x^2 - 1$ (b) $x^2 - 36$ (c) $x^2 - 49$

(d) $x^2 - y^2$ (e) $4x^2 - 9$ (f) $4x^2 - 25y^2$

(g) $x^2 - 5$ (h) $4x^2 - 64$ (i) $81 - x^2$

(j) $100 - 9x^2$ (k) $3x^2 - 5$ (l) $x^4 - y^4$

2.   Factorise the following perfect squares. Watch out for common factors.

(a) $x^2 - 2x + 1$ (b) $x^2 - 4x + 1$ (c) $x^2 - 6x + 9$

(d) $9x^2 + 6x + 1$ (e) $9x^2 + 12xy + 4y^2$ (f) $-x^2 + 6x - 9$

(g) $2x^2 + 32x + 128$ (h) $16x^2 + 40x + 25$ (i) $-16x^2 - 24x - 9$

(j) $3x^2 - 30x + 75$ (k) $5x^2 + 40x + 80$ (l) $x^4 - 4x^2 + 4$

7

Equations and Inequalities

An equation is simply a statement that two quantities are the same.

Hence $\qquad 4 + 5 = 9 \qquad$ is an equation

The two quantities here are $(4 + 5)$ on the left hand side (LHS) and 9 on the right hand side (RHS).

MAINTAINING THE BALANCE

Think of an equation as a pair of perfectly balanced old-fashioned scales.

If we place nine marbles on the RHS, the scales will tilt to the right. If we add four marbles to the LHS, they will still tilt to the right; but adding five more to the LHS will make the scales balance.

So $\qquad\qquad 4 + 5 = 9$

Now add three marbles to the RHS. Clearly we will also have to add three marbles to the LHS to maintain the balance. It doesn't matter how many marbles you add, as long as you add the *same number* of marbles to both sides. The argument works for any objects we care to name, so we could have chosen sausages or nails or sweets etc.

Hence $(4 + 5) + 3 = 9 + 3$ still balances

Now take six marbles away from both sides of the scales. They still balance

and $[(4 + 5) + 3] - 6 = [9 + 3] - 6$ still balances

We see clearly that:

An equation is unchanged by adding or subtracting the same number on both sides.

Now go back to the situation in our first equation. Suppose we had put three times as many marbles on the RHS, i.e. 3×9. Then the balance would only be maintained if we put three times as many on the LHS, i.e. $3(4 + 5)$.

Hence $3(4 + 5) = 3 \times 9$ still balances

and similarly, if we decided to divide both sides by 3:

$$\frac{(4+5)}{3} = \frac{9}{3}$$ still balances

Hence:

An equation is unchanged if we multiply or divide both sides by the same number.

Other Properties of Equations

Four useful properties of equations that we will use often are:

(i)   **An equation can be written with the quantities reversed**

For example, $3 = 4$ is the same statement as $4 = 3$

It may seem rather an obvious point to make, but we cannot do this with all mathematical statements; for example, $4 > 3$ is not the same as $3 > 4$. The second statement is obviously false!

(ii)   **Equations can be added together or can be subtracted from each other**

For example, $3 + 4 = 7$ and $10 = 8 + 2$ can be combined to give

$$3 + 4 + 10 = 7 + 8 + 2$$

and $3 + 4 - 10 = 7 - (8 + 2) = 7 - 8 - 2$

(iii)   **If A = B and B = C then A = C**

So if $(3 + 2) = 5$ and $5 = (7 - 2)$

then we can say that $(3 + 2) = (7 - 2)$

(iv)   **If A = B and A = C then B = C**

So if $5 = (3 + 2)$ and $5 = (7 - 2)$ then $(3 + 2) = (7 - 2)$

At first sight it looks as if (iii) and (iv) are making the same statement but they are not. Consider again the inequality statement:

$5 > 4$ and $4 > 3$ mean that $5 > 4 > 3$ and hence $5 > 3$

so '>' satisfies a rule similar to (ii)

But $5 > 3$ and $5 > 4$ do not imply $3 > 4$

Solving Linear Equations

Often we meet equations involving unknown quantities and it is necessary to find the unknown. A linear equation involves just the first power of the unknown quantity. Finding the value of the unknown is called *solving* the equation. The value found is called the *solution* or *root* of the equation. (Not to be confused with square or cube root.)

Example 7.1   Suppose I have a pair of scales with nineteen marbles on the RHS and seven on the LHS. How many marbles do I have to add to the LHS to balance the equation?

Solution   It is clearly not going to take a genius to solve this problem! The first step is to translate the statement into a shorthand form so that we can concentrate on what we have to do.

Then we need   7 marbles + an unknown number of marbles = 19 marbles

Suppose that the unknown number of marbles is given a shorthand label, say x;

then   7 marbles $+ x$ marbles $= 19$ marbles

or even better, $7 + x = 19$

This is a very simple equation indeed, and of course, we can just guess the answer to be 12 since $7 + 12 = 19$. Hence we need twelve marbles to balance the scales. Guessing the answer is not as easy if we make the problem a little harder.

Example 7.2   Suppose I take a number and multiply it by 5, and then add 4. If I end up with the answer 39, what was the number I started with?

Solution   We need: (5 times the number) $+ 4 = 39$

Now let x stand for the number. This then becomes

$$(5 \text{ times } x) + 4 = 39$$

We can shorten the above even further, remembering that in algebra, we omit the multiplication sign:

$$5x + 4 = 39$$

We know that $35 + 4 = 39$

Therefore $5x$ must be 35

But $5 \times 7 = 35$ so x must be 7

Example 7.3 A group of children are away from home for a week's skiing holiday. Before the holiday, each child gave a teacher the same amount of money for safe keeping. The teacher now has £75 to give back and one of the children, John, already has £7. How many children are there altogether if John ends up with £22?

Solution   Let the number of children be n. Then, talking in pounds sterling, John is given 75 divided by n, which we write as $\dfrac{75}{n}$, but he already has 7 so he now has $\dfrac{75}{n} + 7$.

But this is 22, so $\dfrac{75}{n} + 7 = 22$

Again we can easily find the answer, since we know that $15 + 7 = 22$

Hence $\dfrac{75}{n}$ must be 15. But $\dfrac{75}{5} = 15$ so $n = 5$

The method of solution used in Examples 5.2 and 5.3 is fine if the equation has a fairly easy form but we will not always be able to solve equations this way. For example, it is not too easy to spot the solution of

$$\frac{7x-1}{3} - \frac{5x+1}{2} = \frac{9-2x}{4}$$

The best way to proceed is to consider easy cases first and add more and more complexity. At each stage we will reduce the equation to a type that we have previously solved. This way we can develop a systematic approach to the solution of equations. With experience, you will find many shortcuts, but the methods shown here will be reasonably foolproof! The first two cases are very important as *all* equations can be reduced to either one case or the other.

Type 1.   **Equations where the unknown is part of a product**

Example 7.4   Solve $5x = 35$.

Solution   If we divide both sides by 5, the 5 on the LHS will cancel, leaving the unknown, x:

$$5x = 35$$

i.e.
$$\frac{\cancel{5}x}{\cancel{5}} = \frac{35}{5}$$

Hence
$$x = 7$$

Example 7.5   Solve $4x = 7$.

Solution   Divide both sides by 4:

$$4x = 7$$

i.e.
$$\frac{4x}{4} = \frac{7}{4}$$

So
$$x = \frac{7}{4} = 1.75$$

The above solutions are both equivalent to moving the number *multiplying* x to the other side of the equation and *dividing* by it. We can use this fact in future, to simplify the solution.

Example 7.6 Solve $9x = 3$.

Solution Move the 9 to the RHS by dividing both sides by 9:

$$9x = 3$$

i.e. $$x = \frac{3}{9} = \frac{1}{3}$$

Example 7.7 Solve $8 = 4y$.

Solution We can write an equation either way around, so

$$4y = 8$$

i.e. $$y = \frac{8}{4} = 2$$

Type 2. **Equations where the unknown is part of a quotient**

Example 7.8 Solve $\dfrac{x}{7} = 3$.

Solution This time, multiplying both sides by 7 will leave x on the LHS:

$$\frac{x}{7} = 3$$

Therefore $$\frac{7x}{7} = 7 \times 3$$

i.e. $$x = 21$$

Example 7.9 Solve $\dfrac{x}{3} = \dfrac{5}{6}$.

Solution

$$\frac{x}{3} = \frac{5}{6} \qquad \text{multiply by 3}$$

$$3 \times \frac{x}{3} = 3 \times \frac{5}{6}$$

$$x = \frac{5}{2}$$

Example 7.10 Solve $\dfrac{2x}{3} = 4$.

Solution

$$\frac{2x}{3} = 4 \qquad \text{multiply by 3}$$

$$2x = 3 \times 4 \qquad \text{divide by 2}$$

$$x = \frac{3 \times 4}{2}$$

$$x = 6$$

or, we can do it all in one step:

$$\frac{2x}{3} = 4$$

$$x = \frac{3}{2} \times 4$$

$$x = 6$$

In each case, if the unknown is divided by a number, we can take the number to the other side of the equation and multiply by it.

Example 7.11   Solve $\dfrac{3}{x} = 7$.

Solution   No need to panic just because the unknown is in the denominator!

$$\frac{3}{x} = 7 \qquad \text{multiply both sides by } x$$

$$x \times \frac{3}{x} = 7x \qquad \text{write the other way around}$$

$$7x = 3 \qquad \text{this is now of type 1 ; so divide by 7}$$

$$x = \frac{3}{7}$$

There is an alternative way of solving the above as we will show later.

Type 3.   Equations involving additions and subtractions

Example 7.12   Solve $x - 3 = 5$.

Solution   This time, if we *add 3 to both sides*, the 3 on the LHS cancels by subtraction.

So $x - 3 = 5$ add 3 to both sides

i.e. $x - 3 + 3 = 5 + 3$

Therefore $x = 8$

Example 7.13   Solve $x + 9 = 2$.

Solution $x + 9 = 2$ subtract 9 from both sides

$$x + 9 - 9 = 2 - 9$$

$$x = -7$$

We see that the above examples are equivalent to saying that if a number is added to the unknown on one side, we subtract it from the other side, and vice versa.

Example 7.14   Solve $9 - x = 2$.

Solution   There are two ways we can handle this:

either

$$9 - x = 2$$ add x to both sides
$$9 = 2 + x$$ subtract 2 from both sides
$$9 - 2 = x$$ write the other way around
$$x = 9 - 2$$
$$x = 7$$

or

$$9 - x = 2$$ subtract 9 from both sides
$$-x = 2 - 9$$
$$-x = -7$$ divide by -1
$$x = \frac{-7}{-1} = 7$$

Either method ends up with the same answer but the second is rather more messy!

We must also stress at this stage that, with practice, there is no need to do each step separately. Solutions should be shortened by performing several steps at the same time. For example, the above can be reduced to

$$9 - x = 2$$
$$9 - 2 = x$$
$$x = 7$$

Exercise 7.1

Find x if:

1. $5x = 20$ 2. $20x = 5$ 3. $8x = 24$

4. $8x = 23$ 5. $23x = 8$ 6. $\dfrac{x}{3} = 5$

7. $\dfrac{2x}{3} = 6$ 8. $\dfrac{3x}{4} = 15$ 9. $\dfrac{3x}{2} = \dfrac{2}{3}$

10. $x + 4 = 10$ 11. $x + 10 = 15$ 12 $x - 7 = 13$

13. $3 + x = 9$ 14. $7 - x = 11$ 15. $9 = 3 - x + 2$

16. $\dfrac{5}{x} = 7$ 17. $\dfrac{2}{x} = \dfrac{5}{3}$ 18. $\dfrac{1}{3x} = \dfrac{2}{9}$

19.   If $3x + 5 = 20$, find $3x$ and hence find x.

20.   If $2y - 3 = 15$, find $2y$ and hence find y.

21.   If $\dfrac{a}{2} + 3 = 5$, find $\dfrac{a}{2}$ and hence find a.

Type 4.   Mixing multiplications, additions and subtractions

All equations, no matter how complicated, can be systematically reduced to very easy forms like those above. If the equation contains a mixture of terms which are added, subtracted, multiplied or divided, we need to be very careful about the order in which we perform the movement of terms. The last three problems in Exercise 7.1 give us a clue or how to proceed.

Example 7.15   Solve $2x + 7 = 13$.

Solution   In this case it is best to keep the $2x$ 'in one piece' and move the 7 first:

$$2x + 7 = 13 \qquad \text{subtract 7 from both sides}$$
$$2x = 13 - 7 \qquad \text{simplify the RHS}$$
$$2x = 6 \qquad \text{now divide by 2}$$
$$x = \frac{6}{2} = 3$$

so, once the value of $2x$ was found, the equation was reduced to Type 1, which we know how to solve.

Example 7.16   Solve $7 = 4 - 3y$.

Solution   The $-3y$ could prove a problem, so move it to the other side first, to make it positive.

$$7 = 4 - 3y \qquad \text{add } 3y \text{ to both sides}$$
$$7 + 3y = 4 \qquad \text{subtract 7 from both sides}$$
$$3y = 4 - 7 \qquad \text{simplify RHS}$$
$$3y = -3 \qquad \text{divide by 3}$$

$$y = \frac{-3}{3} = -1$$

Again, we reduced the equation to Type 1 by finding $3y$ and hence y.

We can shorten the above argument as follows:

$$7 = 4 - 3y$$
$$3y = 4 - 7 = -3$$
$$y = \frac{-3}{3} = -1$$

As we have already pointed out, with experience you also should try to keep the steps as brief as possible (without making mistakes!).

Example 7.17   Solve $\frac{y}{2} + 2 = 5$.

Solution

$$\frac{y}{2} + 2 = 5 \qquad\qquad \text{subtract 2}$$

$$\frac{y}{2} = 5 - 2 \qquad\qquad \text{simplify}$$

$$\frac{y}{2} = 3$$

$$y = 6$$

Once $\frac{y}{2}$ was found, the equation was reduced to Type 2 and it was therefore easy to solve.

Example 7.18   Solve $5x - 7 = 3x + 4$.

Solution   This time, the unknown appears on both sides of the equation. Rearrange so that the terms in x all appear on one side and terms not involving x appear on the other.

$$5x - 7 = 3x + 4 \quad \text{subtract } 3x \text{ and add } 7$$

$$5x - 3x = 4 + 7 \quad \text{collect like terms}$$

$$2x = 11 \quad \text{divide by 2}$$

$$x = \frac{11}{2} = 5.5$$

Example 7.19   Solve $2p + 3 = 7p - 5$.

Solution   Since $7p$ is bigger than $2p$, we can avoid negative coefficients of p if we move the $2p$ to join the $7p$.

$$2p + 3 = 7p - 5 \quad \text{subtract } 2p \text{ and add } 5$$

$$3 + 5 = 7p - 2p \quad \text{collect terms}$$

$$8 = 5p \quad \text{write the other way around}$$

$$5p = 8 \quad \text{divide by 5}$$

$$p = \frac{8}{5} = 1.6$$

We would get the same answer if we did the following, but it is more messy!

$$2x + 3 = 7x - 5$$

$$2x - 7x = -5 - 3$$

$$-5x = -8$$

$$x = \frac{-8}{-5} = 1.6$$

Again, it is best to avoid negative coefficients whenever possible!

Exercise 7.2

Find the unknown quantity in the following equations:

1. $3x + 1 = 22$ 2. $5y + 3 = 23$ 3. $7z + 1 = 15$

4. $2x - 1 = 9$ 5. $8y - 3 = 21$ 6. $9z - 5 = 31$

7. $\dfrac{x}{3} - 1 = 6$ 8. $\dfrac{2y}{3} - 1 = 9$ 9. $1 + \dfrac{4y}{7} = 13$

10. $3x - 9 = 0$ 11. $7 - 3y = 4$ 12. $13 - 4z = -5$

13. $2x + 3x = 15$ 14. $2y + 4y = 48$ 15. $z + 7z = 16$

16. $3x + 1 = 15 + x$ 17. $5y - 1 = 14 + 2y$ 18. $7z + 3 = 3z + 27$

19. $9a - 5a + 3a = 28$ 20. $4b - 3 = 2b + 3$ 21. $7c + 1 = 1 + 6c$

22. $-5a + 1 = a - 11$ 23. $4b = b + 9$ 24. $c - 3 = 2c - 14$

Type 5.   **Equations involving brackets**

Example 7.20   Solve $3(x - 5) = 9$.

Solution   We have two alternatives:

$$
\begin{aligned}
3(x - 5) &= 9 && \text{divide by 3}\\
x - 5 &= 3 && \text{add 5}\\
x &= 8
\end{aligned}
$$

or

$$
\begin{aligned}
3(x - 5) &= 9 && \text{multiply out the brackets}\\
3x - 15 &= 9 && \text{add 15}\\
3x &= 24 && \text{divide by 3}\\
x &= 8
\end{aligned}
$$

Example 7.21   Solve $2(x - 5) = 9$.

Solution   Again we could go two ways:

$$2(x-5) = 9 \qquad \text{divide by 2}$$

$$x-5 = \frac{9}{2} \qquad \text{add 5}$$

$$x = \frac{9}{2} + 5 = \frac{9+10}{2} = \frac{19}{2}$$

or

$$2(x-5) = 9 \qquad \text{multiply out the brackets}$$

$$2x - 10 = 9 \qquad \text{add 10}$$

$$2x = 9 + 10 = 19 \qquad \text{divide by 2}$$

$$x = \frac{19}{2}$$

This time, dividing by 2 introduced a denominator and this made the solution rather messier than just multiplying out the brackets. We would advise you, wherever possible, *not* to introduce denominators. If there is not a simple common factor, it is usually best to multiply out the brackets.

Example 7.22   Solve $3(x-7) = 2(x+4)$.

Solution   If you start dividing through by 2 and by 3, you could end up with a real mess here! It is easier to multiply out the brackets:

i.e.

$$3(x-7) = 2(x+4) \qquad \text{multiply out}$$

$$3x - 21 = 2x + 8 \qquad \text{subtract } 2x \text{ and add 21}$$

$$3x - 2x = 8 + 21 \qquad \text{collect like terms}$$

$$x = 29$$

Example 7.23   Solve $3(y+2) - 7(y+1) = 2(y+1) + 6$.

Solution   Multiply out the brackets:

$$3y + 6 - 7y - 7 = 2y + 2 + 6 \qquad \text{simplify}$$

$$-4y - 1 = 2y + 8 \qquad \text{add } 4y \text{ and subtract 8}$$

$$-1 - 8 = 2y + 4y \qquad \text{simplify}$$

$$-9 = 6y \qquad \text{reverse the equation}$$

$$6y = -9 \qquad \text{divide by } 6$$

$$y = \frac{-9}{6} = -\frac{3}{2}$$

Exercise 7.3

Solve the following equations:

1. $5x = 4(x + 1)$

2. $(3x + 1) - (2x + 14) = 13$

3. $5(x - 4) + 2(x - 6) = 8(x - 5)$

4. $(x - 4) - 10(x - 10) = 0$

5. $-1 = 5(x - 10) + 3(x + 3)$

6. $4(x - 16) = 3(x + 4) - 5(1 - 2x) + 1$

7. $7(2z - 1) - 11(z - 2) = 0$

8. $14(y - 1) - 5(y - 4) = 15$

9. $5(z + 6) + 4(2z - 7) - 54 = 0$

10. $7(x - 6) + 2(x - 7) = 5(x - 4)$

11. $2(x - 3) + 3(x + 1) + 2(x - 1) + 4 = 20$

12. $5(x + 1) - 16(4 - x) = 3(3x - 4) - 11$

Type 6.   Equations involving denominators

So far, we have avoided introducing denominators. If they appear in an equation to be solved, it is best to remove them by multiplying by their LCM.

Example 7.24   Solve $\dfrac{x}{2} = \dfrac{x}{3} + 7$.

Solution   The LCM is 6 so we multiply *everything* by 6:

$$\frac{x}{2} = \frac{x}{3} + 7$$

$$6 \times \frac{x}{2} = 6 \times \frac{x}{3} + 6 \times 7$$

$$3x = 2x + 42$$

$$3x - 2x = 42$$

$$x = 42$$

Example 7.25 Solve $\dfrac{2x}{5}+\dfrac{3x}{7}=\dfrac{x}{10}+4$.

Solution The LCM of 5,7 and 10 is 70 so we multiply each term by 70:

$$70 \times \frac{2x}{5}+70 \times \frac{3x}{7}=70 \times \frac{x}{10}+70 \times 4$$

$$28x+30x = 7x+280$$

$$58x = 7x+280$$

$$51x = 280$$

$$x = \frac{280}{51}$$

Example 7.26 Solve $\dfrac{(x+1)}{5}=\dfrac{(x-7)}{4}$.

Solution

$$\frac{(x+1)}{5}=\frac{(x-7)}{4} \qquad \text{multiply by 20}$$

$$\frac{20(x+1)}{5}=\frac{20(x-7)}{4} \qquad \text{cancel}$$

$$4(x+1) = 5(x-7)$$

$$4x+4 = 5x-35$$

giving $\qquad\qquad\qquad\qquad x = 39$

Example 7.27 Solve $\dfrac{x+3}{2}-\dfrac{3x-5}{3}=7$.

Solution

$$\frac{(x+3)}{2}-\frac{(3x-5)}{3}=7 \qquad \text{multiply by 6}$$

$$\frac{6(x+3)}{2}-\frac{6(3x-5)}{3}=42$$

$$3(x+3)-2(3x-5) = 42$$

$$3x+9-6x+10 = 42$$

$$-3x+19 = 42$$

$$-3x = 23$$

$$x = \frac{23}{-3} = -\frac{23}{3}$$

Example 7.28   Solve $\dfrac{3}{(x-1)} = \dfrac{2}{(x+2)}$.

Solution

$$\frac{3}{(x-1)} = \frac{2}{(x+2)} \qquad \text{multiply by } (x-1)(x+2)$$

$$3(x+2) = 2(x-1)$$
$$3x+6 = 2x-2$$
$$x = -8$$

Solution by Cross Multiplying and by Inverting

Example 7.29   Solve $\dfrac{5x}{2} = \dfrac{3}{5}$.

Solution   This equation has just *one* fraction on each side. In a case like this, we can solve by simply transferring the denominators to the other sides and multiplying.

$$\frac{5x}{2} = \frac{3}{5} \qquad \text{'cross multiply' the denominators}$$

$$5 \times 5x = 3 \times 2$$

$$25x = 6$$

$$x = \frac{6}{25} = 0.24$$

It is important to realise that we are not doing anything new here; we are still multiplying by the LCM of the denominators (in this case 10) but the method, known as 'cross multiplication', gives us a convenient shortcut.

Example 7.30   Solve $\dfrac{x-5}{2} = \dfrac{x+5}{7}$.

Solution   There is again just *one* fraction on each side, so we can cross multiply:

$$\frac{(x-5)}{2} = \frac{(x+5)}{7}$$ 'cross multiply' the denominators

$$7(x-5) = 2(x+5)$$
$$7x-35 = 2x+10$$
$$5x = 45$$
$$x = 9$$

Note the introduction of brackets in the first line so we remember to multiply *everything* in the brackets.

Finally, as an alternative to cross multiplication, if just one term appears on both sides of the equation, it sometimes gives a quick solution if both sides are inverted. This is particularly useful if the unknown quantity is in the denominator. Example 7.28 could have been done as follows:

$$\frac{3}{(x-1)} = \frac{2}{(x-2)}$$ invert
$$\frac{(x-1)}{3} = \frac{(x-2)}{2}$$ cross multiply
$$2(x-1) = 3(x-2)$$ and so on

SUMMARY

By manipulation, all equations can be reduced to very basic types. Remember the following points:

• Adding any number to both sides of an equation, or subtracting any number from both sides, does not change the balance.

• Multiplying or dividing both sides of an equation by the same number does not change the balance. Of course, we do not use any old number; we choose numbers that conveniently simplify the equation by a series of stages.

- The order of an equation can be reversed for convenience, without changing the balance.

It is important to have a systematic way of dealing with equations. The following steps do not always lead to the quickest or most elegant solutions, but they do help to minimise mistakes. Once you are familiar with the feel of equation solving, we encourage you to use shortcuts; but don't forget that your solution should read logically and should be understandable by another mathematician.

Step 1. If there are denominators, eliminate them by multiplying by their LCM

Step 2. Multiply out brackets.

Step 3. Collect terms in the unknown quantity on one side.

Step 4. The final step should be a multiplication or division.

Remember to avoid negative coefficients if possible and do not introduce denominators. It is, however, often useful to introduce brackets.

We should now be able to take a stab at solving the equation mentioned earlier.

Example 7.31   Solve $\dfrac{7x-1}{3} - \dfrac{5x+1}{2} = \dfrac{9-2x}{4}$.

Solution

$$\frac{7x-1}{3} - \frac{5x+1}{2} = \frac{9-2x}{4} \qquad \text{introduce brackets}$$

$$\frac{(7x-1)}{3} - \frac{(5x+1)}{2} = \frac{(9-2x)}{4} \qquad \text{multiply by 12}$$

$$\frac{12(7x-1)}{3} - \frac{12(5x+1)}{2} = \frac{12(9-2x)}{4} \qquad \text{cancel denominators}$$

$$4(7x-1) - 6(5x+1) = 3(9-2x) \qquad \text{multiply brackets out}$$

$$28x - 4 - 30x - 6 = 27 - 6x \qquad \text{simplify each side}$$

$$-10 - 2x = 27 + 10 \qquad \text{move unknowns to LHS}$$

$$6x - 2x = 27 + 10 \qquad \text{collect like terms}$$

$$4x = 37 \qquad \text{divide by 4}$$

$$x = \frac{37}{4} = 9.25$$

Exercise 7.4

Solve the following equations:

1. $r + \frac{r}{2} - \frac{r}{3} = \frac{7}{3}$

2. $\frac{x}{2} - \frac{x}{3} = \frac{1}{5}$

3. $4x - \frac{(x+4)}{5} = 41$

4. $\frac{2a-1}{5} = 3 - \frac{3a-1}{4}$

5. $\frac{1+x}{2} = \frac{2x-1}{5} -$

6. $\frac{p+3}{5} = 8 - \frac{p-1}{4}$

7. $\frac{1}{3}(y+1) = \frac{1}{2}(y-1) + 1 + \frac{2y}{3}$

8. $\frac{2x-1}{5} - \frac{3x+1}{2} = \frac{2}{5}$

9. $\frac{3}{2x} - 4 = 3 - \frac{9}{x}$

10. $\frac{a+1}{2} - \frac{a-7}{5} = \frac{a+4}{3}$

Solve the following by 'cross multiplying' the denominators or by inverting:

11. $\frac{x}{2} = \frac{3}{5}$

12. $\frac{3x}{4} = \frac{1}{2}$

13. $\frac{x-1}{5} = \frac{x+4}{2}$

14. $\frac{2a-1}{3} = \frac{1}{4}$

15. $\frac{1}{3x} = \frac{2}{9}$

16. $\frac{4a+3}{a} = \frac{9}{2}$

Construction of Equations from Verbal Data

In the introduction to this chapter we showed that equations arise naturally when solving problems generated verbally. We need to be able to interpret data given in words and write it as a simple equation. The next two examples give further ideas on how to use the technique.

Example 7.32   A man divides £150 among his children so that Alan has £50 less than Betty and Betty has twice as much as Charles. If Charles ends up with £x, how much do Alan, Betty and Charles get?

Solution   If Charles has £x, then Betty has £$2x$ and Alan has £$(2x - 50)$.

Therefore

$$x + 2x + (2x - 50) = 150$$

$$x + 2x + 2x - 50 = 150$$

$$5x - 50 = 150$$

$$5x = 200$$

$$x = 40$$

Hence Alan gets £40, Betty gets £80 and Charles gets £30.

The next problem is rather more involved. In any problem involving speeds at this stage, we are talking about average speeds. Since speed is $\dfrac{\text{distance}}{\text{time}}$, we also use distance $=$ speed $\times$ time and time $= \dfrac{\text{distance}}{\text{speed}}$. (See the next chapter for how to transpose formulae.)

Example 7.33   A man leaves home in the morning from village A and cycles at 8 mph to his friend's house in village B. The friend then drives him to the station in town C through rush hour traffic at 24 mph. The man then catches a train which travels at 48 mph to town D. The whole journey takes 5½ hours. If B is five times as far from C as it is from A and C is half way between A and D, how far is it from A to B?

Solution   Let the distance from A to B be x miles. Then the distance from B to C is $5x$ miles, so from A to C is $6x$ miles and hence from C to D is $6x$ miles.

The time taken on any leg of the journey is given by $\dfrac{\text{distance}}{\text{speed}}$.

Therefore the time from A to B is $\dfrac{x}{8}$ hours, the time from B to C is $\dfrac{5x}{24}$ hours and the time from C to D is $\dfrac{6x}{48}$ hours.

But the total time is 5½ hours, so we have

$$\frac{x}{8} + \frac{5x}{24} + \frac{6x}{48} = \frac{11}{2} \qquad \text{multiply by 48}$$

$$6x + 10x + 6x = 264$$

$$22x = 264$$

$$x = 14 \text{ miles}$$

Exercise 7.5

1.   One number is three times another and their sum is 24. If the smaller number is x, write down an expression for the other number. Hence find the two numbers.

2.   A man travels 28 miles in 3 hours, partly on foot and partly by bike. He walks at 4 mph and cycles at 12 mph. If he walks x miles, write down an expression for the number of miles he cycles. Hence find how far he cycles.

3.   I walked at a certain speed for 3 hours and then cycled for two more hours at twice the speed. If I travelled 35 miles altogether, what was my walking speed?

4.   The sum of two numbers is fourteen and five times the smaller number is twice the larger. Find the numbers.

5.   Divide 39 into two parts so that the first part divided by three is equal to the second part times four.

6.   Divide 85 into two parts so that a quarter of one part is one-sixth of the other.

7.   Divide 20 into two parts so that three times one of the parts and seven times the other part equals 116.

8.   A student gains grades of 60%, 65% and 75% in three algebra tests. What percentage does she need to achieve in the fourth test to end up with an average of 70%?

9.   300 people attend a concert. Tickets are sold at £15 and £12.50 and the total receipts were £3950. How many £15 tickets were sold?

10.   In one week a shop assistant works 35 hours at the normal rate of pay, 6 hours on a Saturday at time and a half and 4 hours on Sunday at double time. Her take-home pay, before stoppages, is £202.80. What is the normal rate of pay?

11.   A rectangle and a square have the same area. If the rectangle is 4 cm longer and 3 cm narrower than the square, find the lengths of the sides of the rectangle.

12. Two rectangles of equal area have widths of 10 cm and 12 cm. The difference in their lengths is 3 cm. Find their areas.

13. A greengrocer bought twelve melons and two went bad before they could be sold. By selling each of the rest at 30p more than she paid for them, she made a profit of £2.10. Find the buying price and selling price.

14. A runs at 7 metres per second and B runs at 6.5 metres per second. If B is given a 10 second start, how soon can A overtake B? If they plan to run 1 km altogether, who wins?

15. In a certain job, the starting salary is £10,000 per year and rises by the same fixed amount each year. In the first 5 years, you would earn £70,000 altogether. Find the annual rise.

16. A legacy was divided between two brothers so that they received $\frac{3}{5}$ and $\frac{2}{5}$ of the total. They each donated £3000 to charity and one brother then had $\frac{3}{5}$ as much as the other. How much was the legacy?

INTERVALS AND INEQUALITIES

In Chapter 1 we discussed how numbers can be pictured as points on a line known as the real line. We also met the inequality symbols '>' and '<'.

Thus $3 < 7$ and $14 > 3$ are inequality statements that show *order* relationships between real numbers.

An inequality can be read both ways, so $3 < 7$ can either be read as '3 is less than 7' or '7 is greater than 3'; and of course, $3 < 7$ is the same statement as $7 > 3$.

Suppose one of the numbers in our inequality is unknown. How can we use the real line to picture a statement such as $x > 3$?

It is important to realise that this statement means that x can be *any* real number larger than 3 itself.

This is shown as:

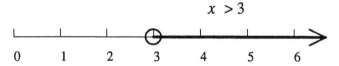

Similarly:

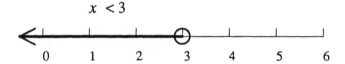

The arrow in both cases indicates that the bold line is continued to infinity (or minus infinity). The small circle around the number 3 shows that 3 itself is *excluded* from the region where the inequality is true. In each case, there are an infinite number of values of x which satisfy the inequality.

We can extend the idea to *include* the value 3. For example:

$$x \geq 3 \text{ means '}x \text{ is greater than or equal to 3'}$$

$$x \leq 3 \text{ means '}x \text{ is less than or equal to 3'}$$

These are shown as:

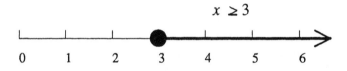

and

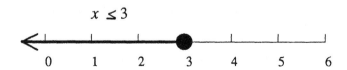

The solid circles indicate that 3 is *included* in the set of numbers.

The set of values of x that satisfy an inequality is called an *interval*. x can be *any* number in the interval. It may well happen that x satisfies two (or more) inequalities simultaneously. In such cases we will need to find the interval where the inequalities are both true.

Example 7.34   Draw a diagram to show the interval $5 \leq x < 7$.

Solution

This time the interval is a finite section of the real line. The number 5 is included in the interval, 7 is not. x lies *between* the two numbers.

Example 7.35   Draw a diagram showing where the following inequalities are true:

$$x \leq 5 \quad \text{and} \quad x > 7$$

Solution

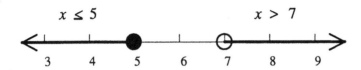

This time, values of x lie in two separate intervals. x lies *outside* the interval from 5 to 7. We again include 5 in the left hand interval but 7 is excluded from the right hand one.

Example 7.36   Find the intervals where the following are true:

$$-3 \leq x \leq 2 \quad \text{and} \quad -1 \leq x \leq 5$$

Solution   In both cases x will lie between the two values given. However, the two intervals overlap so we must be careful. The diagram below shows the two intervals where the inequalities are true and the overlap where they are both true.

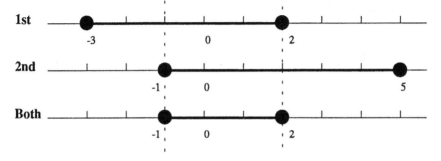

The two sets of inequalities are true simultaneously for $-1 \leq x \leq 2$.

Example 7.37   Find the values of x which satisfy $-3 < x < 3$ and $-1 \leq x < 2$.

Solution

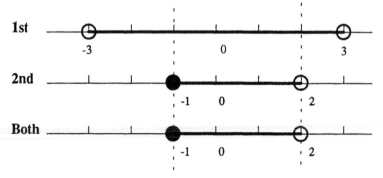

The second interval is included in the first. Both inequalities are true if

$$-1 \leq x < 2$$

Exercise 7.6

1.   Illustrate the following inequalities using diagrams of the real line.

     (a) $x > -3$ (b) $x < -1.5$

     (c) $x \geq 2$ (d) $x \leq 0$

     (e) $1 < x < 3$ (f) $1 \leq x \leq 3$

     (g) $x < 1$ and $x > 3$ (h) $x \leq 1$ and $x \geq 3$

     (i) $-3 \leq x < -1$ (j) $x < -4$ and $x \geq -2$

2.   Find the intervals where the following pairs of inequalities are simultaneously true. Illustrate your answers.

     (a) $-3 \leq x \leq 3$ and $-1 \leq x \leq 1$
     (b) $1 < x < 7$ and $3 < x < 10$
     (c) $-1 \leq x < 1$ and $2 < x \leq 3$
     (d) $-3 \leq x \leq 3$ and $-1 \geq x$ and $x \geq 1$

PROPERTIES OF INEQUALITIES

We illustrate the properties for '<'. They remain true for $\leq$, >, and $\geq$.

(i)   If a, b and c are real numbers such that $a < b$, then

$$a + c < b + c \quad \text{and} \quad a - c < b - c$$

For example, $3 < 7$; so $3 + 4 < 7 + 4$ and $3 - 8 < 7 - 8$ etc. Hence:

> We can add or subtract the same real number on both sides of an inequality without changing the fact that the sides are unequal.

(ii)   If a, b, c and d are real numbers such that $a < b$ and $c < d$, then

$$a + c < b + d$$

For example, $11 > 5$ and $7 > 3$ imply that $11 + 7 > 5 + 3$.

In other words, we can add inequalities together.

(iii)   If a, b and c are real numbers such that $a < b$ and $c > 0$ (i.e. c is positive), then

$$ac < bc \quad \text{and} \quad \frac{a}{c} < \frac{b}{c}$$

> We can therefore multiply or divide both sides of an inequality by a positive number, without changing it.

So far, inequalities seem to be behaving in a similar way to equations. (Of course, we cannot use the 'balancing' analogy.) However, we have stressed in (iii) above that we can multiply or divide both sides of an inequality by a positive real number without changing the inequality. What happens if we try to multiply by a negative number?

It is true that $7 < 16$ but if we multiply both sides by -1, it is *NOT* true that $-7 < -16$. In fact, the opposite is true. Multiplying by -1 *reverses* the inequality.

Similarly $5 < 8 \Rightarrow -2.(5) > -2.(8)$ i.e. $-10 > -16$

(Where the symbol '$\Rightarrow$' means 'implies'.)

and $\qquad 5 < 8 \Rightarrow \dfrac{5}{-2} > \dfrac{8}{-2} \qquad$ i.e. $\quad -2.5 > -4$ and so on

Hence we have a fourth property:

(iv)   If a, b and c are real numbers such that $a < b$ and $c < 0$, i.e. c is negative, then

$$ac > bc \quad \text{and} \quad \frac{a}{c} > \frac{b}{c}$$

> We can multiply or divide both sides of an inequality by a negative number, as long as we reverse the inequality.

Exercise 7.7

1. Starting with the inequality $3 < 5$, write down the two inequalities which are generated when 7 is added to and subtracted from both sides.

2. Write down the two inequalities generated when both sides of the inequality in question 1 are multiplied and divided by 7.

3. Starting with the inequality $-2 < 3$, write down the two inequalities which are generated when -5 is added to and subtracted from both sides.

4. Write down the two inequalities generated when both sides of the inequality in question 3 are multiplied and divided by -5.

5. Starting with the inequality $-6 > -10$, write down the two inequalities which are generated when -2 is added to and subtracted from both sides.

6. Write down the two inequalities generated when both sides of the inequality in question 5 are multiplied and divided by -2.

SOLUTION OF INEQUALITIES

Solving an equation means finding the exact values of the variable for which the equation is true. Solving an inequality means finding the *interval* for which the inequality is true.

We are fortunate in that we can solve inequalities using many of the same techniques that we used to solve equations. However, remember that if an inequality is multiplied or divided by a negative number, the inequality must be reversed. In many cases we must take great care since we may not know whether or not a certain quantity is positive or negative.

Example 7.38   Solve $1 + 2x < 3x + 2$.

Solution $1 + 2x < 3x + 2$ subtract $2x$ from both sides
                                     $1 < x + 2$ subtract 2 from both sides
                                     $-1 < x$

which may also be written $x > -1$ giving the interval below:

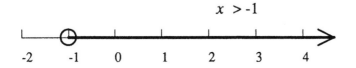

$$x > -1$$

With practice, both subtractions can be done at the same time.

Example 7.39   Solve $3x - 7 > x - 3$.

Solution $3x - 7 > x - 3$ subtract x

                                     $2x - 7 > -3$ add 7

                                     $2x > 4$ divide by 2

                                     $x > 2$

Example 7.40   Solve $\dfrac{3y}{4} > y - 6$.

Solution $\dfrac{3y}{4} > y - 6$ multiply by 4

                                     $3y > 4y - 24$ subtract $3y$

                                     $0 > y - 24$ add 24

                                     $24 > y$

                            or $y < 24$

Example 7.41   Find the interval in which the following is true:

$$5(3x + 1) \leq 2(x - 2)$$

Solution $5(3x + 1) \le 2(x - 2)$ multiply out the brackets

                     $15x + 5 \le 2x - 4$ subtract $2x$

                     $13x + 5 \le - 4$ subtract 5

                     $13x \le -9$

                     $x \le -9/13$

Exercise 7.8

Solve the following inequalities, illustrating your answers.

1. $2x > 6$ 2. $3x < -12$ 3. $9 > 3x$

4. $-3x < -18$ 5. $x - 3 > 4$ 6. $x + 3 > 7$

7. $3x + 8 \ge 12$ 8. $5x - 6 \ge 9$ 9. $1 - x > 4$

10. $-7x + 4 > -2$ 11. $3x - 5 < 7$ 12. $4 \le 7x + 10$

13. $5x + 7 > 3x + 20$ 14. $4x - 3 < 2x - 13$ 15. $3x - 2 \ge 7(x - 4) + 2$

16. $5(x - 3) \le 3(x - 4)$ 17. $2(x - 1) > 3(x - 2)$ 18. $-3x + \dfrac{3}{5} > 6$

19. $\dfrac{3x + 4}{5} > \dfrac{6x + 7}{8}$ 20. $\dfrac{6x - 5}{4} < \dfrac{2x - 1}{2}$ 21. $1 < 2x + 4 < 7$

22. $-1 \le 3 - \frac{1}{2}x \le$ 23. $4 \le 3x + 7 \le 8$ 24. $-5 \le x - 3 < 2$

MODULUS OR ABSOLUTE VALUE

The modulus of a number is the 'size' of the number, ignoring its sign. If the number is x, its modulus is denoted by $|x|$.

A good way of picturing the modulus of a number is as its distance from 0 on the real line.

Hence $|3| = 3$ $|-3| = 3$

                            $\left|\frac{1}{2}\right| = \frac{1}{2}$ $\left|-\frac{1}{2}\right| = \frac{1}{2}$

                            $|\sqrt{2}| = \sqrt{2}$ $|-\sqrt{2}| = \sqrt{2}$

$$|\sqrt{3}-1| = \sqrt{3}-1 \quad |1-\sqrt{3}| = \sqrt{3}-1 \quad \text{etc.}$$

and, of course, $\qquad |0| = 0$

Properties of the Modulus

The following are fairly obvious and will be stated without proof.

For any two real numbers a and b

(i) $\quad |ab| = |a||b|$

(ii) $\quad \left|\dfrac{a}{b}\right| = \dfrac{|a|}{|b|} \qquad$ as long as $b \neq 0$

(iii) $\quad |a^n| = |a|^n \qquad$ which is an extension of (i)

We very often need to solve equations and inequalities involving the modulus. When we do so, the following facts are useful:

If a is a positive number and x is unknown:

(i) $\quad |x| = a \Rightarrow x = +a \text{ or } x = -a, \qquad \text{i.e.} \qquad x = \pm a$

(ii) $\quad |x| < a \Rightarrow -a < x < a \text{ and } x \text{ lies } between -a \text{ and } +a$

(iii) $\quad |x| > a \Rightarrow x > a \text{ or } x < -a \text{ and } x \text{ lies } outside \text{ the interval } -a < x < a$

For example, $\qquad |x| = 5 \qquad$ means $\qquad x = \pm 5$

$\qquad\qquad\qquad\quad |x| < 5 \qquad$ means $\qquad -5 < x < 5$

and $\qquad\qquad\quad |x| > 5 \qquad$ means $\qquad x > 5 \text{ or } x < -5$

In terms of intervals these inequalities are shown as:

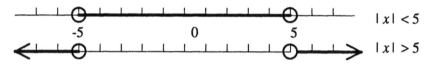

and you can easily adapt them for $|x| \le a$ and $|x| \ge a$.

Example 7.42   Solve the equation $|3x - 7| = 5$.

Solution   This is equivalent to the two separate equations

$$3x - 7 = 5 \qquad \text{or} \qquad 3x - 7 = -5$$
$$3x = 12 \qquad\qquad\qquad 3x = 2$$

i.e. $x = 4$ or $x = \frac{2}{3}$

Example 7.43   Solve the inequality $|2x + 4| < 10$.

Solution   This is the same as

$$2x + 4 < 10 \qquad \text{or} \qquad 2x + 4 > -10$$
$$2x < 6 \qquad\qquad\qquad 2x > -14$$
$$x < 3 \qquad \text{or} \qquad x > -7$$

Hence the inequality is true for $-7 < x < 3$.

Example 7.44   Solve $|3x + 2| \geq 4$.

Solution   Again this is the same as the two separate inequalities

$$3x + 2 \geq 4 \qquad \text{or} \qquad 3x + 2 \leq -4$$
$$3x \geq 2 \qquad\qquad\qquad 3x \leq -6$$
$$x \geq \frac{2}{3} \qquad \text{or} \qquad x \leq -2$$

Exercise 7.9

Solve the following inequalities:

1. $|x - 5| > 3$ 2. $|x - 7| < 4$ 3. $|x + 3| \geq 9$

4. $|2x - 5| < 3$ 5. $|2x - 4| < 1$ 6. $|3 - 2x| < 7$

7. $|5 - 2x| \geq 1$ 8. $|4 - 3x| > 3$ 9. $3|x - 1| > 2$

10. $2|3x - 2| \leq 9$ 11. $\left|x + \frac{1}{2}\right| \geq \frac{3}{2}$ 12. $\left|\dfrac{5x - 7}{3}\right| > 9$

8

Manipulation of Formulae

INTRODUCTION

We introduced formulae in Chapter 3. We will now extend some of the ideas met in that chapter and introduce some new ones.

Remember that a formula is an equation connecting two or more variable quantities. A formula shows the relationship between the variables and gives us a rule for calculating one of them from any combination of the others.

For example, $y = 3x + 2$ is a simple formula connecting two variables x and y.

This is known as an **explicit** formula. It actually *explains* how to calculate y, given any value of x. In an explicit formula, y is known as the **subject**.

A formula does not necessarily have a subject. For example, a much used formula from the theory of lenses is always quoted as

$$\frac{1}{v} + \frac{1}{u} = \frac{1}{f}$$

This is an **implicit** formula. The relationship between the three variables u, v and f is *implied* rather than explained.

SUBSTITUTION

In Chapter 3 we used simple substitution in formulae in order to solve simple problems. Now we have covered solution of equations, we can introduce slightly more difficult problems.

$V = RI$ is a formula needed for analysing electrical circuits, where V is the voltage, R is the electrical resistance of the circuit and I is the current.

You might find it strange that we use I for current rather than C. The reason is historical and I is used by convention. You can use whatever letters you wish. It does, however, illustrate that when letters are introduced in a formula, we should carefully describe what they stand for.

In order to find V from the above formula we will need to know particular values of I and R.

Example 8.1   Use the above formula to find V when $R = 50$ and $I = 4.8$.

Solution

$$V = RI$$

Therefore when $R = 50$ and $I = 4.8$:

$$V = 50 \times 4.8 = 240$$

We can use the same formula to find either I or R provided we know the numerical values of the other two quantities.

Example 8.2   Use the above formula to find (i) I when $V = 100$ and $R = 1000$
                                               (ii) R when $V = 1000$ and $I = 13$

Solution

(i)
$$V = RI$$

When $V = 100$ and $R = 1000$:
$$100 = 1000 \times I$$

This is just a simple equation; therefore

$$I = \frac{100}{1000} = 0.1$$

(ii)   Similarly, when $V = 1000$ and $I = 13$:

$$1000 = R \times 13$$

$$R = \frac{1000}{13} = 76.9 \quad \text{to 1 decimal place}$$

Example 8.3 $\dfrac{1}{v}+\dfrac{1}{u}=\dfrac{1}{f}$ is a formula connecting u, the distance between an object and a lens, v, the distance between the lens and the image produced, and f, the focal length of the lens. Find:

(i)   The focal length when $u = 2$ cm and $v = 3$ cm.

(ii)   The distance of the image from a lens of focal length 5 cm if the object is placed 6 cm away.

Solution

(i)   The first thing to remember is that that the units 'cm' (i.e. centimetres) quoted should not be substituted into the formula!

$$\frac{1}{v}+\frac{1}{u}=\frac{1}{f}$$

When $u = 2$ and $v = 3$ we have, reversing the equation for convenience,

$$\frac{1}{f}=\frac{1}{3}+\frac{1}{2}\qquad\text{common denominator on RHS}$$

$$\frac{1}{f}=\frac{2+3}{6}=\frac{5}{6}\qquad\text{invert}$$

$$f=\frac{6}{5}=1.2\ \text{cm}$$

(ii)   If $u = 10$ and $f = 5$:

$$\frac{1}{v}+\frac{1}{6}=\frac{1}{5}\qquad\text{subtract }\tfrac{1}{6}$$

$$\frac{1}{v}=\frac{1}{5}-\frac{1}{6}=\frac{6-5}{30}=\frac{1}{30}$$

$$v=30\ cm$$

Exercise 8.1

1.   If $P = Tv$, find v when $P = 2000$ and $T = 750$.

2.   If $A = \pi rl$, find l when $A = 750$, $r = 25$ and $\pi = \dfrac{22}{7}$.

3.   If $E = I^2 R$, find I when $E = 12$ and $R = 1000$.

4.   If $y = \dfrac{3x+2}{x+1}$, find x when $y = 15$.

5.   If $v = \sqrt{2gh}$, find h when $v = 16$ and $g = 9.8$.

6.   If $S = \pi r(r+h)$, find h when $S = 175$, $r = 4$ and $\pi = 3.142$.

7.   If $s = ut + \frac{1}{2}at^2$, find u when $s = 200$, $a = 4$ and $t = 7$.

8.   If $T = 2\pi\sqrt{\dfrac{l}{g}}$, find l when $T = 5$, $g = 10$ and $\pi = 3.142$.

TRANSPOSITION

One of the most important skills needed in algebra is the ability to transpose or rearrange a formula. We can build on the skills learned when manipulating simple equations in order to understand the process. There is no real difference between what we are trying to achieve in this section and what we have already mastered in the previous chapter. When we solved simple equations, they contained just one unknown quantity. All the other quantities were numerical. In a formula, we connect two or more unknown quantities.

The two main purposes when transposing formulae are:

- to change the subject of an explicit formula

- to make an implicit formula explicit by making one of the variables the subject.

It is a good idea at this stage to think about why we should wish to perform such tasks.

Suppose $y = 3x + 2$ and we want to find the value of y when $x = 7$. This is a simple substitution problem:

$$y = 3(7) + 2$$
$$= 21 + 2$$
$$= 23$$

Now suppose we wish to find the value of x when $y = 7$. Substituting for y gives us a simple equation which needs to be solved:

$$7 = 3x + 2 \qquad \text{subtract 2}$$
$$5 = 3x \qquad \text{divide by 3 and reverse}$$
$$x = \tfrac{5}{3}$$

This is all very well if there is only one calculation required but suppose we have to find many values of x. We would have to go through the same rearrangement process each time. It makes sense to perform the rearrangement *once only*, before any substitutions are made.

Following the same argument as above, the above formula is transposed as follows:

$$y = 3x + 2 \qquad \text{subtract 2}$$
$$y - 2 = 3x \qquad \text{divide by 3 and reverse}$$
$$x = \frac{(y-2)}{3}$$

We have introduced the brackets to remind ourselves to perform the subtraction before the division. Hence if we now wish to substitute $y = 7$, the hard work has already been done and we get

$$x = \frac{(7-2)}{3}$$
$$= \tfrac{5}{3}$$

Notice that the steps in the process were the same as those used in solving the simple equation. When transposing a formula, once we have decided which variable is to be the subject, we treat all other variables as if they were known numerical quantities.

You should understand that not all formulae can be made explicit. For example, it looks easy to transpose $y = x + x^3$ to make x the subject, but there is no simple way of doing it. It is no good rewriting the above as $x = y - x^3$ since we have merely written x in terms of itself! To transpose the above successfully, it must be rearranged in the form

$$x = \text{an expression involving } y \text{ only}$$

The transposition *can* be done in this case but not by using elementary methods. In many cases, it is impossible to make chosen variables the subject of the formula, in which case we have to be satisfied with the implicit form.

A SYSTEMATIC METHOD FOR TRANSPOSING

Formulae can show relationships between variables in a vast number of ways. We cannot possibly consider every possible combination of variables in a book of *any* size. What we can do however, following the lead of Chapter 5, is to give an idea of how to analyse the structure of a formula and devise a sequence of steps that will normally lead to a successful transposition. There are other routes and many shortcuts when you have gained experience. With practice, you may be able to transpose mentally!

Generally, we may follow similar steps to those used for solving simple equations:

1.   Clear any denominators by multiplying both sides by their LCM.

2.   Multiply out any brackets, particularly those containing the required new subject.

3.   Rearrange the formula so that all like terms containing the required subject are on one side and the rest of the terms are on the other.

4.   After simplifying, you may now have to factorise out the required subject.

5.   The final step should be a multiplication or division.

Always remember that *we are not dealing with letters of the alphabet*. The quantities in a formula are all *numbers* and must behave like numbers! Even though you might be using a systematic method of transposing, you should always understand why certain moves are allowed and others are not.

We will now give a selection of examples that indicate how to proceed with successful transposition of formulae.

Simple Formulae Involving Multiplication or Division

Many important formulae involve three quantities that are connected in this way.

Example 8.4   If an object is moving with constant speed v, and moves a distance s in time t, the three quantities are connected by the formula

$$v = \frac{s}{t}$$

Transpose the formula to make (i) s the subject (ii) t the subject.

Solution   Transposing to make s the subject is a single step. Multiplying both sides by t will cancel t on the RHS:

Therefore $v = \dfrac{s}{t}$ multiply by t

$\Rightarrow$ $vt = \dfrac{st}{t}$ cancel t on the RHS

$vt = s$ reverse the equation

$s = vt$

Transposing for t is now a simple additional step. The *complete* steps for finding t are

$$v = \frac{s}{t} \qquad \text{multiply by } t$$

$$vt = s \qquad \text{divide by } v$$

$$t = \frac{s}{v}$$

There is an alternative way of finding t but it needs more care:

$$v = \frac{s}{t} \qquad \text{divide by } s$$

$$\frac{v}{s} = \frac{s}{st} \qquad \text{cancel } s \text{ on the RHS}$$

$$\frac{v}{s} = \frac{1}{t} \qquad \text{invert the equation}$$

$$\frac{s}{v} = t \qquad \text{reverse the equation}$$

$$t = \frac{s}{v}$$

Notice what happens when s is cancelled. It is a common mistake to leave t on the RHS instead of $\frac{1}{t}$.

Example 8.5   If you borrow a sum of money £P (known as the principal) at a rate of simple interest of $R\%$ per year for time T years, then the interest, £I, you will have to pay is given by the formula

$$I = \frac{PTR}{100}$$

Transpose this formula for P.

Solution   It is a good idea to clear the denominator first, just as we did when solving simple equations. Hence

$$I = \frac{PTR}{100} \qquad \text{multiply by 100}$$

$$100I = PTR \qquad \text{reverse the equation}$$

$$PTR = 100I \qquad \text{divide by } TR$$

$$P = \frac{100I}{TR}$$

Formulae Involving More Than One Term on Each Side

Later in the book, we will study straight line graphs. If two variables x and y produce a graph which is a straight line, then the formula connecting x and y is always of the form $y = mx + c$ where m is the gradient (slope) of the line and c is the value of y where the line crosses the y axis.

Example 8.6   Transpose the formula $y = mx + c$ (i) for c (ii) for m.

Solution

(i) $$y = mx + c$$

There are no denominators to worry about, so to make c the subject, we can subtract mx from both sides:

$$y = mx + c \quad \text{subtract mx}$$

$$y - mx = c \quad \text{reverse the formula}$$

$$c = y - mx$$

(ii)   To transpose for m, go back to the original formula but this time subtract c from both sides:

$$y = mx + c \quad \text{subtract c}$$

$$y - c = mx \quad \text{reverse the formula}$$

$$mx = y - c$$

All we need to do now is to divide both sides by x, to give

$$m = \frac{(y - c)}{x}$$

The brackets are not absolutely necessary but they do remind us that the subtraction must be performed first.

Look at this sequence of numbers:

$$2, \quad 5, \quad 8, \quad 11, \quad 14, \quad 17, \quad 20, ...$$

The sequence is known as an arithmetic progression. Each number in the sequence is generated by adding 3 to the previous number. If we *subtract* any *two consecutive numbers*, we always get the number 3, so 3 is known as the *common difference*. The first term is 2. We notice that

the second term	$= 2 + (1) \times 3 = 2 + (2 - 1) \times 3$
the third term	$= 2 + (2) \times 3 = 2 + (3 - 1) \times 3$
the fourth term	$= 2 + (3) \times 3 = 2 + (4 - 1) \times 3$
the fifth term	$= 2 + (4) \times 3 = 2 + (5 - 1) \times 3$
the sixth term	$= 2 + (5) \times 3 = 2 + (6 - 1) \times 3$
the seventh term	$= 2 + (6) \times 3 = 2 + (7 - 1) \times 3$

If we continue the sequence it is always true that

the last term = the first term + (number of terms – 1) × (the common difference)

This formula is always true for any arithmetic progression. Try to generate the first ten terms of the following: 3,   10,   17, ...

Therefore, in general, for any arithmetic progression

$$l = a + (n - 1)d$$

where each quantity is defined above.

Example 8.7   By transposing the formula $l = a + (n - 1)d$, find an explicit formula for (i) d (ii) n.

Solution

(i)
$$l = a + (n - 1)d \qquad \text{subtract } a$$

$$l - a = (n - 1)d \qquad \text{reverse the formula}$$

$$(n - 1)d = l - a \qquad \text{divide by } (n - 1)$$

$$d = \frac{(l - a)}{(n - 1)}$$

(ii)   To find n, there are two ways we can proceed:

either

$$l = a + (n - 1)d \qquad \text{subtract } a$$

$$l - a = (n - 1)d \qquad \text{reverse the formula}$$

$$(n - 1)d = l - a \qquad\qquad \text{divide by } d$$

$$n - 1 = \frac{(l - a)}{d} \qquad\qquad \text{add 1}$$

$$n = 1 + \frac{(l - a)}{d} \qquad\qquad d \text{ is a common denominator}$$

$$n = \frac{d + l - a}{d}$$

(the last step is not absolutely necessary but it does tidy up the formula)

or

$$l = a + (n - 1)d \qquad\qquad \text{multiply out the brackets}$$

$$l - a = nd - d \qquad\qquad \text{add } d$$

$$l - a + d = nd \qquad\qquad \text{reverse the formula}$$

$$nd = l - a + d \qquad\qquad \text{divide by } d$$

$$n = \frac{l - a + d}{d}$$

I think you will agree, that was quicker and simpler.

Formulae Involving Denominators

Example 8.8 Transpose the formula $y = \dfrac{5x + 3}{x - 2}$ for x.

Solution

$$y = \frac{5x+3}{(x-2)}$$ put brackets around $(x-2)$ and then multiply

$y(x-2) = 5x+3$ multiply out the brackets

$xy - 2y = 5x + 3$ collect terms in x on LHS

$xy - 5x = 2y + 3$ factorise LHS

$x(y-5) = 2y+3$ divide by $(y-5)$

$$x = \frac{2y+3}{y-5}$$

Example 8.9   Transpose the formula $T = 2\pi\sqrt{\dfrac{I}{mgh}}$ for h.

Solution   We must eliminate the square root this time.

$$T = 2\pi\sqrt{\frac{I}{mgh}}$$ square both sides

$$T^2 = 4\pi^2 \cdot \frac{I}{mgh} = \frac{4\pi^2 I}{mgh}$$ multiply by mgh

$$mghT^2 = 4\pi^2 I$$ divide by mgT^2

$$h = \frac{4\pi^2 I}{mgT^2}$$

Example 8.10   Transpose the formula $\dfrac{D}{d} = \sqrt{\dfrac{f+a}{f-a}}$ for f.

Solution

$$\frac{D}{d} = \sqrt{\frac{f+a}{f-a}}$$ square both sides

$$\frac{D^2}{d^2} = \frac{(f+a)}{(f-a)}$$ cross multiply

$$D^2(f-a) = d^2(f+a)$$ multiply out brackets

$$D^2f - D^2a = d^2f + d^2a$$ collect terms in f on the LHS

$$D^2f - d^2f = D^2a + d^2a$$ factorise both sides

$$f(D^2 - d^2) = (D^2 + d^2)a$$ divide by $(D^2 - d^2)$

$$f = \frac{(D^2 + d^2)a}{(D^2 - d^2)}$$

Example 8.11   Transpose $F = mg + \dfrac{mv^2}{r}$ for v.

Solution

$$F = mg + \frac{mv^2}{r}$$ multiply by r

$$Fr = mgr + mv^2$$ subtract mgr

$$Fr - mgr = mv^2$$ divide by m

$$\frac{Fr - mgr}{m} = v^2$$ reverse and take square root

$$v = \pm\sqrt{\frac{(F - mg)r}{m}}$$

Exercise 8.2

1. – 8.   Repeat the Exercise 8.1 by transposing the formulae *before* substituting.

In the following questions, transpose the given formula for the given variable.

9. $A = \pi r l$ for r 10. $\dfrac{v^2}{r} = \sqrt{3g}$ for r

11. $x = \frac{1}{2}gt^2$ for t 12. $s = ut + \frac{1}{2}at^2$ for a

13. $T = 2\pi\sqrt{\dfrac{l}{g}}$ for g 14. $s = \dfrac{v-u}{t}$ for u

15. $F = mg + \dfrac{mv^2}{r}$ for m 16. $x = \dfrac{1-t^2}{1+t^2}$ for t

17. $x = \sqrt{\dfrac{y}{y+z}}$ for z 18. $x = \sqrt{\dfrac{y}{y+z}}$ for y

19. $y = \dfrac{5x-3}{x+4}$ for x 20. $V = \dfrac{R}{R-r}$ for r

21. $V = \dfrac{R}{R-r}$ for R 22. $b = \dfrac{2ac}{a+c}$ for a

23. $\dfrac{D}{d} = \sqrt{\dfrac{f+a}{f-a}}$ for a 24. $v^2 = 2k\left(\dfrac{1}{x} - \dfrac{1}{a}\right)$ for a

25. $\dfrac{1}{v} - \dfrac{1}{u} = \dfrac{\mu-1}{r}$ for r 26. $\dfrac{1}{v} - \dfrac{1}{u} = \dfrac{\mu-1}{r}$ for μ

ELIMINATION

Often we know how two variables are related to a third variable and want to know how they are related to each other. We then need to eliminate the third variable, which is often called a **parameter**.

Example 8.12   The circumference of a circle of radius r is given by $C = 2\pi r$ and the area of the same circle is given by $A = \pi r^2$. Find the relationship between C and A.

Solution   To find this relationship, we need to eliminate the radius, r.

The equation $C = 2\pi r$ may be transposed to give $r = \dfrac{C}{2\pi}$

We can then substitute this into the formula for A:

i.e.
$$A = \pi r^2 = \pi \left(\frac{C}{2\pi}\right)^2$$

Therefore
$$A = \pi.\frac{C^2}{4\pi^2} = \frac{\pi C^2}{4\pi^2} = \frac{C^2}{4\pi}$$

Example 8.13   If $x = ct$ and $y = \dfrac{c}{t}$, find the relationship between y and x.

Solution   We need to eliminate t.

The relationship between x and t gives $t = \dfrac{x}{c}$

The relationship between y and t gives $t = \dfrac{c}{y}$

Hence
$$\frac{x}{c} = \frac{c}{y}$$

We now cross multiply, to give $xy = c^2$ which is an implicit formula and the neatest way of writing this relationship.

Example 8.14   The volume (V) and surface area (A) of a sphere of radius r are given by the formulae

$$V = \frac{4}{3}\pi r^3 \qquad \text{and} \qquad A = 4\pi r^2$$

Find an implicit formula connecting V and A.

Solution

$$= \frac{4}{3}\pi r^3 \qquad \text{multiply by 3 and then square both sides}$$

$$9V^2 = 16\pi^2 r^6$$

Hence

$$r^6 = \frac{9V^2}{16\pi^2}$$

Similarly

$$A = 4\pi r^2 \qquad \text{cube both sides}$$

$$A^3 = 64\pi^3 r^6$$

So

$$r^6 = \frac{A^3}{64\pi^3}$$

We can now eliminate r^6:

$$\frac{9V^2}{16\pi^2} = \frac{A^3}{64\pi^3}$$

which simplifies to $\qquad 36\pi V^2 = A^3$

Exercise 8.3

1. Eliminate t from $\qquad x = at^2$ and $y = 2at$

2. Eliminate v from $\qquad x = 1 + v$ and $y = 1 + \dfrac{1}{v}$

3. Eliminate k from $\qquad p = h + \dfrac{1}{k}$ and $q = k + \dfrac{1}{h}$

4. Eliminate t from $\qquad v = u + at$ and $s = ut + \frac{1}{2}at^2$

5.   Eliminate T from $T = 2\pi\sqrt{\dfrac{l}{g}}$ and $T = 2\pi\sqrt{\dfrac{I}{mgh}}$

Hence find an explicit formula for h.

6.   Eliminate R from $V = \dfrac{R}{R-1}$ and $C = \dfrac{2}{R+1}$

Hence find V in terms of C.

9

Quadratic Equations

ZERO PRODUCTS

Before we begin, you need to be aware of a basic mathematical fact:

> If the product of two numbers is zero at least one of them must be zero.

So $4 \times 0 = 0$ $0 \times 7 = 0$ and, of course, $0 \times 0 = 0$

Hence:

> If $ab = 0$ then $a = 0$ or $b = 0$ or both a and $b = 0$.

Remember that this only works for zero! If $ab = 1$, this does not mean that $a = 1$ or $b = 1$. There are an infinite number of possibilities; for example, $a = 0.5$ and $b = 2$ or $a = 0.8$ and $b = 1.25$ etc.

The idea can be extended to as many numbers as we like: if $abcd = 0$ then at least one of a, b, c or d must be zero.

SOLUTION OF QUADRATIC EQUATIONS

We very often have to solve equations involving quadratic expressions. These equations are usually written in the form $ax^2 + bx + c = 0$ where a, b and c are known constants and we have to find the values of x which satisfy the equation. These solutions are normally called the *roots* of the equation.

Special Case: if the constant term is missing

If the constant term is missing, i.e. if $c = 0$, x is always a common factor, so we can solve as follows:

Example 9.1   Solve the equation $x^2 - 3x = 0$.

Solution

$$x^2 - 3x = 0$$

Hence $$x(x - 3) = 0$$

We now have two numbers whose product is zero, so *either* $x = 0$ *or* $(x - 3) = 0$

and so $x = 0$ and $x = 3$ are the two solutions

Example 9.2   Solve the equation $2x^2 - 5x = 0$.

Solution

$$2x^2 - 5x \quad = 0$$
$$x(2x - 5) \quad = 0$$
$$x = 0 \ \text{ or } \ 2x = 5$$

i.e. $$x = 0 \ \text{ or } \ x = 2.5$$

Example 9.3   Solve $3x^2 + 6x = 0$.

Solution

$$3x^2 + 6x = 0$$
$$3x(x + 2) = 0$$

We now have three factors but 3 cannot be zero, so we can cancel it. Again, be aware; we can only do this if the RHS is zero.

Therefore $x = 0 \text{ or } x = -2$

Example 9.4   We end this section with a warning by solving the equation

$$x^2 = x$$

There is a great temptation here to cancel an x, leaving the solution $x = 1$. However, we know that so far quadratic equations have all had two solutions. If we cancel, we will therefore lose one of them! We should really do the following:

Solution

$$x^2 = x$$

i.e. $$x^2 - x = 0$$

so $$x(x - 1) = 0$$

giving $$x = 0 \text{ or } x = 1$$

The above is *very* important. In trigonometry we will often meet quadratic equations. If we cancel a term that might be zero, we will not lose just one solution but *an infinite number* of them!

Special Case: the term in x is missing

Example 9.5   Solve $x^2 - 9 = 0$.

There are two possible ways of solving this:

First Solution

$$x^2 - 9 = 0$$

i.e. $$x^2 = 9 \quad \text{transposing}$$

Take the square root of both sides and note carefully that we must always include two solutions for this type of equation. In this particular case, both 3^2 and $(-3)^2$ give 9; so the solutions are $x = 3$ and $x = -3$.

Second Solution

$$x^2 - 9 = 0$$

i.e. $$(x - 3)(x + 3) = 0 \quad\quad\quad \text{difference of two squares}$$

so $$x = 3 \text{ or } x = -3$$

N.B.   When solving the above equation, $x^2 = 9$, the solutions are $x = \pm 3$.

Another way of putting this is to say that *the modulus of x is 3,*

so $x = \pm 3$ and $|x| = 3$ are equivalent statements

Example 9.6 Solve $2x^2 - 6 = 0$.

Solution

$$2x^2 - 6 = 0$$
$$2(x^2 - 3) = 0 \qquad \text{we may cancel the 2}$$
$$x^2 = 3$$

i.e. $$x = \pm\sqrt{3}$$

Example 9.7 Solve $3x^2 - 4 = 0$.

Solution

$$3x^2 - 4 = 0$$

i.e. $$3x^2 = 4$$

$$x^2 = \frac{4}{3}$$

$$x = \pm\sqrt{\frac{4}{3}} = \pm1.155 \quad \text{to 3 decimal places}$$

Example 9.8 Solve $x^2 + 9 = 0$.

Solution

$$x^2 + 9 = 0$$

$$x^2 = -9$$

$$x = \pm\sqrt{-9}$$

But we do not know of any numbers (certainly not at this stage) whose square is -9. We know that $3^2 = 9$ and that $(-3)^2 = 9$ also. For the moment we will say that such an equation has imaginary roots since we know of no real numbers which satisfy the equation.

Example 9.9   Solve $3x^2 + 4 = 0$.

Solution

$$3x^2 + 4 = 0$$
$$x^2 = -\tfrac{4}{3}$$

again the roots are imaginary.

Exercise 9.1

Solve the following equations. If the roots are imaginary, say so.

1.	$x^2 - 7x = 0$	2.	$x^2 + 4x = 0$	3.	$3x^2 - 6x = 0$
4.	$x^2 = 3x$	5.	$x^2 = -4x$	6.	$3x^2 - 5x = 0$
7.	$x^2 = 49$	8.	$x^2 = 17$	9.	$x^2 - 25 = 0$
10.	$x^2 - 26 = 0$	11.	$3x^2 - 147 = 0$	12.	$x^2 - y^2 = 0$
13.	$4x^2 - 9 = 0$	14.	$4x^2 - 64 = 0$	15.	$4x^2 - 25y^2 = 0$
16.	$100 - 9x^2 = 0$	17.	$x^2 = -4$	18.	$x^2 + 16 = 0$
19.	$x^2 + 51 = 0$	20.	$7x^2 + 19 = 0$	21.	$81 + x^2 = 0$

Solution by Factorising

If the quadratic expression factorises, the solution is easy. If you have forgotten how to factorise quadratic expressions, look back to Chapter 4.

Example 9.10   Solve $x^2 + 3x + 2 = 0$.

Solution

$$x^2 + 3x + 2 = 0$$
$$(x + 1)(x + 2) = 0$$
$$x = -1 \text{ or } x = -2$$

Example 9.11   Solve $x^2 + 2x - 35 = 0$.

Solution

$$x^2 + 2x - 35 = 0$$

$$(x-5)(x+7) = 0$$

$$x = 5 \ or \ x = -7$$

Example 9.12   Solve $3x^2 + 10x - 8 = 0$.

Solution

$$3x^2 + 10x - 8 = 0$$
$$3x^2 + 12x - 2x - 8 = 0$$
$$3x(x+4) - 2(x+4) = 0$$
$$(x+4)(3x-2) = 0$$

Hence $x = -4 \ or \ x = \frac{2}{3}$

Example 9.13   Solve $x^2 + 14x + 49 = 0$.

Solution

$$x^2 + 14x + 49 = 0$$

$$(x+7)^2 = 0 \qquad \text{since this is a perfect square}$$
$$x = -7 \qquad \text{is the only solution}$$

Exercise 9.2

Solve the following quadratic equations by factorising. Watch out for perfect squares and common factors that can be cancelled.

1. $x^2 + 5x + 6 = 0$ 2. $x^2 + 4x + 3 = 0$ 3. $x^2 + 6x + 8 = 0$

4. $x^2 + 10x + 24 = 0$ 5. $x^2 + 15x + 54 = 0$ 6. $x^2 + 20x + 51 = 0$

7. $x^2 - 4x + 3 = 0$ 8. $x^2 - 12x + 32 = 0$ 9. $x^2 - 3x - 40 = 0$

10. $x^2 - 7x + 10 = 0$ 11. $x^2 - 17x + 60 = 0$ 12. $y^2 + 5y + 6 = 0$

13. $y^2 + 5y + 4 = 0$ 14. $3a^2 - 6a - 9 = 0$ 15. $a^2 + 3a - 4 = 0$

16. $2p^2 - 12p + 18 = 0$ 17. $z^2 - z - 12 = 0$ 18. $5x^2 - 5xy - 30y^2 = 0$

19. $x^2 - 2x + 1 = 0$ 20. $x^2 - 4x + 4 = 0$ 21. $x^2 - 6x + 9 = 0$

22. $x^2 + 6xy - 7y^2 = 0$ 23. $x^2 - 10xy + 9y^2 = 0$ 24. $p^2 + 5pq - 14q^2 = 0$

25. $h^2 + 9hk + 14k^2 = 0$ 26. $3x^2 + 10x + 3 = 0$ 27. $6x^2 + 5x + 1 = 0$

Solution by Formula

All quadratic equations can be solved by using your calculator and the formula we are about to develop. In future you will have the choice of always using this formula or using the factorisation methods. We recommend a sensible mixture of both techniques which can be developed with experience.

Firstly, we will need a technique known as 'completing the square'.

$$x^2 + 14x + 49 \qquad \text{is a perfect square}$$
$$x^2 + 14x \qquad \text{is } \textit{not} \text{ a perfect square}$$

How do we make the second expression into a perfect square? The answer is pretty obvious: we add 49! Is there a way of working this out without knowing the answer?

Look again at the first expression:

$$x^2 + 14x + 49 = x^2 + 2(7)(x) + (7)^2$$

$$\uparrow \qquad \uparrow$$

7 is half the coefficient of x

so, to make $x^2 + 14$ into a perfect square we divide 14 by 2 to get 7. This is then squared and added on.

In other words:

> To complete the square, square half the coefficient of x and add it on.

Example 9.14   Complete the square $x^2 + 10x + \ldots$.

Solution

$$x^2 + 10x + \ldots \qquad\qquad \text{half the coefficient of } x \text{ is } 5$$

Hence $\qquad\qquad x^2 + 10x + (5)^2 = (x + 5)^2 \quad$ is a perfect square

Example 9.15 $\qquad$ Complete the square $x^2 - 12x + \ldots$.

Solution

$$x^2 - 12x + \ldots \qquad\qquad \text{half the coefficient of } x \text{ is } -6$$

$$x^2 - 12x + (-6)^2 = (x - 6)^2 \quad \text{is a perfect square}$$

Example 9.16 $\qquad$ Complete the square $x^2 + 9x + \ldots$.

Solution

$$x^2 + 9x + \ldots \qquad\qquad \text{half the coefficient of } x \text{ is } \tfrac{9}{2}$$

$$x^2 + 9x + \left(\frac{9}{2}\right)^2 = \left(x + \frac{9}{2}\right)^2 \quad \text{is a perfect square}$$

Why on Earth are we doing this? Because it gives us a rather clever way of solving quadratic equations!

Example 9.17 $\qquad$ Solve the equation $x^2 + 7x + 3 = 0$ by completing the square.

Solution $\qquad$ This expression does not have simple factors, since $b^2 - 4ac = 37$, which is not an integer so we cannot solve by factorising.

Solving a quadratic equation is not like solving a simple equation. There is no simple way of rearranging the equation in the form $x =$ 'an expression without x in it'. However, we can do the following:

$$x^2 + 7x \qquad = -3 \qquad\qquad \text{add } \left(\frac{7}{2}\right)^2 \text{ to complete the square}$$

$$x + 7x + \left(\frac{7}{2}\right)^2 = -3 + \left(\frac{7}{2}\right)^2 \qquad \text{we add to both sides to keep the balance}$$

$$\left(x+\frac{7}{2}\right)^2 = -3+\frac{49}{4} = \frac{37}{4}$$

$$\left(x+\frac{7}{2}\right) = \pm\sqrt{\frac{37}{4}} = \pm\frac{\sqrt{37}}{2}$$

$$x = \frac{-7\pm\sqrt{37}}{2}$$

giving us two solutions:

$$\frac{-7+\sqrt{37}}{2} = -0.459 \quad \text{to 3 decimal places}$$

$$\text{and} \quad \frac{-7-\sqrt{37}}{2} = -6.541 \quad \text{to 3 decimal places}$$

The Quadratic Formula

The above method, although impressive, can become tedious to do every time; so we will do it just once more with general values of a, b and c. We will then develop a formula which we can just plug numbers into and use our calculators.

If you find the following argument difficult at a first reading, come back to it at a later date. It is worth the effort of understanding each step.

To solve the equation $\qquad\qquad ax^2 + bx + c = 0$

it is easier to continue if the coefficient of x^2 is one:

dividing by a $\qquad\qquad\qquad x^2 + \left(\frac{b}{a}\right)x + \left(\frac{c}{a}\right) = 0$

i.e. $\qquad\qquad\qquad\qquad x^2 + \left(\frac{b}{a}\right)x = -\left(\frac{c}{a}\right)$

Complete the square $\qquad x^2 + \left(\frac{b}{a}\right)x + \left(\frac{b}{2a}\right)^2 = \left(\frac{b}{2a}\right)^2 - \left(\frac{c}{a}\right)$

$$\left(x+\frac{b}{2a}\right)^2 = \frac{b^2}{4a^2} - \frac{c}{a} = \frac{b^2-4ac}{4a^2}$$

Take the square root of both sides
$$x+\frac{b}{2a} = \pm\frac{\sqrt{b^2-4ac}}{2a}$$

$$x = \frac{-b\pm\sqrt{b^2-4ac}}{2a}$$

This is not as bad as it looks!

You will recognise our old friend b^2-4ac which we used to test for simple linear factors. We can now explain why this works.

Example 9.18   Solve $x^2 + 3x - 10 = 0$.

Solution   In this case $a = 1$, $b = 3$ and $c = -10$, so $b^2-4ac = 9 + 40 = 49$.

Hence, using the formula:

$$x = \frac{-3\pm\sqrt{49}}{2(1)} = \frac{-3\pm 7}{2} = -\frac{10}{2} \text{ or } \frac{4}{2}$$

So $x = -5$ or $x = 2$

This means that the quadratic expression had simple linear factors $(x - 2)$ and $(x + 5)$. If b^2-4ac is a perfect square, then $\sqrt{b^2-4ac}$ will always be a positive integer and the quadratic expression will have simple factors.

b^2-4ac is called the *discriminant* of the equation and it has further properties as the following shows:

Properties of the Discriminant

Look again at the formula for the roots:

$$\boxed{x = \frac{-b\pm\sqrt{b^2-4ac}}{2a}}$$

Case 1:   if $b^2 - 4ac > 0$ the equation will have *two, distinct, real roots.*

Case 2:   if $b^2 - 4ac = 0$ the square root term will disappear and we will be left
               with just one root. (We normally say that the equation has *two equal
               roots.*)

Case 3:   if $b^2 - 4ac < 0$ the square root term will be imaginary and the quadratic
               will have *no real roots.* (We usually still think of having two roots; they
               are called *complex roots* and have a real part and an imaginary part.)

Example 9.19   Solve $2x^2 + 5x - 4 = 0$.

Solution $a = 2, \ b = 5, \ c = -4$ so $b^2 - 4ac = 25 - 4(2)(-4) = 57$

$\therefore$ $x = \dfrac{-5 \pm \sqrt{57}}{2(2)} = \dfrac{-5 \pm 7.5498}{4}$ to 4 d.p.

               $x = 0.637$ or $x = -3.137$ to 3 d.p. using a calculator

Example 9.20   Solve $x^2 - 6x + 9 = 0$.

Solution $a = 1, \ b = -6, \ c = 9$ so $b^2 - 4ac = (-6)^2 - 4(1)(9) = 36 - 36 = 0$

Hence the solutions are $x = \dfrac{-(-6) \pm 0}{2(1)} = \dfrac{6}{2} = 3$ only but it is convenient still to think of
the equation of having two equal roots, $x = 3$ (twice). Note that the quadratic
expression is a perfect square, so we could have factorised.

Example 9.21   Solve $3x^2 - 4x + 5 = 0$.

Solution

This time $a = 3, \ b = -4, \ c = 5$ and $b^2 - 4ac = (-4)^2 - 4(3)(5) = 16 - 60 = -44$

We cannot find the square root of this number so the equation has complex roots.

Exercise 9.3

1. Add the appropriate term to the following expressions in order to complete the square. In each case, state what perfect square has been generated.

    (a) $x^2 + 12x + ...$ (b) $x^2 - 18x + ...$ (c) $x^2 + 7x + ...$

    (d) $x^2 - 5x + ...$ (e) $3x^2 + 6x + ...$ (f) $2x^2 + 5x + ...$

2. For each of the following, work out the discriminant.

    If the discriminant is a perfect square or zero, solve by factorising.

    If the discriminant is positive and not a perfect square, use the formula.

    If the discriminant is negative what should you say about the roots?

    (a) $x^2 - 15x + 56 = 0$ (b) $x^2 - x - 1 = 0$

    (c) $x^2 + 7x + 10 = 0$ (d) $x^2 - 6x + 12 = 0$

    (e) $x^2 + 4x + 1 = 0$ (f) $5x^2 - 4x + 1 = 0$

    (g) $5x^2 - 4x - 1 = 0$ (h) $4x^2 - 3x - 2 = 0$

    (i) $3x^2 + 5x + 8 = 0$ (j) $2x^2 - 7x + 4 = 0$

    (k) $5x^2 - 7x + 4 = 0$ (l) $5x^2 - 9x + 1 = 0$

    (m) $4x^2 + 3x - 7 = 0$ (n) $3x^2 + 5x - 4 = 0$

Problems Leading to Quadratic Equations

Often a quadratic equation will be generated when we try to solve real problems. One important point here is to realise that some of the solutions generated *will not be solutions to the problem*! You should always be aware of what the question is about and discard any solutions which are unreasonable. Sometimes the additional root will be the solution to a related real problem.

Example 9.22   A stone is thrown down a mineshaft with an initial speed of 5 metres/second. The formula for the distance dropped after t seconds is given by $s = ut + 5t^2$ where u is the initial speed. How long does it take to drop 140 metres?

Solution　　　　Since $u = 5$ and $s = 140$, we have $140 = 5t + 5t^2$

Therefore　　　　　　　　　　　　　$0 = 5t^2 + 5t - 140$

i.e.　　　　　　　　　　　　　　　$0 = t^2 + t - 28$

$$t = \frac{-1 \pm \sqrt{113}}{2} = \frac{-1 \pm 10.6}{2} \quad \text{to 1 d. p.}$$

$$t = \frac{9.6}{2} = 4.8 \text{ seconds approx. since } t \text{ cannot be negative}$$

Example 9.23　　　　Two people cycle a distance of 45 kilometres at different speeds. One of them cycles at 1 kilometre per hour faster than the other and arrives at the destination half an hour earlier. Find their speeds.

Solution　　　　Let the speed of the slower cyclist be u km/h. Then the other speed is $(u + 1)$ km/h. Let the slower cyclist take t hours. Then the other cyclist takes $(t - 0.5)$ hours

Since speed $= \dfrac{\text{distance}}{\text{time}}$

then　　　　　　　　　$u = \dfrac{45}{t}$　and　$(u+1) = \dfrac{45}{(t-0.5)}$

Hence　　　　　　　$ut = 45$　　and　$(u+1)(t-0.5) = 45$

eliminate t　　　　　　　　$(u+1)\left(\dfrac{45}{u} - 0.5\right) = 45$

multiply by u　　　　　　　$(u+1)(45 - 0.5u) = 45u$

So　　　　　　　　　　$45u + 45 - 0.5u^2 - 0.5u = 45u$

$$0.5u^2 + 0.5u - 45 = 0$$

multiply by 2　　　　　　　　　$u^2 + u - 90 = 0$

This factorises to give $u = 9$ or $u = -10$ and since u is a speed and cannot be negative, the two speeds are 9 km/h and 10 km/h.

Example 9.24 Two manufacturers make circular saw blades of different qualities and the higher quality blade sells for £4 more than the other. As a hardware shopkeeper, if you bought the cheaper type you could get five more for £120. What are the prices of the two blades?

Solution Let £x be the price of the cheaper blade. Then the dearer blade would cost £$(x + 4)$.

$$\text{number of cheaper blades that can be bought for } £120 \ = \frac{120}{x}$$

$$\text{number of dearer blades that can be bought for } £120 = \frac{120}{(x+4)}$$

But number of cheapter blades = number of dearer blades + 5

Hence
$$\frac{120}{x} = \frac{120}{(x+4)} + 5$$

i.e. $\quad 120(x+4) = 120x + 5x(x+4)$

$$24(x+4) = 24x + x(x+4)$$

$$24x + 96 = 24x + x^2 + 4x$$

so $\quad 0 = x^2 + 4x - 96$

$$0 = (x-8)(x+12)$$

$x = 8$ is the only realistic solution and the two blades cost £8 and £12

Exercise 9.4

1. A rectangular field is 6 metres longer than it is wide. Its area is 160 square metres. Find its length and width.

2. Two cyclists travel from Exeter to Plymouth, a distance of 45 miles. One of the cyclists travels an average of 1 mph faster than the other and takes 30 minutes less to complete the journey. What were their average speeds?

3. A car travels 288 miles on country roads. If it could increase its average speed by 4 mph, it would take 1 hour less. Find its average speed.

4. The product of two consecutive odd integers is 1443. What are the integers?

5. Find three consecutive integers whose product is 40 times their sum. (Hint: let the integers be $(x - 1)$, x and $(x + 1)$.)

6. A small garden has a square lawn with sides of length x metres. There is a path 1 metre wide all the way around the lawn. If the area of the lawn is $\frac{3}{5}$ of the total area, find x.

7. A farmer has 70 metres of fencing to enclose three sides of a rectangular pen. The fourth side is a wall. The area of the pen is 600 square metres. Find the length of the sides.

8. The sum of the first n positive integers (i.e. $1 + 2 + 3 + 4 + ... + n$) is given by the formula

$$S = \frac{n(n+1)}{2}$$

Find n if   (a) $S = 120$ (b) $S = 231$.

9. If a polygon has n sides, it has D diagonals, where

$$D = \frac{n(n-3)}{2}$$

Find n if (a) $D = 20$ (b) $D = 35$ (c) $D = 40$.

What can you deduce from your answer in (c) above?

10.   When a stone is thrown vertically upwards, its height h after t seconds is given by the formula

$$h = 10t - 5t^2$$

How long does it take to reach heights of

(a)   4 metres   (b)   5 metres   (c)   6 metres?

Try to give physical explanations for each of your answers.

11.   The final bill for a recent party was £400. There were two guests of honour who did not pay anything so the cost was spread equally amongst the rest. If they each paid £10 more than they would otherwise have done, how many people were at the party altogether?

10

Simultaneous Equations

INTRODUCTION

The simple equations met so far have involved one unknown quantity. Even when we transposed formulae involving several variables, we always considered only one of them as unknown. Suppose we wish to solve a problem where there are two unknowns and we want to find both of them.

Example 10.1   We are told that the average of two numbers is 36. Is this enough information to find the two numbers?

Solution   If the two numbers are x and y we are given that

$$\frac{x+y}{2} = 36$$

i.e.
$$x + y = 72$$

Hence $y = 72 - x$ and to find y, we will need to know x (and vice versa); so it is not possible to find x and y from this single piece of information.

Suppose we also know that the difference between the two numbers is 16; then

$$x - y = 16$$

We now have two equations that must be solved simultaneously. Writing them down again:

$$x + y = 72 \qquad\qquad (1)$$
$$x - y = 16 \qquad\qquad (2)$$

If we add them together, y is eliminated therefore

$$2x = 88 \qquad \text{and so} \qquad x = 44$$

If we subtract, x is eliminated:

i.e. $\qquad\qquad\qquad\qquad x + y - (x - y) = 56$

so $\qquad\qquad\qquad\qquad x + y - x + y = 56$

Therefore $\qquad\qquad\qquad 2y = 56 \qquad \text{and} \quad y = 28$

Note that there is only *one* solution for x and *one* solution for y. Equations (1) and (2) are called *linear equations* and what we have done is find the single point where two straight lines intersect.

We need two equations if we need to find two unknown quantities. Pairs of equations of this type are called **simultaneous equations** since they must both be true at the same time. Simultaneous equations occur frequently in mathematics and we need to know how to deal with them.

SOLUTION BY ELIMINATION

The above method is known as the **method of elimination** and is possibly the quickest method in simple cases. Here is another example:

Example 10.2 Solve the following simultaneous equations:

$$a + 2b = 7 \qquad\qquad (1)$$
$$3a - 2b = 4 \qquad\qquad (2)$$

Solution '$2b$' occurs in both equations with *opposite* signs. We can therefore *add* the equations to give

$$4a = 11 \qquad \text{so} \qquad a = \frac{11}{4}$$

We can now substitute this value into (1):

$$\frac{11}{4} + 2b = 7$$

This is just a simple equation with one unknown:

$$11 + 8b = 28 \qquad \text{multiplying by 4}$$

$$8b = 17$$

$$b = \frac{17}{8}$$

It is a good idea to check these values in (2):

$$3a - 2b = \frac{33}{4} - \frac{17}{4} = 4 \qquad \text{correct}$$

We can extend the idea slightly to solve equations where the coefficients are different.

Example 10.3   Solve

$$3a + 2b = 12 \qquad (1)$$

$$a + 3b = 11 \qquad (2)$$

Solution   This time we cannot eliminate a or b directly but we can make the coefficients of b the same in both equations if we multiply equation (2) by 3:

$$3a + 9b = 33 \qquad (3)$$

$$3a + 2b = 12 \qquad (1)$$

Subtracting the equations gives $7b = 21$ and $b = 3$

and substituting this in (2) gives $a = 2$

This should now be checked by substituting a and b back into either (1) or (3).

Exercise 10.1

Solve the following pairs of equations. In each case decide which unknown can be most easily eliminated and whether to add or subtract.

1. $a + b = 3$
    $a - b = 1$

2. $a + b = 7$
    $a - b = 4.5$

3. $2a + b = 3$
    $a + b = 2$

4. $3a - 4b = 6$
    $5a + 4b = 2$

5. $2a - 3b = 2$
    $2a + 7b = 16$

6. $3a + 2b = 5$
    $a + 2b = 7$

7. $3x + y = 4$
    $x + 2y = 1$

8. $3y - z = 11$
    $2y - 3z = 5$

9. $2p + 3q = 4$
    $4p - 7q = -18$

SOLUTION BY SUBSTITUTION

In all but the simplest cases demonstrated above, it is usually better to use this method. If one of the equations is not linear, it is the *only* method.

Example 10.4   Solve

$$7x - 6y = 20 \qquad (1)$$
$$2x - 5y = 9 \qquad (2)$$

Solution   The idea is to transpose the equations for either x or y. We can transpose one equation and substitute in the other or transpose both of them. It is just a matter of preference. In this case it is easiest to rearrange both equations to make x the subject:

rearranging (1): $7x = 20 + 6y$ giving $x = \dfrac{20 + 6y}{7}$

similarly for (2): $2x = 9 + 5y$ giving $x = \dfrac{9 + 5y}{2}$

Hence

$$\frac{20 + 6y}{7} = \frac{9 + 5y}{2}$$

$$2(20 + 6y) = 7(9 + 5y)$$

$$40 + 12y = 63 + 35y$$

$$23y = -23$$

$$y = -1$$

Now substitute this into one of the *transposed* equations:

i.e. $$x = \frac{(20 + 6y)}{7} = \frac{(20 + 6(-1))}{7} = \frac{14}{7} = 2$$

is answer should be checked by substituting $y = -1$ into the other equation.

Example 10.5   Solve the equations

$$\frac{3x}{2} + 2y = -\frac{11}{2} \qquad\qquad (1)$$

$$5x + 6y = -7 \qquad\qquad (2)$$

Solution   Multiplying (1) by 2, to clear the denominators gives $3x + 4y = -11$ which can be transposed to give

$$y = \frac{-11 - 3x}{4} \qquad\qquad (3)$$

Transposing (2) gives $y = \dfrac{-7 - 5x}{6}$; so, equating these two expressions:

$$\frac{-11 - 3x}{4} = \frac{-7 - 5x}{6}$$

$$\frac{11 + 3x}{4} = \frac{7 + 5x}{6} \qquad \text{multiplying both sides by } -1$$

$$\frac{11 + 3x}{2} = \frac{7 + 5x}{3} \qquad \text{cancelling factors in denominators}$$

$$33 + 9x = 14 + 10x$$
$$x = 19$$

and $y = \dfrac{-11 - 3(19)}{4} = -\dfrac{68}{4} = -17$ from equation (3)

Exercise 10.2

Solve the following pairs of simultaneous equations:

1. $\begin{aligned} 4x + 3y &= 1 \\ 5x + 4y &= 2 \end{aligned}$

2. $\begin{aligned} 2x + 5y &= 8 \\ 3x + 4y &= 5 \end{aligned}$

3. $\begin{aligned} 6s - 5t &= 21 \\ 5s - 4t &= \frac{35}{2} \end{aligned}$

4. $\begin{aligned} 6a - 5b &= 24 \\ 9a - 4b &= 22 \end{aligned}$

5. $\begin{aligned} \frac{5}{x} + \frac{3}{y} &= 9 \\ \frac{2}{x} - \frac{5}{y} &= 16 \end{aligned}$

6. $\begin{aligned} \frac{3x}{2} + 2y &= -\frac{11}{2} \\ 5x + 6y &= -7 \end{aligned}$

SOLUTION IF ONE OF THE EQUATIONS IS NOT LINEAR

This time the *only* method is substitution. We find that there is usually more than one pair of solutions. The problem is equivalent to finding where a straight line crosses a curve.

Example 10.6   Solve the equations

$$x + y = 2 \qquad\qquad (1)$$
$$x^2 + y^2 = 10 \qquad\qquad (2)$$

Solution   Be careful! The LHS of (2) is not just the LHS of (1) squared!

$$y = 2 - x \qquad\qquad (3)$$

Substitute *directly* in (2):

$$x^2 + (2 - x)^2 = 10$$
$$x^2 + 4 - 4x + x^2 = 10$$
$$2x^2 - 4x - 6 = 0$$

i.e.
$$x^2 - 2x - 3 = 0$$
$$(x + 1)(x - 3) = 0$$

so
$$x = -1 \text{ or } x = 3$$

and from (3)
$$y = 3 \text{ or } y = -1$$

Hence there are two pairs of solutions: $x = -1, y = 3$ and $x = 3, y = -1$

Example 10.7   Solve

$$x - 3y = 7 \qquad\qquad (1)$$
$$xy = 9 \qquad\qquad (2)$$

Solution

from (1):
$$x = 7 + 3y \qquad\qquad (3)$$
$$(7 + 3y)y = 9$$

Substitute in (2): $\qquad\qquad\qquad 7y + 3y^2 = 9$

$$3y^2 + 7y - 9 = 0$$

Solving by using the quadratic formula gives $y = -3.25$ and $y = 0.92$ to 2 d.p.

Substituting these values into (3) gives $x = -2.76$ and $x = 9.76$ to 2 d.p.

Example 10.8   Solve

$$x + y = 4 \qquad\qquad\qquad (1)$$
$$x^2 - y^2 = 40 \qquad\qquad\qquad (2)$$

Solution   This time we can use our knowledge of factors to help us.

Since the LHS of (2) is a difference of two squares, it can be factorised:

$$(x + y)(x - y) = 40 \qquad\qquad\qquad (3)$$

Substitute from (1):

$$4(x - y) = 40$$

i.e. $x - y = 10 \qquad\qquad\qquad (4)$

(1) and (4) can now be solved easily. Adding gives $x = 7$ and subtracting gives $y = -3$.

Exercise 10.3

Solve the following pairs of simultaneous equations:

1. $\begin{aligned} x^2 + y^2 &= 13 \\ x + y &= 5 \end{aligned}$ 2. $\begin{aligned} x^2 + y^2 &= 25 \\ x + 3y &= 13 \end{aligned}$ 3. $\begin{aligned} 2(x - y) &= 11 \\ xy &= 20 \end{aligned}$

4. $\begin{aligned} 2y + 3x &= 8 \\ 3y^2 + 2x^2 &= 11 \end{aligned}$ 5. $\begin{aligned} x + y &= 1 \\ x^2 - xy &= 153 \end{aligned}$ 6. $\begin{aligned} 2x - y &= 5 \\ 3x^2 - y^2 + 2x + y &= 50 \end{aligned}$

PROBLEMS LEADING TO SIMULTANEOUS EQUATIONS

We began this chapter with a simple problem that generated two simultaneous equations. It is often necessary to be able to convert verbal problems into mathematical shorthand. Often when we do so, we see that the problem cannot be solved without the use of simultaneous equations. In these problems it is *essential* that you begin by introducing variables of your own, as the following examples show. At the end of the problem you should give your solutions in a form that refers back to the original question.

Example 10.9   Howard was 30 years old when his daughter Kate was born and the sum of their ages is now 58. What are their current ages?

Solution   Let the present ages of man and daughter be m and d.

We know that the difference in their ages is 30; hence

$$m - d = 30 \qquad\qquad (1)$$

But the sum of their ages is 58, so

$$m + d = 58 \qquad\qquad (2)$$

Adding (1) and (2) gives $m = 44$,   subtracting gives $d = 14$

Hence the man's age is 44 and his daughter is 14.

Example 10.10   A family decides to go on holiday to France using the new Channel Tunnel shuttle service. The basic cost of a family holiday is made up of the shuttle fare plus a certain amount per night for the hotel. If they stay for seven nights, the cost is £580 and if they stay for ten nights it costs £700. Find the price of the shuttle fare, the cost of one night's hotel bill and the price of a fourteen day holiday.

Solution   Let the shuttle fare be s and the cost per night be c.

Then for a seven night stay,

$$s + 7c = 580 \qquad\qquad (1)$$

and for a ten night stay,

$$s + 10c = 700 \qquad\qquad (2)$$

Hence, subtracting (1) from (2) gives $3c = 120$ and so $c = 40$. Substituting this into (1) leads to $s = 300$.

The cost of the shuttle is therefore £300 and the cost per night in the hotel is £40.

A fourteen night stay would cost $£(300 + 14 \times 40) = £860$

Example 10.11   Last year a small local firm employed five full-time people and three part-timers. The total weekly wage bill was £1240. This year the number of full-time workers has been reduced to three and three extra part-timers have been taken on. There have been no pay rises. This makes the new weekly wage bill £1080. Find the weekly full-time wage and the weekly part-time wage.

Solution   Let the full-time and part-time wages be f and p pounds per week.

Then last year's weekly wage bill was given by

$$5f + 3p = 1240 \tag{1}$$

and this year's weekly bill satisfies the equation

$$3f + 6p = 1080 \tag{2}$$

Multiply (1) by 2 to give

$$10f + 6p = 2480 \tag{3}$$

$$3f + 6p = 1080 \tag{2}$$

Subtracting (2) from (3) gives $7f = 1400$ and so $f = 200$. Substituting back into (1) or (2) or (3) leads to $p = 80$.

Hence the full-timers earn £200 per week and the part-timers earn £80 per week.

Example 10.12   A keen mountain biker averages 25 km/h on roads and 15 km/h across country. A recent 75 km journey took 4 hours. How much of this time was spent cycling on roads?

Solution   Let the time spent on roads be t_1 and the time across country be t_2; then

$$t_1 + t_2 = 4 \tag{1}$$

Since average speed = distance/time, then distance = speed × time.

Hence distance travelled on the road = $25t_1$ and distance across country = $15t_2$, so the total distance satisfies

$$25t_1 + 15t_2 = 75 \tag{2}$$

Dividing (2) by 5

$$5t_1 + 3t_2 = 15 \tag{3}$$

It is t_1 that we want; so substitute for t_2 from (1) into (3):

$$5t_1 + 3(4 - t_1) = 15$$
$$5t_1 + 12 - 3t_1 = 15$$
$$2t_1 = 3$$
$$t_1 = 1.5$$

Therefore the cyclist spends 1.5 hours cycling on roads.

Exercise 10.4

Use the information given in the following problems to write down pairs of simultaneous equations. Hence solve the equations.

1.   Find two numbers whose sum is 35 and whose difference is 5.

2.   Find two numbers whose sum is 20 and whose product is 91.

3.   A bill of £125 was paid for with a mixture of £10 and £5 notes. If a total of eighteen notes was used altogether, how many of each note were used?

4.   In response to the popularity of the new Channel Tunnel Family Saver holidays (see Example 10.10) a cross channel ferry company decides to offer a similar scheme in competition. They would also like to encourage longer breaks so with their scheme, ferry fare plus seven nights' stay costs £600 but ferry plus a ten night stay only costs £81 more. Find the family fare, the cost per night at the hotel and the cost of a fourteen day break.

5.   Two motorists start at noon from York and London, a distance of 220 miles. They both agree to average 60 mph and meet half way but the driver leaving London is caught in heavy traffic and can only average 50 mph. They are in contact by car 'phone so the driver travelling from York keeps on travelling. How far from London do they meet and at what time of day?

6.   A party of seven Morris Men went to a pub which sells two different Real Ales at different prices. Feeling generous, Lionel bought the first round: 4 pints of 'Xtra' and 3 pints of 'Yukkybrew'. This came to £11.50.

      When Nigel was buying the second round, the 'Xtra' ran out after 2 pints had been pulled, so he had to buy 5 pints of 'Yukkybrew'. His round came to £11.70. What was the price per pint of each beer?

11

Indices and Logarithms

In Chapter 4, we introduced the idea of indices and deduced three laws which these indices must obey.

Suppose a is the base and m and n are indices. The laws of indices are:

Law 1. Multiplication $\qquad a^m.a^n = a^{m+n} \qquad$ add indices

Law 2. Division $\qquad \dfrac{a^m}{a^n} = a^{m-n} \qquad$ subtract indices

Law 3. Powers $\qquad (a^m)^n = a^{mn} \qquad$ multiply indices

In addition, we have two rules for finding powers of products and quotients. These are not really new rules but simply applications of the basic laws of multiplication. If a and b are real numbers and n is an index, then

$$(ab)^n = a^n b^n \quad and \quad \left(\frac{a}{b}\right)^n = \frac{a^n}{b^n}$$

The indices we have met so far arise naturally from a desire to use shorthand notation and have all been positive integers. We now see how our basic ideas can be extended to include other real numbers. If you do not feel confident with the early material, now is a good time to go back and revise the work.

FURTHER INDICES

The positive integer indices we have worked with so far are just a special case of something much deeper. In the following, we make the assumption that the laws of indices can be extended to other numbers. It is particularly law 2 that is the key to much of the following.

Zero Index

This was introduced in Chapter 4 but is worth going through again: 'What happens if we divide a number in power form by itself?'

For example: divide a^4 by a^4

$$\frac{a^4}{a^4} = 1 \quad \text{(since anything divided by itself is one)}$$

But using law 2 of indices gives $\quad \frac{a^4}{a^4} = a^{4-4} = a^0 \quad$ and so we can extend our original ideas about indices to include zero index as long as we define:

$$\boxed{a^0 = 1}$$ 'anything to the power zero is one'

Negative Indices

We can extend the above idea even further:

Suppose we divide a^3 by a^2: then $\dfrac{a^3}{a^2} = \dfrac{a.a.a}{a.a} = a$ as we have already seen.

However, if we try the division the other way around:

$$\frac{a^2}{a^3} = \frac{a.a}{a.a.a} = \frac{1}{a}$$

But, from law 2, assuming it still works

$$\frac{a^2}{a^3} = a^{2-3} = a^{-1}$$

Hence we have found a meaning for a negative index:

$$a^{-1} = \frac{1}{a}, \text{ the reciprocal of } a$$

Hence, for example,

$$2^{-1} = \frac{1}{2} \qquad 3^{-1} = \frac{1}{3} \qquad 10^{-1} = \frac{1}{10} \qquad \text{and so on}$$

Note also that, since an index of -1 gives a reciprocal, then

$$\left(\frac{1}{2}\right)^{-1} = 2 \qquad \left(\frac{5}{6}\right)^{-1} = \frac{6}{5} \qquad\qquad \text{and in general}$$

$$\left(\frac{a}{b}\right)^{-1} = \frac{b}{a}$$

Of course, this idea can be taken further still:

By cancellation, $\qquad \dfrac{a^3}{a^5} = \dfrac{a.a.a}{a.a.a.a.a} = \dfrac{1}{a.a} = \dfrac{1}{a^2}$

and by using law 2, $\qquad \dfrac{a^3}{a^5} = a^{(3-5)} = a^{-2}$

Hence $\qquad\qquad\qquad a^{-2} = \dfrac{1}{a^2}$

Therefore, $\quad 3^{-2} = \dfrac{1}{3^2} = \dfrac{1}{9} \quad 7^{-2} = \dfrac{1}{7^2} = \dfrac{1}{49} \qquad$ and so on

We can use similar reasoning to give a meaning to any negative integer index, so

$$a^{-3} = \frac{1}{a^3} \qquad a^{-5} = \frac{1}{a^5} \qquad a^{-17} = \frac{1}{a^{17}}$$

and in general
$$a^{-n} = \frac{1}{a^n}$$
where n is a positive integer

Hence *any* negative integer index gives a reciprocal quantity. Note therefore that

$$\frac{1}{a^{-n}} = a^n$$

Exercise 11.1

1. Write down the values of the following:
    (a) 6^0 (b) $\left(\frac{1}{2}\right)^0$ (c) $(-3)^0$ (d) 6247.5^0
    (e) 6^{-1} (f) $\left(\frac{1}{3}\right)^{-1}$ (g) $\left(\frac{4}{5}\right)^{-1}$ (h) $(-1)^{-1}$

2. Write down the values of the following:
    (a) 2^{-2} (b) 3^{-2} (c) $(-1)^{-2}$ (d) $\left(\frac{1}{2}\right)^{-2}$
    (e) 2^{-3} (f) 3^{-3} (g) $(-1)^{-3}$ (h) $\left(\frac{1}{2}\right)^{-3}$

3. Simplify:
    (a) $\dfrac{1}{2^{-1}}$ (b) $\dfrac{1}{10^{-1}}$ (c) $\dfrac{1}{5^{-2}}$ (d) $\dfrac{1}{7^{-3}}$
    (e) $\left(\frac{1}{4}\right)^{-2}$ (f) $\left(\frac{1}{6}\right)^{-3}$ (g) $\left(\frac{2}{3}\right)^{-4}$ (h) $\dfrac{1}{\left(\frac{3}{4}\right)^{-2}}$

Fractional (Rational) Indices

Now that we know it is possible to define indices that are not positive integers, let us try to find a meaning for $a^{\frac{1}{2}}$. The easiest way to see what it means is to multiply it by itself. Therefore,

$$a^{\frac{1}{2}} \cdot a^{\frac{1}{2}} = a^{(\frac{1}{2}+\frac{1}{2})} \quad \text{(law 1)}$$
$$= a^1$$
$$= a$$

But we know that the number which gives a when multiplied by itself is the square root of a. In other words

$$\sqrt{a}.\sqrt{a} = a$$

So by comparison, we see that if $a^{\frac{1}{2}}$ has a meaning, it must mean the square root of a,

i.e.
$$\boxed{a^{\frac{1}{2}} = \sqrt{a}}$$

By similar reasoning, $a^{\frac{1}{3}} = \sqrt[3]{a}$, $a^{\frac{1}{4}} = \sqrt[4]{a}$ and in general

$$\boxed{a^{\frac{1}{n}} = \sqrt[n]{a}}$$

Hence, for example,

$$3^{\frac{1}{4}} = \sqrt[4]{3} \qquad 7^{\frac{1}{5}} = \sqrt[5]{7} \qquad 125^{\frac{1}{3}} = \sqrt[3]{125} = 5 \text{ and } 81^{\frac{1}{4}} = \sqrt[4]{81} = 3$$

Other Positive Fractional Indices

We have found that an index of the form $\frac{1}{n}$ means an nth root We now investigate whether indices can be extended to include other fractions. Can we find a meaning for $a^{\frac{2}{3}}$? Law 3 can help us here!

$$a^{\frac{2}{3}} = a^{2 \times \frac{1}{3}} = (a^2)^{\frac{1}{3}} \qquad \text{(using law 3 'backwards')}$$

Therefore
$$\boxed{a^{\frac{2}{3}} = \sqrt[3]{a^2}}$$

(Note that the root denominator of the index gives the root.)

We could equally well have done the following:

$$a^{\frac{2}{3}} = a^{\frac{1}{3} \times 2} = (a^{\frac{1}{3}})^2$$

giving:
$$\boxed{a^{\frac{2}{3}} = (\sqrt[3]{a})^2}$$

Both expressions lead to the same answer but in numerical work, it is best to take the root first. For example:

$$125^{2/3} = (\sqrt[3]{125})^2 = 5^2 = 25 \qquad \text{(Easy, no need for calculators!)}$$

but $\quad 125^{2/3} = \sqrt[3]{(125)^2} = \sqrt[3]{15625} = 25 \quad$ (Calculator definitely essential)

As usual, by similar reasoning, we can show that $a^{3/4} =$ either $(\sqrt[4]{a})^3$ or $\sqrt[4]{a^3}$ and in general:

$$a^{m/n} = \text{ either } (\sqrt[n]{a})^m \text{ or } \sqrt[n]{a^m}$$

Exercise 11.2

1. Write as powers of x:

   (a) $\sqrt[5]{x}$ (b) $\sqrt[3]{x^2}$ (c) $\sqrt[7]{x^4}$ (d) $\sqrt{x^6}$

2. Write the following in index form:

   (a) $\sqrt{3}$ (b) $\sqrt{100}$ (c) $\sqrt[3]{2}$ (d) $\sqrt[5]{16}$

   (e) $\sqrt[6]{4}$ (f) $\sqrt[7]{49}$ (g) $\sqrt[9]{81}$ (h) $\sqrt[11]{11}$

   (i) $\sqrt{17^3}$ (j) $\sqrt[3]{17^2}$ (k) $\sqrt{5^6}$ (l) $\sqrt[4]{13^8}$

3. Write the following in a form involving roots:

   (a) $4^{1/2}$ (b) $5^{1/2}$ (c) $7^{1/3}$ (d) $11^{1/4}$

   (e) $64^{1/2}$ (f) $64^{1/3}$ (g) $\left(\frac{1}{6}\right)^{1/5}$ (h) $\left(\frac{3}{4}\right)^{1/2}$

4. Evaluate:

   (a) $8^{1/3}$ (b) $125^{1/3}$ (c) $64^{2/3}$ (d) $64^{1/2}$

   (e) $32^{2/5}$ (f) $16^{3/4}$ (g) $\left(\frac{1}{27}\right)^{2/3}$ (h) $\left(\frac{1}{16}\right)^{5/4}$

5.   Evaluate:

(a) $\left(\frac{16}{49}\right)^{1/2}$ (b) $\left(\frac{25}{36}\right)^{1/2}$ (c) $\left(\frac{8}{27}\right)^{2/3}$ (d) $\left(\frac{32}{243}\right)^{2/5}$

(e) $\dfrac{9^{1/2}.16^{1/2}}{25^{1/2}}$ (f) $\dfrac{9^{1/2}.2^{1/2}}{8^{1/2}}$ (g) $\dfrac{4^{1/3}.4^{0}.16^{1/3}}{64^{1/3}}$ (h) $\dfrac{5^{1/2}.25}{125^{1/2}}$

Negative Fractional Indices

We can find a meaning for $a^{-2/3}$ by using what we have already learned. A negative index means a reciprocal; therefore

$$a^{-2/3} = \frac{1}{a^{2/3}}$$

and we know that $a^{2/3} = $ either $(\sqrt[3]{a})^2$ or $\sqrt[3]{a^2}$

Hence, $a^{-2/3} = $ either $\dfrac{1}{(\sqrt[3]{a})^2}$ or $\dfrac{1}{\sqrt[3]{a^2}}$

and, in general: $\boxed{a^{-m/n} = \text{ either } \dfrac{1}{(\sqrt[n]{a})^m} \text{ or } \dfrac{1}{\sqrt[n]{a^m}}}$

To sum up our findings:

$$a^0 = 1 \qquad \text{(anything to the power zero is one)}$$

$$a^{-n} = \frac{1}{a^n} \qquad \text{(negative indices are reciprocals)}$$

$$a^{1/n} = \sqrt[n]{a} \qquad \text{(fractional indices are roots)}$$

$$a^{m/n} = \text{ either } (\sqrt[n]{a})^m \text{ or } \sqrt[n]{a^m}$$

$$a^{-m/n} = \text{ either } \frac{1}{(\sqrt[n]{a})^m} \text{ or } \frac{1}{\sqrt[n]{a^m}}$$

These are not new laws of indices; they are meanings given to powers that are not positive integers and have been deduced from the three laws introduced earlier. We have found a meaning for zero index, negative indices and fractional indices. We will not try to find a meaning here for irrational indices, as they rarely crop up, but you may assume that they still obey the three laws.

Indices are not difficult to understand but you need to learn the meanings of all the different types of indices in order to deal with them successfully.

Exercise 11.3

Find the values of:

1. $9^{-\frac{1}{2}}$ 2. $27^{-\frac{1}{3}}$ 3. $81^{-\frac{1}{4}}$

4. $125^{-\frac{1}{3}}$ 5. $256^{-\frac{1}{4}}$ 6. $27^{-\frac{4}{3}}$

7. $\left(\frac{1}{64}\right)^{-\frac{2}{3}}$ 8. $\left(\frac{1}{64}\right)^{-\frac{3}{2}}$ 9. $\left(\frac{9}{100}\right)^{-\frac{3}{2}}$

LOGARITHMS

Logarithm is merely another word for a power or index. If y is written as 2^x, then 2 is known as the base and x is called 'the logarithm of y to base 2'.

We write: if $y = 2^x$ then $x = \log_2 y$

We have not introduced anything new here, merely a change of emphasis. The second equation above is just a transposed form of the first, with the subject changed from y to x.

As numerical examples, we have

$$9^2 = 81 \qquad \text{and so} \qquad 2 = \log_9 81$$

Similarly, $\qquad 5^{-2} = \frac{1}{25} \qquad \text{and so} \qquad -2 = \log_5\left(\frac{1}{25}\right)$

Example 11.1   Change the following into a form involving logarithms:

(a) $5^3 = 125$ (b) $49^{\frac{1}{2}} = 7$ (c) $\left(\dfrac{1}{3}\right)^{-3} = 27$

Solution

(a) $5^3 = 125$ $\Rightarrow$ $\log_5 125 = 3$

(b) $49^{\frac{1}{2}} = 7$ $\Rightarrow$ $\log_{49} 7 = \frac{1}{2}$

(c) $\left(\dfrac{1}{3}\right)^{-3} = 27$ $\Rightarrow$ $\log_{\frac{1}{3}} 27 = -3$

Example 11.2   Change the following into a form involving indices (also known as exponential form):

(a) $\log_8 64 = 2$ (b) $\log_{10} 0.1 = -1$ (c) $\log_2 0.25 = -2$

Solution   (a) $\log_8 64 = 2$ $\Rightarrow$ $64 = 8^2$

(b) $\log_{10} 0.1 = -1$ $\Rightarrow$ $0.1 = 10^{-1}$

(c) $\log_2 0.25 = -2$ $\Rightarrow$ $0.25 = 2^{-2}$

Example 11.3   Rewrite the two equations $3^2 = 9$ and $9^{\frac{1}{2}} = 3$ in terms of logarithms and deduce a relationship between the two logs.

Solution

$$3^2 = 9 \;\Rightarrow\; \log_3 9 = 2 \;\text{ and }\; 9^{\frac{1}{2}} = 3 \;\Rightarrow\; \log_9 3 = \tfrac{1}{2}$$

giving $$\log_9 3 = \dfrac{1}{\log_3 9}$$

The above example is an illustration of a general rule. If a and b are numbers such that $a = b^n$ then $b = a^{\frac{1}{n}}$. Hence the log of a to base b is n and the log of b to base a is $\frac{1}{n}$ and so it is always true that:

$$\log_a b = \frac{1}{\log_b a}$$

Exercise 11.4

1. Change the following to logarithmic form:

    (a) $3^4 = 81$ (b) $2^3 = 8$ (c) $7^4 = 2401$

    (d) $10^{0.301} = 2$ (e) $3^{-3} = \dfrac{1}{27}$ (f) $\left(\dfrac{1}{2}\right)^{-2} = 4$

    (g) $4^{5/2} = 32$ (h) $216^{-2/3} = \frac{1}{36}$ (i) $7^0 = 1$

    (j) $43^0 = 1$ (k) $17^1 = 17$ (l) $0.5^1 = 0.5$

2. Change to exponential form:

    (a) $\log_4 16 = 2$ (b) $\log_{16} 4 = \frac{1}{2}$ (c) $\log_2 32 = 5$

    (d) $\log_{1/2} 8 = -3$ (e) $\log_4 \left(\dfrac{1}{4}\right) = -1$ (f) $\log_{10} 1 = 0$

    (g) $\log_3 1 = 0$ (h) $\log_{10} 100 = 2$ (i) $\log_{100} 10 = \frac{1}{2}$

    (j) $\log_{27} 9 = \frac{2}{3}$ (k) $\log_{10} 10 = 1$ (l) $\log_{17} 17 = 1$

3. Find $\log_2 64$ and hence write down $\log_{64} 2$.

4. Find $\log_{11} 121$ and hence write down $\log_{121} 11$.

5. Find $\log_{49} 343$ and hence write down $\log_{343} 49$.

6. Write down the logs to base 3 of:

    (a) 27   (b) $\sqrt{3}$ (c) $\dfrac{1}{3}$ (d) $\dfrac{1}{3\sqrt{3}}$

7.   Evaluate without using a calculator:

     (a) $\log_{10} 100,000$ (b) $\log_2 4$ (c) $\log_{1/2} 4$ (d) $\log_{81} 3$

     (e) $\log_9 9$ (f) $\log_2 2^3$ (g) $10^{\log_{10} 10}$ (h) $5^{\log_5 25}$

Common Logarithms

These are logarithms to base 10. Up to about 20 years ago, when pocket calculators became common, these were very important in calculations although they are used much less nowadays. However, there are still important formulae which require their use and you should find them available on your calculator.

When using common logs, we usually write them as simply 'log', leaving out the base. Your calculator should therefore have a button labelled 'log'. Similarly, if you see log in a modern textbook, it will normally be referring to logs to base 10.

Properties of Logarithms

The previous exercise highlighted two important facts which we will state here generally. If a is any base of logarithms, then

         Since $a^1 = a$ then

i.e.   the log of any number to base itself equals one

Also   since $a^0 = 1$ then

i.e.   the log of 1 to any base is zero

Two other important facts that can be helpful in manipulating logarithms are:

Let a be any base; then $y = a^x$ and $x = \log_a y$ mean the same thing. If we eliminate y:

$$\log_a a^x = x$$

Similarly, since $x = a^y$ and $y = \log_a x$ mean the same, we can again eliminate y, giving:

$$a^{\log_a x} = x$$

These results are merely statements about performing a mathematical process and then doing its inverse. If you choose a number, put it in exponential form to base a and then find the logarithm to base a, you are back where you started. Similarly, if you find the logarithm of a number to base a and then put this into exponential form to base a, you will end up with your original number.

This is a similar idea to squaring a square root or finding the square root of a perfect square. For example,

$$\left(\sqrt{7}\right)^2 = 7 \quad \text{and} \quad \sqrt{7^2} = 7$$

The Three Laws of Logarithms

Since logarithms are only indices seen from a different viewpoint and there are three basic laws of indices, then there must be three laws of logarithms. If a is any base and x and y are positive, the laws state

$$\log_a(xy) = \log_a x + \log_a y$$

$$\log_a\left(\frac{x}{y}\right) = \log_a x - \log_a y$$

$$\log_a x^y = y \log_a x$$

We will prove the first law and leave the others as an exercise. The proofs are very similar.

Let $x = a^b$ and $y = a^c$. This means that $b = \log_a x$ and $c = \log_a y$.

If we now multiply x and y together:

$$xy = a^b . a^c$$

$$= a^{(b+c)} \qquad \text{by the first law of indices}$$

This is the same as $b + c = \log_a xy$

Hence $$\log_a xy = b + c = \log_a x + \log_a y$$

These laws are used extensively. They can be used from left to right as written above, in order to 'expand' the log of a product or quotient into a sum or difference of logs. In such cases, laws 1 and 2 are used first and then law 3.

Example 11.4   'Expand' $\log\left(\dfrac{x^3 y}{z^2}\right)$.

Solution   Note that the first two steps below could be performed together.

$$\log\left(\frac{x^3 y}{z^2}\right) = \log x^3 y - \log z^2 \qquad \text{(law 2)}$$

$$= \log x^3 + \log y - \log z^2 \quad \text{(law 1)}$$

$$= 3\log x + \log y - 2\log z \quad \text{(law 3)}$$

Example 11.5   Expand   (a) $\log\dfrac{1}{x}$ (b) $\log_2\dfrac{1}{x}$

Solution

(a)
$$\log\frac{1}{x} = \log 1 - \log x \qquad \text{by law 2}$$

$$= -\log x \qquad \text{since } \log 1 = 0$$

Similarly   (b)

$$\log_2\frac{1}{x} = \log_2 1 - \log_2 x$$

$$= -\log_2 x \qquad \text{since } \log_2 1 = 0$$

The above example shows two instances of an important general result that is worth memorising, i.e.:

$$\log_a \frac{1}{x} = -\log_a x$$

In other applications, we need to 'reduce' a string of logarithmic terms to a single logarithm. In such cases, coefficients may be present and these must be 'tucked out of the way' by using law 3 before laws 1 and 2.

Example 11.6   Write as a single logarithm

$$3\log x - 2\log z + \tfrac{1}{2}\log y$$

Solution

$$3\log x - 2\log z + \tfrac{1}{2}\log y = \log x^3 - \log z^2 + \log y^{1/2} \quad \text{(law 3)}$$

$$= \log\left(\frac{x^3 y^{1/2}}{z^2}\right)$$

WARNING: It is a common mistake to try to simplify terms such as $\log(x+y)$ and $\log(x-y)$. There is no law of logarithms that deals with such cases. They cannot be further simplified and should be left as they are.

Exercise 11.5

1.   Prove the second and third laws of logarithms.

2.   By calculating each side separately, using your calculator, verify the logarithm laws for common logs, in the following cases:

   (a) $\log 2 + \log 3 = \log 6$ (b) $\log 5 + \log 11 = \log 55$

   (c) $\log 7 - \log 2 = \log 3.5$ (d) $\log 35 - \log 5 = \log 7$

   (e) $\log 3^2 = 2\log 3$ (f) $\log \sqrt{11} = \tfrac{1}{2}\log 1$

3.   Given that $\log 2 = 0.3010$ and $\log 3 = 0.4771$, to 4 decimal places, calculate the following logarithms *without using a calculator*.

     (a) $\log 6$ (b) $\log 1.5$ (c) $\log 8$ (d) $\log 9$

     (e) $\log 20$ (f) $\log \frac{2}{3}$ (g) $\log 5$ (h) $\log 15$

4.   'Expand' the following common logs:

     (a) $\log xy$ (b) $\log xyz$ (c) $\log x^2 y^3 z^4$ (d) $\log \sqrt{xy}$

     (e) $\log \dfrac{x}{y}$ (f) $\log \dfrac{xz}{y}$ (g) $\log \dfrac{x}{yz}$ (h) $\log \dfrac{1}{x}$

     (i) $\log \sqrt{\dfrac{x}{y}}$ (j) $\log 10x^2$ (k) $\log \dfrac{x^3}{10}$ (l) $\log \dfrac{10}{\sqrt{x}}$

5.   Express as a single logarithm:

     (a) $2\log 5 - 3\log 7$ (b) $\frac{1}{2}\log 4 + \frac{1}{3}\log 27$

     (c) $\frac{1}{2}\log 8 + 2\log \frac{1}{\sqrt{2}}$ (d) $2\log 7 - \log \frac{3}{7} + \frac{1}{2}\log \frac{9}{16}$

     (e) $\log_a x + 2\log_a y$ (f) $\frac{1}{2}\log_a x + 4\log_a y - 3\log_a z$

     (g) $\log_a x - \frac{1}{2}\log_a y$ (h) $2\log_a x + 3\log_a y^2 - 6\log_a z^{\frac{1}{2}}$

12

Functions and Graphs

FUNCTIONS

'Choose a number, double it, then add 5.'

The outcome of this statement will be different for each number that is chosen. For example, the following table shows various applications:

chosen number (input)	'double it'	'add 5' (output)
3	6	11
7	14	19
−3	−6	−1
0	0	5

Table 12.1

This activity demonstrates all the ingredients of a quantity in mathematics called *a function*. Essentially a function consists of three parts:

(i) a set of inputs called its *domain*;
(ii) a *rule*;
(iii) a set of outputs called its *range* or *codomain*.

Consider the example above and denote the input numbers as x and the output numbers as y. Then the rule between x and y is

$$y = 2x + 5$$

x and y are called *variables*. The *rule* provides a set of processing instructions for getting from the *independent variable x* to the *dependent variable y*.

For example:

1.   The area of a circle is a function of its radius, $A = \pi R^2$

        independent variable,   radius R

        dependent variable,   area A

2.   The volume of a cylinder is a function of its height and radius, $V = \pi R^2 h$

        independent variables,   radius R and height h

        dependent variable,   volume V

Exercise 12.1

In the following examples, state the independent and dependent variables:

1.   The Fahrenheit and Centigrade temperature scales are connected by the equation $C = 5(F - 32)/9$.

2.   The interest on a sum of money invested in a Building Society is a function of the time invested and also of the rate of interest.

3.   The circumference of a circle is a function of its radius, $C = 2\pi R$.

Diagrammatically the three elements that make up a function can be represented as in Figure 12.1. Mathematical modelling, using functions, is about choosing appropriate variables and defining the function rule that best describes a physical situation.

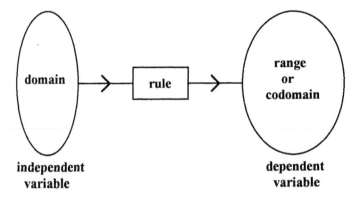

Figure 12.1 The three elements of a function

When talking about functions in general where no definite rule is given explicitly, the following notation is used;

$$y = f(x)$$

where x is the independent variable and y is the dependent variable. We say that 'y *is a function of x*'. One important requirement of the rule is that it must be unambiguous, i.e. given any value of the independent variable from the domain, the rule must specify a unique value of the dependent variable in the range.

For instance, the rule $y = 2x + 5$ is unambiguous, so that for each value of x there is a unique value of y. However, we must be careful when defining $y = x^{\frac{1}{2}}$ as a function. The possible values of $x^{\frac{1}{2}}$ are $+\sqrt{x}$ and $-\sqrt{x}$; for example, $4^{\frac{1}{2}}$ takes the value $+2$ and -2. So strictly speaking $y = x^{\frac{1}{2}}$ is not a function because the rule does not give a unique value in the range.

The general functional notation $y = f(x)$ provides the following precise mathematical notation:

The value of $f(x)$ when $x = 2$ is denoted by $f(2)$
The value of $f(x)$ when $x = 0$ is denoted by $f(0)$
The value of $f(x)$ when $x = a$ is denoted by $f(a)$

and so on.

Example 12.1 For the rule $f(x) = 3x^2 - 1$ find the values of $f(1)$ and $f(3)$. Find the range of the function $f(x)$ if the domain is $1 \le x \le 3$.

Solution To evaluate $f(1)$ we substitute for $x = 1$ in the rule:

$$f(1) = 3(1)^2 - 1 = 2$$

Similarly

$$f(3) = 3(3)^2 - 1 = 26$$

For values of x between 1 and 3 (the domain) the values of $f(x)$ lie between 2 and 26 (the range). The range of the function is $2 \le y \le 26$.

COMPOSITE FUNCTIONS

'Choose a number, add 3, then square.'

This statement consists of two rules: first we add the number 3 and then we square the result. For example, if we choose the number 5, then adding 3 gives 8, and squaring gives 64. Writing x for the input number, then the first part of the rule can be written as $f(x) = x + 3$. The second part of the rule can be written as $g(y) = y^2$. Putting the two rules together gives

$$g(f(x)) = (x + 3)^2$$

Such a combination is called a *composite function* or a *function of a function*. Figure 12.2 shows diagrammatically the process $g(f(x))$.

$$x \quad\quad \rightarrow \quad \boxed{\text{rule for } f} \quad \rightarrow \quad \boxed{\text{rule for } g} \quad \rightarrow \quad g(f(x))$$
input   output

Figure 12.2 Forming a composite function

Example 12.2   For the two rules $f(x) = +\sqrt{x}$ and $g(x) = x - 3$ find the values of $f(g(4))$ and $g(f(4))$.

Solution   For $f(g(4))$ we begin with the 'inner function' $g(x)$ and then substitute into $f(x)$:

$$g(4) = 4 - 3 = 1$$
$$f(g(4)) = f(1) = +\sqrt{1} = 1$$

Consider now $g(f(4))$; here we evaluate $f(4)$ first.

$$f(4) = +\sqrt{4} = 2$$
$$g(f(4)) = g(2) = 2 - 3 = -1$$

Note that in general $f(g(x))$ and $g(f(x))$ give different outputs for the same input.

Exercise 12.2

1. For the two rules $f(x) = x^3 - 3$ and $g(x) = -\sqrt{x}$,

    (a) find the values of $f(0), f(1), f(5), f(-4), f(16), g(0), g(1), g(-4), g(16)$;

    (b) find the range of each function for the domain $0 \le x \le 25$.

2. Find the range for each of the following functions which are defined over the given domain D.

    (a) $f(x) = 3x$ $D\{-1, 0, 1, 2\}$

    (b) $f(x) = 1 + x$ $D = -2 \le x \le 3$

    (c) $g(y) = \dfrac{1}{y}$ $D = \{1, 2, 3, 4, 5\}$

    (d) $h(u) = u^2 + 1$ $D = 1 \le u \le 4$

3. If $f(x) = 3x$ and $g(x) = x + 3$ find $g(f(2))$ and $f(g(2))$.

4. If $f(x) = x^2$ and $g(x) = -2x$ find $g(f(3))$ and $f(g(3))$.

5. If $f(x) = +\sqrt{x}$ and $g(x) = 4x + 3$ find the values of $g(f(x))$ and $f(g(x))$ for $x = 1, 2, 3, 4, 5$. Write down general rules for $f(g(x))$ and $g(f(x))$ and state whether there are any restrictions on the values of their domains.

6. If $f(x) = x^2$ and $g(x) = \dfrac{1}{x}$ show that $g(f(x)) = f(g(x))$.

7. If $f(x) = x^2$ and $g(x) = +\sqrt{x}$ show that $g(f(x)) = f(g(x)) = x$.

GRAPHS OF FUNCTIONS

In mathematics, a rule connecting two variables x and y can be described pictorially by drawing a graph. This shows the behaviour of the relation between x and y; for example

(i)   a graph can show easily how y changes as x changes,

(ii)   a graph can show where the largest or smallest values of a function occur.

To draw a graph of a relation the first step is to construct two perpendicular axes, labelling the horizontal axis with the independent variable (often denoted by x) and labelling the vertical axis with the dependent variable (often denoted by y). The point where the axes intersect is called the origin and is usually denoted by 'O'.

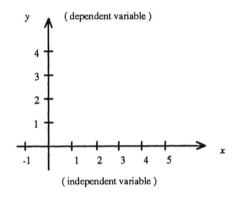

Figure 12.3   Clearly labelled axes for a graph

(The arrows show the direction of increasing values of x and y.)

1.   Linear Functions

The function $f(x) = mx + c$ where m and c are constants is called *a linear function of x*. It is not obvious why this is called a linear function, until we draw its graph, as the following examples show.

Example 12.3   Draw the graph of $y = 3x + 2$.

Solution   We make a table as follows taking several sensible values of x and working out the corresponding values of y.

x	0	1	2	3	−1	−2
$3x$	0	3	6	9	−3	−6
$y = 3x + 2$	2	5	8	11	−1	−4

Notice that the values of y increase as the values of x increase.

Plotting the values of y against the values of x, we obtain the graph of Figure 12.4.

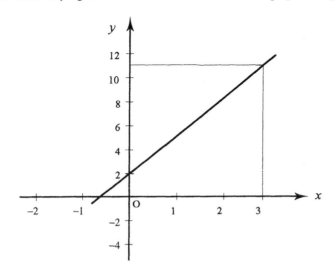

Figure 12.4

As we can see, the graph of the y values plotted against the x values gives a *sloping straight line*. This is why we call $f(x) = 3x + 2$ a linear function. The graph slopes 'upwards' in the sense that the y values increase as the x values increase.

Example 12.4   Draw or sketch the graph of $2y = 3 - x$. (Note: 'sketch' means that it is not necessary to plot points accurately.)

Solution   We rewrite this equation in the form $y = -\dfrac{1}{2}x + \dfrac{3}{2} = -0.5x + 1.5$

Choosing integer values of x from −1 to +4 gives the table

x	-1	0	1	2	3	4
$-0.5x$	0.5	0	-0.5	-1	-1.5	-2
$y = -0.5x + 1.5$	2	1.5	1	0.5	0	-0.5

This time, values of y *decrease* as x increases. Figure 12.5 shows that we again generate a straight line graph but this time it slopes 'down' from left to right, reflecting the fact that y is decreasing.

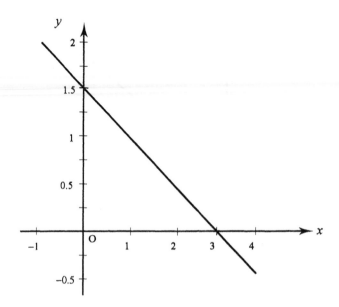

Figure 12.5

Hence we will state, without further proof, that:the function $f(x) = mx + c$, where m and c are constants, **always gives a graph which is a straight line.**

The meaning of c

A glance at the two graphs drawn in Figures 12.4 and 12.5 clearly shows that the graph of $y = 3x + 2$ crosses the y axis at $y = 2$ and the graph of $y = -0.5x + 1.5$ crosses the y axis at $y = 1.5$. We can also see this algebraically, since these are the values taken by y when $x = 0$.

Hence, the constant c tells us where the graph crosses the y axis. It is known as the *intercept* on the y axis or simply the *y intercept.*

The meaning of m

A further example will point the way to the meaning of m.

Example 12.5   Draw sketch graphs of the following linear functions:

(a) $y = 2x + 5$ (b) $y = 3x + 5$

(c) $y = 0.5x + 5$ (d) $y = -2x + 5$

Solution   The functions have graphs as shown below (Figure 12.6):

(a)   (b)

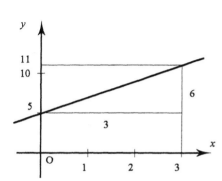   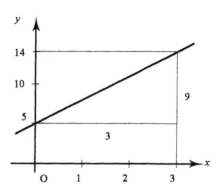

(c)   (d)

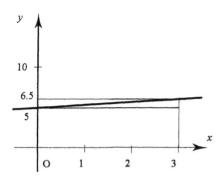   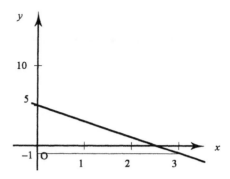

Figure 12.6

All the graphs have the same y intercept. They all cross the y axis at $y = 5$. What is clearly different about the four graphs is their *slope* or *gradient*.

In everyday language we speak of a gradient of 1 in 10 or 10%. What we mean by this is that for every 10 horizontal units travelled, we rise (or drop) 1 unit depending on whether we are travelling uphill or downhill.

We define the gradient of a straight line as follows:

If A and B are two points on a straight line (Figure 12.7)

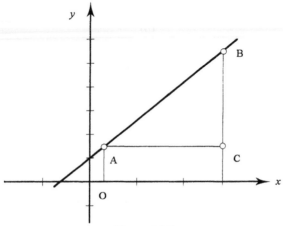

Figure 12.7

$$\text{Gradient of line AB} = \frac{\text{Difference in vertical height (BC)}}{\text{Difference in horizontal distance (AC)}}$$

If we apply this definition to the graphs in Example 12.5, we see that:

(a)　　$y = 2x + 5$ rises 6 units vertically over a horizontal distance of 3 units. This is the same as 2 units vertically for each horizontal unit. Its gradient is 2.

(b)　　$y = 3x + 5$ rises 9 units vertically for a change of 3 horizontal units. This is the same as 3 units vertically for each horizontal unit. Its gradient is 3.

(c)　　For each horizontal step of 1 unit, $y = 0.5x + 5$ only rises by half a unit and so its gradient is 0.5.

(d)　　Finally, $y = -2x + 5$ *falls* by 2 units for each horizontal unit travelled. The graph hence has a *negative* gradient of -2.

In each case, the calculated gradient is identical to the coefficient of x in the equation of the line. Therefore, the constant m always tells us the gradient of the line.

Hence, to summarise:

The function $f(x) = mx + c$ always has a linear graph of **gradient** m and **intercept** c (see Figure 12.8).

If the gradient is positive, y increases as x increases and the graph 'slopes up' from left to right.

If the gradient is negative, y decreases as x increases and the graph 'slopes down' from left to right.

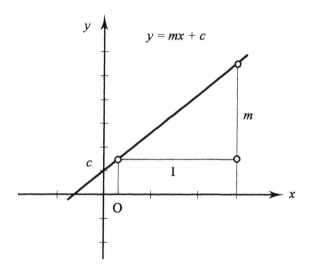

Figure 12.8 The graph of a linear function

Exercise 12.3

1. Draw sketch graphs of the following three 'special cases':

    (a) $c = 0$ (b) $m = 0$ (c) $m = 1$

2. Draw the graphs of the following linear functions. Write down the gradient and intercept in each case.

    (a) $y = 2x + 1$ (b) $y = 2x - 1$

(c) $y = -2x + 1$ (d) $y = -2x - 1$

(e) $x + y = 2$ (f) $2y = 3x + 5$

(g) $y = 6 - 3x$ (h) $3x + 4y = 12$

(i) $y = 3$ (j) $x = 7$

2.   Quadratic Functions

The function $y = ax^2 + bx + c$ where a, b and c are constants is called a quadratic function of x. The graph of this type of function is similar in shape to $y = x^2$. It is called a *parabola*.

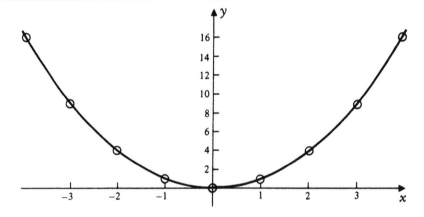

Figure 12.9   The graph of $y = x^2$ is a parabola

Example 12.6   Draw the graph of $y = 2x^2 + 3x - 5$.

Solution   Put $y = 0$ to find the intersections with the x axis.

$$0 = 2x^2 + 3x - 5$$

$$0 = (x - 1)(2x + 5)$$

Hence $x - 1 = 0$ or $2x + 5 = 0$

Solving for x,

$$x = 1 \quad \text{or} \quad x = -\frac{5}{2}$$

How do we decide which range of values of x to choose? We must include both $x = 1$ and $x = -2\frac{1}{2}$.

Suppose we choose an extra $\frac{1}{2}$ or 1 unit at the ends of this range. (We could choose more than this, or less, but we must include the range $-2\frac{1}{2}$ to $+1$, not merely this range alone.) Thus we make a table of values to plot the graph between $x = -3$ and $x = +2$, as below:

$$y = 2x^2 + 3x - 5$$

x	-3	$-2\frac{1}{2}$	-2	-1	0	1	2	$-\frac{3}{4}*$	
x^2	$+9$	$+\dfrac{25}{4}$	$+4$	$+1$	0	1	4	$+\dfrac{9}{16}$	
$2x^2$	18	$\dfrac{25}{2}$	8	2	0	2	8	$\dfrac{9}{8}$	Add these values
$+3x$	-9	$-\dfrac{15}{2}$	-6	-3	0	3	6	$-\dfrac{9}{4}$	to give y
-5	-5	-5	-5	-5	-5	-5	-5	-5	
y	4	0	-3	-6	-5	0	9	$-6\frac{1}{8}$	

The graph is shown in Figure 12.10.

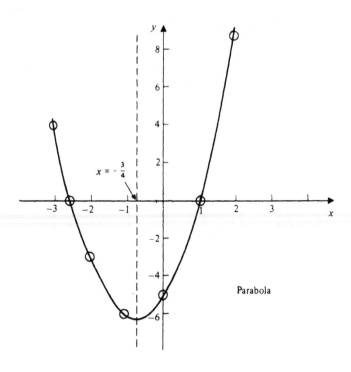

Figure 12.10 The graph of $y = 2x^2 + 3x - 5$

Note that if the coefficient of x^2 is negative, for example $y = 4 - x^2$, in which $a = -1$, then the curve is inverted as in Figure 12.11.

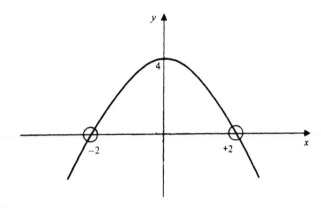

Figure 12.11 The graph of $y = 4 - x^2$

Where the minimum and maximum values occur are called *turning points*.

For $y = 2x^2 + 3x - 5$ there is a turning point at $\left(-\frac{3}{4}, -\frac{49}{8}\right)$ and for $y = 4 - x^2$ there is a turning point at $(0,4)$.

Exercise 12.4

Draw the graphs of the following quadratic functions:

1. $y = x^2 - 2x$
2. $y = 2x - x^2$

3. $y = x^2 + x - 1$
4. $y = x^2 - 3x - 1$

5. $y = 1 - 2x - 3x^2$
6. $y = 4x^2 - 2x + 1$

7. Draw the graph of $y = x^2 - 4x + 4$ between $x = 0$ and $x = 4$.
   Use your graph to find the minimum value of y.
   Use your graph to solve the equation $x^2 - 4x + 4 = 0$.

8. Draw the graph of $y = 2 - x - x^2$ between $x = -3$ and $x = +3$.
   Use your graph to find the maximum value of y.
   Use your graph to solve the equation $x^2 + x - 2 = 0$.

Summary so far

The highest power of x determines the type of graph which is obtained.

Highest power of x	Example	Type of graph
Zero	$y = -2$	Straight line parallel to x axis at distance 2 units below the x axis
First	$y = 4 - 2x$	Straight line gradient -2 through point $(0,4)$
Second	$y = x^2 - 3x + 2$ $= (x - 2)(x - 1)$	Parabola, passing through points $(1,0)$, $(2,0)$ and $(0,2)$

3.   Cubic Functions

The function $y = ax^3 + bx^2 + cx + d$ where a, b, c, d are constants is called a cubic function.

Example 12.7   Draw the graph of the cubic function $y = (x - 1)(x + 2)(3x - 7)$.

Solution   The highest power of x, on working out the brackets, would be $3x^3$ obtained by multiplying all the x terms together so that it is a cubic curve.

If we put $y = 0$,

i.e. $0 = (x - 1)(x + 2)(3x - 7)$

we obtain the solutions $x = 1$ or -2 or $\dfrac{7}{3}$ so that the curve cuts the x axis at these points.

It therefore looks like one of the curves of Figure 12.12.

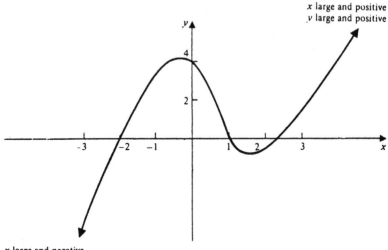

Figure 12.12 (a)

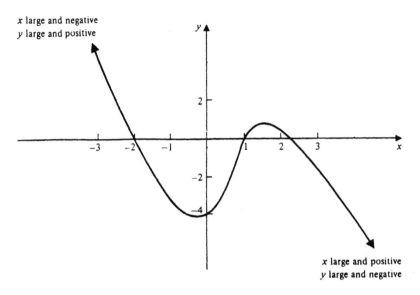

Figure 12.12(b)

We can decide which of these is correct by asking ourselves the following questions:

1. If x is large and positive, what do we know about y?
2. If x is large and negative, what do we know about y?

If x is large and positive,

using
$$y = (x - 1)(x + 2)(3x - 7)$$

we get

y = (a large positive number) × (a large positive number) × (a large positive number)

y = a large positive number

If x is large and negative

y = (a large negative number) × (a large negative number) × (a large negative number)

y = a large negative number

These two facts show that diagram (a) is the only possibility.

N.B.   **A cubic curve has two bends.**

Summary

$f(x)$	Type of graph	Type of curve
$y = ax + b$	Straight line, gradient a, no turning points	*a +ve* or *a — ve*
$y = ax^2 + bx + c$	Parabola, one turning point	*a +ve* or *a −ve*
$y = ax^3 + bx^2...$	Cubic, two turning points	or
$y = ax^4 +...$	Quartic, three turning points*	or

* We are making a reasonable guess here, but an example will check.

Note that in some special curves, two or more turning points may coalesce as in $y = x^3$ (Figure 12.13).

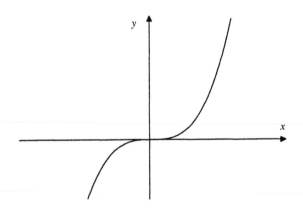

Figure 12.13 The graph of $y = x^3$

In general, the number of turning points is one less than the highest power of the polynomial.

Expressions such as $1 + 2x + 5x^2$ and $3 + 4x - 2x^3 + x^4$ consisting of the sum of a number of positive whole number powers of the variable are called *polynomials*. The highest power of the variable which occurs is called the *degree of the polynomial*.

Exercise 12.5

Draw the graph of each of the following functions. From your graph find the coordinates of the turning points.

1. $y = x^3 - x$ 2. $y = 2 + x - 2x^2 - x^3$

3. $y = (1 - x)(x - 2)(x - 3)$ 4. $y = x^4$

5. $y = 4 - 3x^2 + x^4$ 6. $y = x^4 - 2x^3 - x^2 + 2x$

7.   Draw the graph of $y = \dfrac{1}{x}$ for values of x between 0.5 and 5.

8.   Draw the graph of $y = \dfrac{4}{1+2x}$ for values of x between 0 and 3.

13

Linear Graphs and Their Use in Experimental Work

Many students view graphs with apprehension. Yet graphs are a part of everyday life. One very rarely opens a newspaper without being presented with a graph illustrating what is happening to such things as unemployment, the birth rate, profits, etc. We need graphs to give us a quick picture of trends, to find out whether a certain item is increasing, decreasing or levelling off. No set of figures could do this.

In the same way, scientists, economists, statisticians and many others need to study the relationship between different variables to see if there is any connection, so that they are able to predict what will happen when changes are made. For example, how will a small increase in productivity per machine affect profits, bearing in mind extra bonuses paid, and a bigger sales drive required to sell the extra goods?

In our study of different types of graph, there is one graph which is very easily drawn: the straight line. Conversely, if we are given a set of points on graph paper, the only one which is easily identified, even allowing for experimental errors, is the straight line.

Often, a practical problem can be solved to sufficient accuracy, by assuming a linear law between the variables. In this case we talk about *modelling* the problem using a linear law.

This, then, is our first task: to find the law which governs the set of points which seem to fall into a straight line, always bearing in mind the relationship between any graph and its equation, that is:

- The (x,y) values at every point *on* the graph satisfy the equation of the graph.

- The (x,y) values at every point *not on* the graph *do not* satisfy the equation of the graph.

Example 13.1   A man on a motor cycle passes a certain point on a motorway at 09.00 hours while travelling at a constant speed of 60 km/h. Draw a graph illustrating his distance from the given point and find out how far he will have travelled at 09.25, 10.15, 10.35, and 10.55. Also find at what time he reaches a point A, 35 km along the road.

Solution   We know that the distance increases steadily with time, so that we have a straight line graph. We also know that to draw a straight line we only need two points. Let us choose:

Time	Distance from start in km
09.00	0
11.00	120

Plotting these on a graph, with a time scale chosen to show 5 minute intervals, we obtain the graph below.

GRAPH OF MOTOR CYCLIST'S JOURNEY

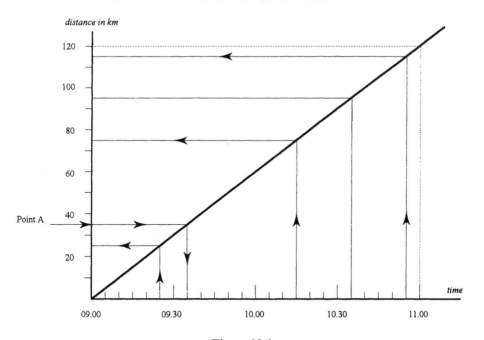

Figure 13.1

Reading off results:

Time	Distance from start in km
09.25	25
10.15	75
10.35	95
10.55	115

He reaches point A at 09.35.

We can check these easily since we know that the motorcyclist travels 1 km per minute and we can see that each point on the graph represents the time and place at some point of the journey. Also, any point off the graph represents a point of time and place where he will not be.

This is what we mean when we say this represents the graph of the journey, and if he continues to obey the law of the graph, i.e. that he continues to travel at 60 km/h, we can predict where he will be at any time.

This is an obvious case, but the same principles apply to more complicated graphs.

THE STRAIGHT LINE LAW

We have already seen in Chapter 12 that the straight line has a special kind of equation of the form

$$y = mx + c$$

where m = gradient and c = intercept on the y axis.

We can easily draw such a graph when the equation is known, by calculating two points on the line, and using a third point as a check.

Now we need to do the reverse process. Given a set of points on a graph, we need to find its equation. This equation is often required for use in predicting other results in combination with other equations. We proceed as below.

Example 13.2   Suppose we wish to lift a certain load by using a simple pulley system. To do so, we will have to use a certain amount of effort. The efforts used (E) to lift certain loads (W) are connected by the following table.

E	270	295	320	345	370	395
W	1000	1100	1200	1300	1400	1500

(i)   Draw the graph showing load, W, horizontally and find the law connecting E and W.

(ii)   Find E when $W = 50$, assuming the law holds for this load.

Solution

(i)   **Graph showing effort against load for a simple pulley**

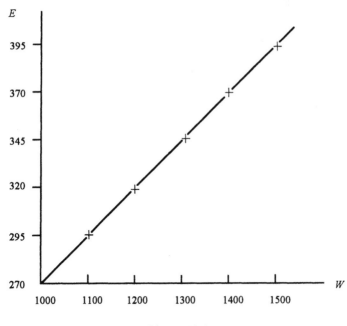

Figure 13.2

Note that in this case, the effort we must use depends on the load we need to lift, not the other way round. Hence E is the dependent variable and is plotted as the 'y axis'.

The graph is a straight line and is therefore of the form

$$E = mW + c$$

where m is the gradient and c the intercept on the vertical E axis.

Now, it could be assumed that the easiest method of finding m and c is merely to read off the gradient and the intercept on the vertical axis. But the intercept must be the *intercept on the vertical axis through the origin* and in this case (and many cases) it does not appear on the graph. Therefore we can use the following method.

Take two points on the line, well separated, and having convenient readings horizontally. (The vertical readings which follow have to take care of themselves.) Suppose we take the first and last readings in our table which are obviously on the line.

$$\text{When} \qquad E = 270 \qquad W = 1000$$
$$E = 395 \qquad W = 1500$$
$$\text{But} \qquad E = mW + c$$

Substituting the above values in turn:

$$270 = 1000m + c \qquad (1)$$
$$\text{and} \qquad 395 = 1500m + c \qquad (2)$$

Subtract (1) from (2) to eliminate c:

$$125 = 500m$$

Therefore

$$m = \frac{125}{500} = \tfrac{1}{4}$$

Substituting $m = \tfrac{1}{4}$ in (1)

$$270 = 1000 \times \tfrac{1}{4} + c$$

i.e.

$$270 = 250 + c$$

so

$$c = 20$$

Hence the law of the straight line is $E = \tfrac{1}{4}W + 20$.

(ii)   When $W = 50$

$$E = \tfrac{1}{4} \times 50 + 20$$

so

$$E = 32.5$$

This method is usually easier to complete than other methods.

Exercise 13.1

1. The velocity v cm/s of a body after travelling for t seconds is found by a series of experiments to be given by the following table:

t	2.5	4	6	9	10
v	33.5	38	44	53	56

If v and t are connected by a law $v = u + ft$, find u and f.

2. The volume V of a gas was measured in an experiment with increasing temperatures, T, as shown in the following table:

T	10	20	30	40	50	60	70
V	94.9	98.3	101.7	105.1	109.0	111.9	115.3

(i)   Show by drawing a graph of V against T that one of these readings was not sound, and say which it is.

(ii) Find the equation connecting V and T in the form $V = aT + b$ where a and b are constants.

(iii) What is the temperature when $V = 100$?

In the following question:

- Since all of the given values are experimental all of them may be slightly inaccurate. In other words none of the given pairs of values need be on the required straight line, which should be drawn evenly among them, i.e. with as many points above the line as below.

- When evaluating a and b make sure that the values of R and T used are taken from points actually on the line since none of the given values may be accurate enough to use.

3.   The resistance R ohms of a length of copper wire at different temperatures $T\,°C$ was found experimentally to be as follows:

$T\,°C$	0	40	75	105	160	200
R ohms	39.8	46.4	52.0	56.6	65.4	71.8

Plot R vertically against T horizontally, and see if they suggest a law of the linear form $R = aT + b$ where a and b are constants. Draw the probable line between the points and find the values of a and b. Hence find T when $R = 48$ and R when $T = 90$.

TO FIND LINEAR LAWS IN PRACTICAL CASES

In experimental work, the task of fitting an equation to a *curve* (rather than a straight line) may be very tedious and time consuming. If a set of points obtained by experiment seem to fit on a reasonable curve, how do we find out what type of equation will fit? It might be part of a parabola, a cubic, or one of the curves we will meet in later chapters. How can we find the equation of this curve? The only set of points which is easily recognised is that which gives a straight line.

Hence, if we can 'straighten out the curve' into a straight line we are well on the way to solving our problem. This is essentially a trial and error method, but some knowledge of the methods available is useful. The skill here is to make an intelligent guess about the possible equation of the curve and it is therefore important to recognise the shapes of basic curves.

Firstly we consider a basic list of equations whose graphs can be reduced to a straight line. There are many more 'tricks' which may be used, beyond the scope of this book.

Types of Laws Which Can Be Reduced to Straight Line Graphs

(i)   Algebraic types:

(a) $$y = a + \frac{b}{x}$$

and $$y = \frac{b}{x}$$

(b) $$y = \frac{b}{x^2}$$ and similarly for higher powers of x

(c)
$$y = a + b\sqrt{x}$$
$$y = a + \frac{b}{\sqrt{x}}$$

(d)
$$y = \frac{a}{1 + bx}$$

(e)
$$xy + ax + by = 0$$

(f)
$$y = \frac{ax}{1 + bx}$$

(g)
$$y = ax + bx^2$$

(In all the above, a and b represent constants.)

(ii)   The law $y = ax^n$ where a and n are constants.

(iii) The 'exponential law' $y = ae^{cx}$ where a and c are constants.

N.B.   Types (ii) and (iii) involve using logarithms and will be dealt with in a later chapter. Note that type (ii) passes through the origin.

Algebraic Types Not Involving Logarithms

(a)
$$y = a + \frac{b}{x}$$

Substitute $\dfrac{1}{x} = X$ to give $y = a + bX$

This we can compare with our standard form

$$y = c + mX$$

Plotting y against X, i.e. $\dfrac{1}{x}$, reduces the curve to a straight line whose gradient is b and intercept on the y axis is a.

Example 13.3   It is suspected that the following data obeys a law of the form

$$y = a + \frac{b}{x}$$

x	1	2	3	4	5	6
y	7	5.5	5	4.75	4.6	4.5

(i)   Draw a graph of y against x. Verify that it is *not* linear.

(ii)   Draw a graph of y against $\frac{1}{x}$, verify that it is linear and from it find a and b to the nearest whole number.

Solution

(i)

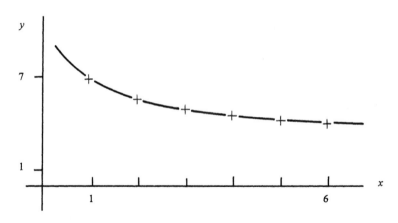

Figure 13.3

It would be very difficult to find a and b from this non-linear graph. What do you think happens to the graph as x gets closer to 0? Use your calculator and try $x = 0.9$, 0.5, 0.1, 0.01, 0.001, etc.

This type of curve is called a rectangular hyperbola.

(ii)   We can now find the reciprocals of the x values.

$\frac{1}{x}$	1	0.5	0.33	0.25	0.2	0.17
y	7	5.5	5	4.75	4.6	4.5

Note that there is no point quoting $\dfrac{1}{x}$ to more than 2 d.p. Why?

The graph is:

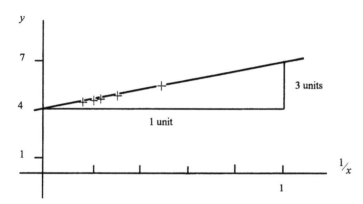

Figure 13.4

This graph is linear. The y intercept is 4 and the gradient 3.

Now compare $$y = a + b\left(\frac{1}{x}\right)$$

with $$y = c + mX$$

b is the gradient $= 3$

a is the y intercept $= 4$

Therefore $$y = 4 + \frac{3}{x}$$

(b) $$y = a + \frac{b}{x^2}$$

Let $X = \dfrac{1}{x^2}$. Then $y = a + bX$.

We can produce a straight line graph by plotting y against $\dfrac{1}{x^2}$.

The gradient is b, and a is the intercept on the y axis through (0,0).

Exercise 13.2

You can try one of this type for yourselves.

We suspect the following experimental data to follow a law of the form $y = a + \dfrac{b}{x^2}$.
Confirm this by drawing up a new table, plotting y against $\dfrac{1}{x^2}$ and using your graph
to find a and b.

x	0.5	1	1.5	2	2.5
y	1.0	–4.9	–6.1	–6.5	–6.7

Your resulting graph should be linear and lie mainly *below* the x axis. From it, you should be able to show that b = gradient = 2 and a = y intercept = –7, both to the nearest whole number. (Remember, you cannot read this number off if you don't start the y axis from zero.)

(c) $y = a + b\sqrt{x}$ and $y = a + \dfrac{b}{\sqrt{x}}$

                 Let $X = \sqrt{x}$ Let $X = \dfrac{1}{\sqrt{x}}$

            Plot y against $\sqrt{x}$ Plot y against $\dfrac{1}{\sqrt{x}}$

The gradient is b, and a is the intercept on the y axis through (0,0).

Similarly for any power of x, positive, negative or fractional.

Exercise 13.3

Here is another one for you to try

Confirm that the following data fits a law of the form $s = a + b\sqrt{t}$ by plotting s against $\sqrt{t}$. Show that to 1 d.p., $a = 3.5$ and $b = 1.2$.

t	1	2	3	4	5
s	4.7	5.2	5.6	5.9	6.2

We will now look at two harder examples, which show the general sneakiness we have to possess to squeeze a straight line out of non-linear data!

(d) $$y = \frac{a}{1+bx}$$

Cross multiply: $y + bxy = a$

$\therefore$ $y = -bxy + a$

Plotting y against (xy) gives a straight line. The gradient is $-b$, and a is the intercept on the y axis through $(0,0)$.

Example 13.4   The following data obeys the law $p = \dfrac{a}{1+bq}$. Find a and b.

q	0	1	2	3	4	5
p	5	1.67	1	0.71	0.56	0.45

Solution   Using the above theory, rearranging gives

$$p = -b(pq) + a$$

Therefore devise a new table showing pq against p:

pq	0	1.67	2	21.3	2.24	2.25
p	5	1.67	1	0.71	0.56	0.45

giving the graph:

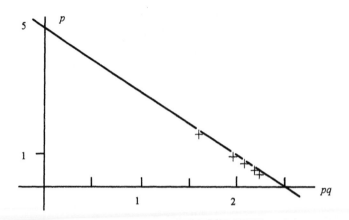

Figure 13.5

$\therefore$ intercept $= 5 \Rightarrow a = 5$ and   gradient $= -\dfrac{4}{2} = -b \Rightarrow b = 2$

$\therefore$ the law is $p = \dfrac{5}{1+2q}$

(e) $xy + ax + by = 0$

Divide by xy: $1 + \dfrac{a}{y} + \dfrac{b}{x} = 0$

$\therefore$ $\dfrac{1}{y} = -\dfrac{b}{a}\cdot\dfrac{1}{x} - \dfrac{1}{a}$

Plotting $\dfrac{1}{y}$ against $\dfrac{1}{x}$ gives a straight line.

The gradient is $-\dfrac{b}{a}$ and $-\dfrac{1}{a}$ is the intercept on the $\dfrac{1}{y}$ axis through $(0,0)$.

So this time we have to do a little more work! Note that the transposition has to be done so that *only two* terms contain a variable. One term must end up as a constant which will then be the 'y intercept'.

An alternative way of rearranging the above would be $xy = -by - ax$ therefore

$$y = -b\left(\frac{y}{x}\right) - a$$

A plot of y against y/x gives a straight line. Its gradient will give us $-b$ and the y intercept will give $-a$.

(f) $$y = \frac{ax}{1+bx}$$

Cross multiply: $$y + bxy = ax$$

Divide by y: $$1 + bx = a.\frac{x}{y}$$

$$\therefore \qquad \frac{x}{y} = \frac{1}{a} + \left(\frac{b}{a}\right).x$$

Plotting $\frac{x}{y}$ against x gives a linear graph. The gradient is $\frac{b}{a}$ and $\frac{1}{a}$ is the intercept on the $\frac{x}{y}$ axis through (0,0).

(g) $$y = ax + bx^2 \qquad \text{(a parabola through the origin).}$$

Divide by x: $$\frac{y}{x} = a + bx$$

Plotting $\frac{y}{x}$ against x gives a straight line whose is gradient b, and intercept on the $\frac{y}{x}$ axis through (0,0) is a.

In all these cases there are only two constants at most in the relationship, and reduction to a straight line will yield these two constants and no more.

If the relationship to be verified contains more than two constants, one of them must be found by some other means, e.g. by moving the origin or finding an intercept on the y axis directly from the graph.

WARNING: Having obtained the probable law of the curve, **it is extremely dangerous to extrapolate**, i.e. continue the curve and predict values outside the range. Conditions beyond the observed values may fundamentally change the tendency of the curve.

Exercise 13.4

1. Devise strategies in order to plot linear graphs from the following formulae. In each case a and b are constants. State how to find a and b once the graph is drawn.

   (a) $\dfrac{1}{y} = ax + b$　　　　　(b) $y^2 = ax + b$

   (c) $\dfrac{a}{x} + \dfrac{b}{y} = 1$　　　　　(d) $y = ax^2 + bx$

   (e) $y = ax - bx^3$　　　　　(f) $y = \dfrac{b}{x(x-a)}$

2. Show that the following values of two observed quantities, x and y, satisfy a law of the form $y = a + bx^2$, and find a and b.

x	1.1	1.8	2.5	2.9	3.6	4.3	4.8	5.4
y	1.91	2.13	2.40	2.65	3.10	3.66	4.09	4.73

3. In an experiment, values for the quantities x and y were found as follows:

x	1	2	4	6	8	9	10
y	0.39	0.667	1.045	1.27	1.43	1.5	1.55

   Show that these values are satisfied by a relation of the form $y = \dfrac{ax}{1+bx}$ and find a and b.

4. The following variables are thought to be connected by a law of the form $y = ax^2 + bx$.

x	10	20	30	40	50	60	70	80
y	150	700	1550	3000	4750	6900	9400	12,400

   Owing to experimental error, it is suspected that two of these readings are wrong. By drawing a suitable graph, isolate these wrong results and draw a linear graph through the other points. Hence find a and b.

14

Calculus – the Mathematics of Change

LIMITS – THE BACKGROUND TO CALCULUS

Many of the terrors of so-called 'advanced' mathematics are associated with a study of calculus, probably because it is so different in approach from the study of 'ordinary' mathematics. The language of mathematics is a shorthand way of saying something which is usually quite simple. In this chapter we shall talk about δx, δy, δt, etc. This symbol δ (the Greek letter delta) merely means 'a little bit of', and we say 'delta x' or 'delta y' or 'delta t' meaning 'a little bit of x', 'a little bit of y', 'a little bit of t'. We refer to it as 'an element of...' or 'an increment in...'.

For instance, if we were thinking of x as a distance in miles, then δx could be an inch or $\frac{1}{2}$ inch or even $\frac{1}{100}$ of an inch; but it will always be something small in relation to the whole. However, *the sum of all these little δx's makes a whole x*. We write this as: $\sum \delta x = x$ (read: sigma delta $x = x$). The sign sigma, $\sum$, is the shorthand way of saying 'the sum of'.

What Is Smallness?

In calculus we often have to deal with different degrees of smallness. We all know that smallness is relative. To a millionaire a tip of £5 for a service would be small, but to you or me it would not be small.

Suppose we think of some other examples. Relative to a week, a minute is small, i.e. it is minute, hence its name, but relative to the time humans have been on earth, a week is small, and a minute is then a small part of a small part. It is thus of a second degree of smallness.

We could say that a penny is small in value compared with a £5 note, but a £5 note is small in comparison with the latest win of £1 million on the pools, so that relative to this sum a penny is negligible.

However, we must always remember that small bits, if multiplied by a large enough number, can be large. A few thousand small coins amount to a considerable sum of money.

THREE IMPORTANT 'FRACTIONS'

Before we can proceed, we will need to think more carefully about the meanings of the three fractions introduced in Chapter 1:

(a) $\dfrac{0}{\text{something}}$ (b) $\dfrac{\text{something}}{0}$ (c) $\dfrac{0}{0}$

We might get several answers if we ask several different people to interpret their meaning. We will consider each one separately:

(a) Suppose we want to divide zero by 4 (say). If we were asked the question: 'If we divide nothing into 4 equal parts, what is the value of each part?', I think we would all say 'nothing'.

Hence $\dfrac{0}{4} = 0$ and generally $\dfrac{0}{\text{something}} = 0$

(b) Now suppose we want to divide 4 (say) by zero. It is difficult to work this out directly. We approach the problem by asking: 'what is 4 divided by a small fraction?'

For example, divide 4 by $\dfrac{1}{10}$:

$$4 \div \frac{1}{10} = 4 \times \frac{10}{1} = 40$$

Now divide 4 by $\dfrac{1}{100}$:

$$4 \div \frac{1}{100} = 4 \times \frac{100}{1} = 400$$

Similarly $4 \div \dfrac{1}{1000} = 4000$ and $4 \div \dfrac{1}{1,000,000} = 4,000,000$ etc.

As we divide by smaller and smaller fractions, our answer becomes larger and larger. We still have not reduced our denominator to zero, but we have seen that the smaller the fraction we divide by, the bigger the answer, so that by dividing by a fraction small enough we can get an answer with as many noughts on the end as we like. This means that however big a number we can think of, 4 divided by a very small number can be made bigger than this.

We have a symbol for this thing which is always beyond any number we can think of. We call it *infinity* and write it '∞' like an 8 on its side.

We say that a finite number n divided by another number x which is approaching zero, gives a result which approaches infinity, and we write

$$\text{The limit of } \left(\frac{n}{x}\right) \text{ as } x \to 0 \text{ is } \infty$$

(Read: 'The limit of n over x as x approaches zero is infinity'.)

or, in shorthand:

$$\underset{x \to 0}{Lim}\left(\frac{n}{x}\right) = \infty$$

provided x is never allowed actually to become zero.

Note that you should *never* write $\dfrac{\text{something}}{0} = 0$. The statement is meaningless!

(c)   What about $\dfrac{0}{0}$? Is it 0? Is it 1? Is it ∞?

In the same way that we were unable to answer the previous question directly we have to look at this from another angle.

(i)   Let us consider $\dfrac{x^2 - 4}{x - 2}$ as x approaches 2.

If we put $x = 2$ in the above we get $\dfrac{4-4}{2-2} = \dfrac{0}{0}$, so we get no further.

However, if we factorise the numerator:

$$\frac{x^2 - 4}{x - 2} = \frac{(x-2)(x+2)}{(x-2)} = (x+2)$$

We can always cancel by $(x - 2)$ *provided x is not actually equal to* 2.

Now what is the value of $(x + 2)$ as $x \to 2$? (*But does not quite equal* 2.)

This value approaches 4; so the ratio of two quantities both approaching zero works out *in this case* to be approaching 4, and can be made as close as we like to the value 4.

We write
$$\underset{x \to 2}{Lim} \left(\frac{x^2 - 4}{x - 2} \right) = 4$$

(Read: the limit of $\dfrac{x^2 - 4}{x - 2}$ as x approaches 2 is 4.)

(ii)   Take another case:

$$\left[\frac{x^2 - 9}{x - 3} \right] = \frac{(x+3)(x-3)}{(x-3)}$$

$$= (x+3) \text{ provided } x \neq 3$$

so:
$$\underset{x \to 3}{Lim} \left[\frac{x^2 - 9}{x - 3} \right] = 3 + 3 = 6$$

This time, we get a *different answer* for the ratio of two things both approaching zero.

(iii)   One more example. I am sure you can make others up for yourself.

$$\frac{x^2 + 3x}{x} = \frac{x(x+3)}{x}$$

$$= x + 3$$

$$\to 0 + 3 \quad \text{as } x \to 0$$

and this time the limit is 3. We see that the question 'What is zero divided by zero?' is *meaningless!* There is no *single* answer.

We can only ask 'What is the limit of $\dfrac{a}{b}$ as both a and b approach zero?' The answer depends on the values of a and b. We say that the limit is *indeterminate*.

It is limits of this kind that we will study in *differential calculus*.

It is limits of this kind that we will study in *differential calculus*.

Function Notation and Limits

As we have seen above, a function of x may not have a value when $x = a$ (i.e. its value at $x = a$ is indeterminate). It may nevertheless approach a limiting value when x approaches the value a, and we write this as

$$\underset{x \to a}{Lim}\ f(x)$$

which we read as 'the limit of $f(x)$ as x approaches a'. We can now incorporate this notation.

Example 14.1 If $f(x) = \dfrac{x^2 - 100}{x + 10}$ find $\underset{x \to -10}{Lim}\ f(x)$.

Solution
$$f(x) = \frac{x^2 - 100}{x + 10}$$
$$= \frac{(x - 10)(x + 10)}{(x + 10)}$$
$$= (x - 10)$$

(cancelling $(x + 10)$, provided $x \neq -10$).

$\therefore$
$$\underset{x \to -10}{Lim}\ f(x) = -10 - 10 = -20$$

We now realise that $\dfrac{0}{0}$ is indeterminate but that the fraction $\dfrac{A}{B}$ can approach a definite limit as both A and B approach zero. There is no single numerical value of this limit; it will depend on the problem.

Exercise 14.1

Find the limits of the given functions as x approaches the given values:

1. $\dfrac{x^2 - 1}{x - 1}$ $(x \to 1)$ 2. $\dfrac{x^2 + x}{x}$ $(x \to 0)$ 3. $\dfrac{x^2 - 9}{x + 3}$ $(x \to -3)$

4. $\dfrac{x^2 - a^2}{x - a}$ $(x \to a)$ 5. $\dfrac{10x}{x^2 + 3x}$ $(x \to 0)$

THE BEGINNINGS OF RELATIVE GROWTH

We are now equipped with the basic tools which will help us to study how one quantity changes when we change another, i.e. how things grow relative to each other. For example, we may need to find out how the cost of producing a certain type of article in a factory changes if we increase the number of employees, or if the total number of machines engaged is increased. The idea of relative change is very important, and this is the basic study of differential calculus.

Gradients and Rates of Change

We have met the idea of the gradient of a straight line in earlier chapters. For example, we speak of the gradient of a road as being *1 in 10, or 10%*, or *1:10*. Remember that this we means that, for every 10 units we travel along the road, we rise 1 unit (or drop 1 unit, according to whether we are travelling uphill or downhill). How can we apply our knowledge to finding the gradient of a *curve* at a point?

The Gradient at a Point on a Curve

Consider any point on the curve $y = f(x)$.

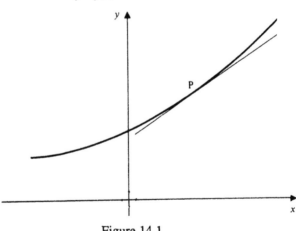

Figure 14.1

To calculate gradients, we need the coordinates of two points on a *straight* line but we do not have a straight line! However, at any point P on the curve, there is an important straight line which just touches the curve. There is only one such line for each point P and it is called the ***tangent*** at P as shown in Figure 14.1.

> When we speak of the gradient of the curve at the point P,
> we mean the gradient of the tangent to the curve at that point.

If we take a ruler and draw in the line we think is the tangent at P we are only getting an approximation. It is not necessarily accurate and in any case, we may need to know a number of gradients at various points and this method would become very cumbersome to use.

The following method is much more useful.

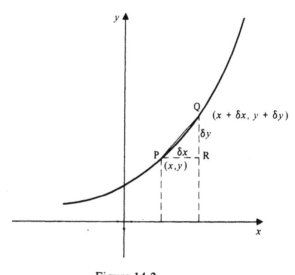

Figure 14.2

Let P be the point (x,y) and let Q be a point $(x + \delta x, y + \delta y)$ where δx *and* δy are very small distances.

Hence, Q is very, very close to P and our diagram is necessarily exaggerated so that we can see what is happening.

Thus in the $\triangle$ PQR in Figure 14.2

$$PR = \delta x, \quad QR = \delta y.$$

In $\triangle$ PQR the gradient of chord $PQ = \dfrac{\delta y}{\delta x} = \dfrac{\text{vertical step}}{\text{horizontal step}}$

Now let the point Q approach the point P, so that δx and δy get smaller, and the triangle PQR shrinks to a point at P.

What happens to the chord PQ? By putting in various positions for Q you will see that the chord approaches the position of the tangent at P, and in the limit when P and Q coincide, the chord becomes a tangent.

We say that the gradient of the tangent at P is the limit of $\dfrac{\delta y}{\delta x}$ as both δy and $\delta x \to 0$. This is why we had to make a study of the ratio of very small quantities, both approaching zero, and we define

$$\boxed{\underset{\delta x \to 0}{Lim}\frac{\delta y}{\delta x} = \frac{dy}{dx}}$$ (pronounced 'dee-y by dee-x')

We call this method *differentiation*.

N.B. $\dfrac{dy}{dx}$ does *not* mean $\dfrac{d \times y}{d \times x}$ and we cannot cancel the d's.

$\dfrac{dy}{dx}$ is called the **differential coefficient** or **derivative** of y with respect to x.

Let us apply the method to several curves.

To find the gradient at any point of the curve $y = x^2$

Let $P \equiv (x, y)$ and $Q \equiv (x + \delta x, y + \delta y)$. (See Figure 14.3.)

That is, let x and y grow by small amounts δx, δy.

Complete the $\triangle$ PQR. Then

$$PR = \delta x \quad QR = \delta y$$

therefore   gradient of $PQ = \dfrac{\delta y}{\delta x}$

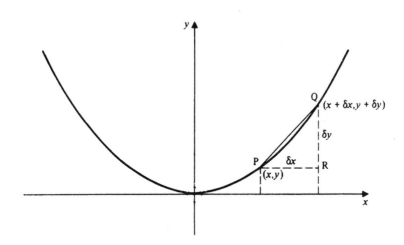

Figure 14.3

Note that δx, δy are greatly enlarged for easier viewing.

We evaluate this as follows:

Since the point (x,y) lies on the curve then

$$y = x^2 \qquad (1)$$

Since the point $(x+\delta x, y+\delta y)$ also lies on the curve

$$y+\delta y = (x+\delta x)^2 \qquad (2)$$

N.B. The square of the x value must always equal the y value for any point on this curve $y = x^2$, i.e. the y value $= (x$ value$)^2$.

From (2) $\qquad y+\delta y = x^2 + 2x.\delta x + (\delta x)^2$ (using $(a+b)^2 = a^2 + 2ab + b^2$)

From (1) $\qquad\qquad y = x^2$

Subtracting (1) from (2):

Divide by δx: $\qquad\qquad \dfrac{\delta y}{\delta x} = 2x + \delta x$

Therefore   Gradient of PQ $= 2x + \delta x$.

As $\delta x \to 0$ gradient of PQ $\to 2x$.

Hence the gradient of the tangent at $P = 2x$,

i.e. $\dfrac{dy}{dx} = 2x$ for the curve $y = x^2$

For example,   where $x = 1$ the gradient of the tangent $= 2 \times 1 = 2$

Similarly: where $x = 2$ the gradient of the tangent $= 4$

where $x = 3$ the gradient of the tangent $= 6$

where $x = 3.1$ the gradient of the tangent $= 6.2$

where $x = -\frac{1}{2}$ the gradient of the tangent $= -1$ etc.

The gradient of the curve $y = x^2$ at any point is the same as the gradient of the tangent and is always twice the x value at that point.

To find the gradient at any point on the curve $y = x^3$

Following the same procedure as before:

Let x and y both grow by small amounts δx and δy:

$$\text{Gradient of PQ} = \frac{\delta y}{\delta x}$$

But since P and Q both lie on the curve

$$y = x^3 \qquad\qquad\qquad (1)$$

$$y + \delta y = (x + \delta x)^3 \qquad\qquad (2)$$

so $y + \delta y = x^3 + 3x^2.\delta x + 3x(\delta x)^2 + (\delta x)^3 \qquad (3)$

(using $(a + b)^3 = a^3 + 3a^2 b + 3ab^2 + b^3$.)

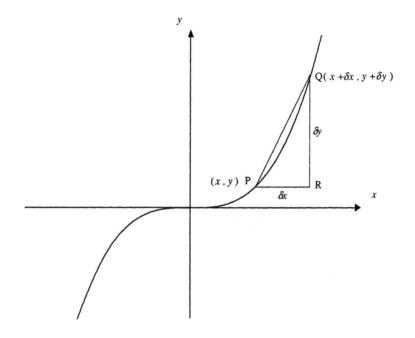

Figure 14.4

Subtracting (1) from (3):

$$\delta y = 3x^2.\delta x + 3x(\delta x)^2 + (\delta x)^3$$

Divide through by δx:

$$\frac{\delta y}{\delta x} = 3x^2 + 3x.\delta x + (\delta x)^2$$

Now we know that δx is small; therefore, $(\delta x)^2$ is of the second degree of smallness, and is very small indeed.

For example if $\quad$ $\delta x = 0.001$ $\quad$ $(\delta x)^2 = 0.001 \times 0.001 = 0.000001$

then $\quad$ $\delta x = 0.0001$ $\quad$ $(\delta x)^2 = 0.0001 \times 0.0001 = 0.00000001$

It is therefore negligible in comparison with the other terms.

Let δx and $\delta y \to 0$. Then the gradient of the tangent at P is

$$\frac{dy}{dx} = 3x^2$$

Thus we can calculate the gradient of the tangent at any point on the curve $y = x^3$ by squaring the x value of the point and then multiplying by 3:

e.g.   the gradient of tangent where $x = 2$ is $3 \times 2^2 = 12$

the gradient of tangent where $x = -1\frac{1}{2}$ is $3x\left(-\frac{3}{2}\right) \times \left(-\frac{3}{2}\right) = +\frac{27}{4}$

To find the gradient of the tangent to the curve $y = x^4$

Here we shall need the expansion

$$(a+b)^4 = a^4 + 4a^3b + 6a^2b^2 + 4ab^3 + b^4$$

Proceed as in the previous examples.

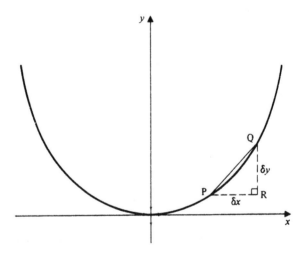

Figure 14.5

Since P lies on the curve:

$$y = x^4 \qquad\qquad (1)$$

Since Q lies on the curve:

$$y + \delta y = (x + \delta x)^4 \qquad\qquad (2)$$

Expanding:

$$y + \delta y = x^4 + 4x^3.\delta x + 6x^2 (\delta x)^2 + 4x(\delta x)^3 + (\delta x)^4$$

From (1): $$y = x^4$$

Subtracting:

$$\delta y = 4x^3.\delta x + 6x^2 (\delta x)^2 + 4x(\delta x)^3 + (\delta x)^4$$

$$\therefore \quad \frac{\delta y}{\delta x} = 4x^3 + 6x^2 \delta x + 4x(\delta x)^2 + (\delta x)^3$$

As δy and $\delta x \to 0$, $\lim \dfrac{\delta y}{\delta x} = 4x^3$.

$$\therefore \quad \frac{dy}{dx} = 4x^3 = \text{the gradient of the tangent at P for the curve } y = x^4$$

The simple case of $y = x$

This obviously has a gradient of 1, by considering the horizontal and vertical steps.

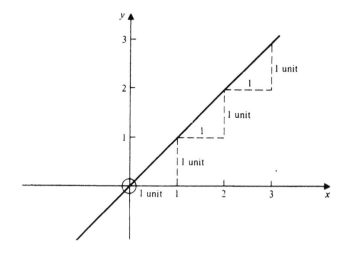

Figure 14.6

$$\frac{dy}{dx} = 1 \text{ for the straight line } y = x$$

The case of $y = c$, where c is a constant

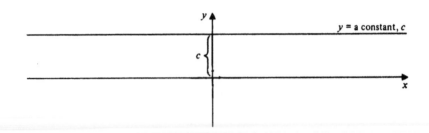

Figure 14.7

This also is a simple case. Whatever the value of the constant, the graph is a line parallel to the x axis, and therefore has zero gradient.

$$\frac{dy}{dx} = 0 \text{ for the straight line } y = c \qquad (c \text{ a constant})$$

We can now tabulate our results:

y	$\dfrac{dy}{dx}$	
x [or x^1]	1 [or $1x^0$]	*since* $x^0 = 1$
x^2	2x	
x^3	$3x^2$	
x^4	$4x^3$	
x^5	$5x^4$	*and so on.*
x^n	nx^{n-1}	

Guessing the next: (row x^5)

In general: (row x^n)

So that:

$$\text{If } y = x^n \text{ then } \frac{dy}{dx} = nx^{n-1}.$$

At this stage we do not prove this formally. It is the use of this formula which is of most importance. It is true for all values of n, positive, negative and fractional.

Notation

The gradient of the tangent may be written in the following forms:

	1.	$\dfrac{dy}{dx}$	(Read: 'dee-y by dee-x'.)
or	2.	$\dfrac{d}{dx}(y)$	(Read: 'dee by dee-x of y'.)
or	3.	$\dfrac{d}{dx}(f(x))$ where $y = f(x)$	(Read: 'dee by dee-x of eff-x'.)
or	4.	$f'(x)$	(Read: 'eff-dash x'.)
or	5.	the rate of change of y with respect to x.	

Negative Gradients: Zero Gradients

We note that just as the gradient of a straight line can be positive or negative, so can the gradient of a curve (Figure 14.8).

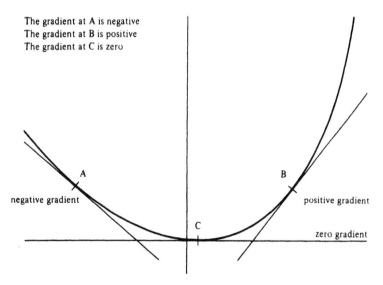

The gradient at A is negative
The gradient at B is positive
The gradient at C is zero

Figure 14.8

A negative gradient implies that y *decreases* as x increases from left to right.

How to Deal with Constants

There are two different types:

(i)   Constants added to (or subtracted from) other terms involving a variable.

(ii)   Constants which multiply a variable (i.e. *coefficients*).

Added constants

An example will illustrate the way to deal with these.

Let $$y = x^2 - 4 \qquad\qquad (1)$$

In exactly the same way as previously, suppose x increases to $x + \delta x$, while y increases to $y + \delta y$.

Then from (1)

$$y + \delta y = (x + \delta x)^2 - 4$$
$$\therefore \qquad y + \delta y = x^2 + 2x\delta x + (\delta x)^2 - 4$$

Subtracting the original $y = x^2 - 4$ we get

$$\delta y = 2x\delta x + (\delta x)^2$$

Divide by δx:

$$\frac{\delta y}{\delta x} = 2x + \delta x$$

As δx and δy approach zero:

$$\underset{\delta x \to 0}{Lim} \frac{\delta y}{\delta x} = 2x$$

$$\therefore \qquad \frac{dy}{dx} = 2x \text{ for } y = x^2 - 4$$

In other words, the –4 has disappeared from the result. It has not affected the change in y. In the same way, whatever number we add to or subtract from x^2, the same thing will happen, whether it is a whole number or a fraction, positive or negative. Hence we say 'the derivative of a simple constant is zero'.

We could also see this result by drawing the graphs of $y = x^2$ and $y = x^2 - 4$; they are of the same shape, only the second one is 4 units vertically below the first. So for a given value of x they have the same gradient.

Multiplying constants (*coefficients*)

As a simple example consider

$$y = 5x^2$$

As before

$$y + \delta y = 5(x + \delta x)^2$$
$$= 5(x^2 + 2x\delta x + (\delta x)^2)$$

$$\therefore \quad y + \delta y = 5x^2 + 10x\delta x + 5(\delta x)^2$$

Subtract $y = 5x^2$. This gives

$$\delta y = 10x\delta x + 5(\delta x)^2$$

Divide by δx:

$$\frac{\delta y}{\delta x} = 10x + 5\delta x$$

As δx and δy approach zero

$$\underset{\delta x \to 0}{Lim} \frac{\delta y}{\delta x} = 10x \qquad \text{since } 5\delta x \to 0$$

$$\therefore \qquad \frac{dy}{dx} = 10x \qquad \text{for } y = 5x^2$$

Thus to differentiate $5x^2$ we multiply the derivative of x^2, which is $2x$, by 5 and we obtain $10x$.

In the same way if we differentiate $y = ax^2$ where a is any constant we obtain

$$\frac{dy}{dx} = a \text{ multiplied by } 2x$$

$\therefore$ $$\frac{dy}{dx} = 2ax$$

This method holds for any power of x, but the general proof is beyond the scope of this book and we have the general result:

$$\text{If } y = ax^n, \text{ then } \frac{dy}{dx} = anx^{n-1}.$$

Sums and Differences

How do we deal with several terms added or subtracted? The answer is quite simple. We treat each term separately. For example, consider

$$y = x^4 + 3x + 6$$

We know that if we increase x, x^4 will increase by a certain amount and $3x$ will increase by a different amount. The constant 6 will not change. Therefore it will not affect the change in y. Hence the total increase in y due to a change in x is the sum of the increases in x^4 and $3x$. This means that the rate of increase of y is equal to the sum of the rates of increase of x^4 and $3x$. A formal proof is as follows:

Let $y = u + v$ (1)

where u and v are both functions of x (u corresponds to x^4 and v to $3x$ in the example above).

If x increases by δx

then u will increase by δu

v will increase by δv

and y will increase by δy

From (1): $y + \delta y = (u + \delta u) + (v + \delta v)$

Subtract (1): $y = u + v$

and we obtain $\delta y = \delta u + \delta v$

Divide by δx:
$$\frac{\delta y}{\delta x} = \frac{\delta u}{\delta x} + \frac{\delta v}{\delta x}$$

Let δx approach zero. Then δu, δv and δy will approach zero. Hence in the limit:

$$\text{If } y = u + v, \text{ then } \frac{dy}{dx} = \frac{du}{dx} + \frac{dv}{dx}.$$

This method applies however many terms we have, whether positive or negative. To differentiate a function consisting of sums and differences of terms, differentiate each term separately.

Example 14.2   If $y = x^6$, find $\frac{dy}{dx}$.

Solution $n = 6$

Therefore
$$\frac{dy}{dx} = 6x^{6-1}$$
$$= 6x^5$$

Example 14.3   If $y = 3x^2$ find $\frac{dy}{dx}$.

Solution
$$\frac{dy}{dx} = 3 \times \text{derivative of } x^2$$
$$= 3 \times 2x$$
$$= 6x$$

Example 14.4   If $y = 4x^3 - x + 6$ find $\frac{dy}{dx}$ and the gradients of the tangents at $x = 2, 1, 0, -1$.

Solution
$$\frac{dy}{dx} = 4.3x^2 - 1 + 0 \text{ since } \frac{d}{dx}(6) = 0$$
$$\frac{dy}{dx} = 12x^2 - 1$$

When $x = 2$
$$\frac{dy}{dx} = 12.2^2 - 1$$

$\therefore$ Gradient of tangent $= 47$ at $x = 2$

When $x = 1$ $\dfrac{dy}{dx} = 12 - 1$

$\therefore$ Gradient of tangent $= 11$ at $x = 1$

When $x = 0$ $\dfrac{dy}{dx} = -1$

$\therefore$ Gradient of tangent $= -1$ at $x = 0$

When $x = -1$ $\dfrac{dy}{dx} = 12(-1)^2 - 1$

$\therefore$ Gradient of tangent $= 11$ at $x = -1$

Thus the gradients at $x = 2, 1, 0, -1$ are 47, 11, -1, 11 respectively.

Example 14.5   Differentiate the following functions by rule:

(i) $(x + 3)^2$ (ii) $ax(a - x)$ (iii) $\dfrac{x^n + 1}{x}$ (iv) $\dfrac{1}{x^2}$ (v) $\dfrac{1}{x}$

Solution

(i)   Let $y = (x + 3)^2 = x^2 + 6x + 9$

$$\dfrac{dy}{dx} = 2x + 6$$

(ii)   Let $y = ax(a - x) = a^2 x - ax^2$

$$\dfrac{dy}{dx} = a^2 - a(2x)$$
$$= a^2 - 2ax$$

(iii)   Let $y = \dfrac{x^{n-1} + 1}{x}$

i.e. $y = x^{n-1} + \dfrac{1}{x} = x^{n-1} + x^{-1}$

$$\dfrac{dy}{dx} = (n-1)x^{n-2} + (-1x^{-1-1})$$
$$= (n-1)x^{n-2} - \dfrac{1}{x^2}$$

(iv)   Let $y = \dfrac{1}{x^2} = x^{-2}$

$\therefore$
$$\frac{dy}{dx} = -2x^{-2-1}$$
$$= -2x^{-3}$$
$$= \frac{-2}{x^3}$$

(v) $y = \dfrac{1}{\sqrt{x}} = \dfrac{1}{x^{\frac{1}{2}}} = x^{-\frac{1}{2}}$

$$\frac{dy}{dx} = -\frac{1}{2}x^{-\frac{1}{2}-1}$$
$$= -\frac{1}{2}x^{-\frac{3}{2}}$$
$$= -\frac{1}{2\sqrt[2]{x^3}}$$

Now you know how to differentiate powers of x. It is all very easy, but if necessary, before doing the next exercise, revise fractional and negative indices in Chapter 9.

Exercise 14.2

In the following, use $\dfrac{d}{dx}(x^n) = nx^{n-1}$ to find $\dfrac{dy}{dx}$.

1.	$y = x^5$	2.	$y = 3x^3$	3.	$y = 5x$
4.	$y = -x$	5.	$y = \frac{1}{4}x^4$	6.	$y = 6x^{\frac{1}{2}}$
7.	$y = \sqrt[3]{x}$	8.	$y = 2x^2 - 4$	9.	$y = \dfrac{2}{x}$
10.	$y = \dfrac{1}{2x}$	11.	$y = x + \dfrac{1}{x}$	12.	$y = \dfrac{3}{x} - \sqrt{x}$
13.	$y = \dfrac{x^6}{6}$	14.	$y = \dfrac{x^7}{7}$	15.	$y = \dfrac{x^8}{8}$
16.	$y = \dfrac{x^{n+1}}{n+1}$	17.	$y = \dfrac{x^4 - 4}{x}$	18.	$y = \dfrac{x^4 - 4}{x^5}$
19.	$y = (x + 2)^2$	20.	$y = 3x^2 + 2x - 1$	21.	$y = ax^2(a - x^2)$
22.	$y = ax^2 + bx + c$				

15

Applications of Differentiation

RATES OF CHANGE WITH RESPECT TO TIME

Linear Velocity

Suppose an object is moving along the x axis.

Figure 15.1

Let O be the fixed origin and P a variable point on the line showing the position of the object at time t. Let x be the distance of P from O at a given time t.

Then if P travels to an adjacent point P' where $PP' = \delta x$, in a time δt, then

$$\text{average speed} = \frac{\text{distance}}{\text{time}} = \frac{\delta x}{\delta t} \text{ during the interval } \delta t$$

Then if δx and $\delta t \to 0$, the limit is the speed at the point P. It is, however, rather more than just the speed. The result will be positive if P moves in the positive x direction and negative otherwise. This combination of speed and direction is known as **velocity**,

i.e. $$\frac{dx}{dt} = \text{velocity at } P = v \text{ (say)}$$

The velocity at P is the rate of change of distance with respect to time.

N.B. 1 $\dfrac{dx}{dt}$ is positive when x is increasing with time, i.e. when P is moving in the positive x direction.

$\dfrac{dx}{dt}$ is negative when x is decreasing with time, i.e. when P is moving in a negative x direction.

N.B. 2 $v = \dfrac{dx}{dt}$ means that on the graph of distance plotted against time, the velocity is the gradient of the (x, t) graph.

Linear Acceleration

Let v denote the velocity at time t.

Then $\dfrac{dv}{dt}$ is the rate at which the velocity is changing at the instant t. This is known as the *acceleration* at time t. If a denotes the acceleration

$$a = \frac{dv}{dt}$$

i.e. the acceleration, a, is the gradient of the (v, t) graph.

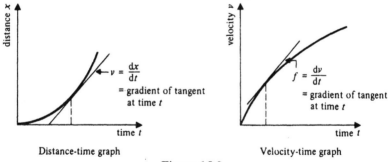

Distance-time graph   Velocity-time graph

Figure 15.2

Example 15.1   If x is measured in metres and t in seconds, and if $x = \frac{1}{4}t^4 + 3t^2 + 2t$, find the velocity and acceleration when $t = 3$.

Solution

$$v = \frac{dx}{dt} = \frac{1}{4}.4t^3 - 3.2t + 2 \qquad (1)$$

$$= t^3 - 6t + 2$$

$\therefore$ when $t = 3$: $v = 3^3 - 6(3) + 2$

$$= 27 - 18 + 2$$

$$v = 11 \text{ metres/second}$$

Also $a = \dfrac{dv}{dt} = 3t^2 - 6$ (by differentiating (1))

When $t = 3$: $a = 3(3)^2 - 6$

$$= 27 - 6$$

$$a = 21 \text{ metres/second}^2$$

Note that metres/second^2 means 'metres per second per second', i.e. the increase in velocity (metres per second) every second.

Example 15.2   If an object moves such that $x = t^3 - 9t^2 + 24t$ find the values of t at which the object is momentarily at rest.

Solution

$$v = \frac{dx}{dt} = 3t^2 - 18t + 24$$

When at rest, $v = 0$

i.e. $0 = 3t^2 - 18t + 24$

$$0 = 3(t^2 - 6t + 8)$$

$$0 = 3(t - 4)(t - 2)$$

$$t = 4 \text{ or } 2$$

Hence P is momentarily at rest after 2 seconds and 4 seconds.

Exercise 15.1

1.   The distance, x metres, travelled by a car t seconds after the brakes are applied, is given by $x = 44t - 6t^2$.

     (i)   What is the speed when the brakes are applied (i.e. $t = 0$)?
     (ii)   What is the retardation (i.e. the negative acceleration)?

(iii)   How long does it take to come to rest?

(iv)   How far does it travel before it comes to rest?

2. With the usual notation, find the velocity and acceleration at the time for the following functions. Also find values of t when P is momentarily at rest.

(i) $x = 2t^3 - 9t^2 + 12t$ when $t = 2$

(ii) $x = t(3 - t)^2$ when $t = 2$

(iii) $x = t^2(4 - t^2)$ when $t = 1$

3. The displacement of an object at time t is given by $x = t^2 + 27\sqrt{t}$. Show that, as long as the object obeys this law, the velocity is never zero but the acceleration is zero when $t = 2.25$.

STATIONARY VALUES

We know that the graphs of many functions have turning points. It is often useful to be able to locate these without drawing an accurate graph. We now proceed to find a method for doing this.

Consider the following sketch graph (Figure 15.3).

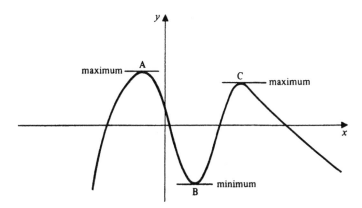

Figure 15.3

Note that this graph is *continuous*. There are no breaks in it and we can draw the whole curve from left to right, *without lifting pen from paper*.

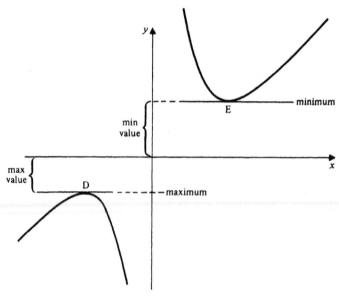

Figure 15.4

The graph in Figure 15.4 is *very* different. There are two separate parts and we cannot draw the whole curve 'in one go'. The graph (and the function it represents) are *discontinuous*.

Nevertheless, the graphs do have certain points in common:

At A, C and D we have *maximum turning points*.

At B and E we have *minimum turning points*.

Points A, C and D are called *local maxima*, because they are 'higher up' than all points in their *local* neighbourhood. Notice that we are talking about the y *values* of these points, i.e. the actual values of the function at A, C, D are bigger than the values nearby.

Similarly, B and E are called *local minima*.

All these points have one thing in common: the tangents are horizontal and therefore the gradients of the curves at these points are all zero.

$$\text{Thus } \frac{dy}{dx} = 0 \text{ is a condition for maximum or minimum point.}$$

Let us see how we can make use of this fact.

Consider the curve

$$y = 2x^3 + 3x^2 - 36x + 6 \qquad (1)$$

Differentiating:

$$\frac{dy}{dx} = 6x^2 + 6x - 36$$

For a turning point $\frac{dy}{dx} = 0$

i.e.
$$6x^2 + 6x - 36 = 0$$
$$6(x^2 + x - 6) = 0$$
$$6(x + 3)(x - 2) = 0$$
$$x = -3 \quad \text{or} \quad x = +2$$
$$y = 87 \quad \text{or} \quad y = -38$$

We suspect that since we are studying a *cubic* curve, there will be *a maximum and a minimum* value (two turning points).

Can we assume, because $y = +87$ at one of these and $y = -38$ at the other, that the first of these gives the maximum value of y and the second gives the minimum value? This is a dangerous assumption in many cases, although in this case it could be justified by reference to a sketch graph. In many cases, a minimum turning point can give a higher value of y than a maximum turning point, as in Figure 15.4.

In fact, for *discontinuous* curves this can often happen. We can, however, safely say that if a *continuous* curve has two turning points, the maximum turning point is always higher than the minimum. (Try it by scribbling!)

It is good practice to draw sketch graphs whenever possible and it is certainly useful to *learn* the shapes of common curves as it can save much time when finding turning points. However, we will find it useful to have a method that will help us to distinguish between local maxima and local minima, *without* having to draw the sketch.

Tests to Distinguish Maximum and Minimum Points

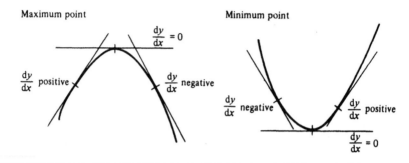

Figure 15.5

For a maximum point:

On the LHS $\quad\quad\quad\quad\quad \dfrac{dy}{dx}$ is positive (+ve)

On the RHS $\quad\quad\quad\quad\quad \dfrac{dy}{dx}$ is negative (–ve)

so that $\dfrac{dy}{dx}$ changes from +ve $\rightarrow$ zero $\rightarrow$ –ve as the tangent rolls around the curve.

This means that $\dfrac{dy}{dx}$ is decreasing (+ve, zero, –ve).

For a minimum point:

On the LHS $\quad\quad\quad\quad\quad \dfrac{dy}{dx}$ is negative (–ve)

On the RHS $\quad\quad\quad\quad\quad \dfrac{dy}{dx}$ is positive (+ve)

so that $\dfrac{dy}{dx}$ changes from –ve $\rightarrow$ zero $\rightarrow$ +ve as the tangent rolls round the curve.

This means that $\dfrac{dy}{dx}$ is increasing (–ve, zero, +ve).

How do we express the fact that $\dfrac{dy}{dx}$ is decreasing (for a maximum point) or increasing (for a minimum point)?

If any function is decreasing (or increasing) we know that its gradient, i.e. *its* derivative, is negative (or positive). Hence the derivative of $\dfrac{dy}{dx}$, which is itself a function of x, must be negative for a maximum point and positive for a minimum point.

Figure 15.6 illustrates what happens to $\frac{dy}{dx}$ at a maximum point and at a minimum point.

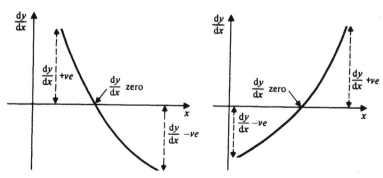

Figure 15.6

In the first diagram $\frac{dy}{dx}$ is decreasing, and therefore has a negative gradient or derivative. In the second, $\frac{dy}{dx}$ is increasing, and therefore has a positive gradient or derivative. How do we write the derivative of $\frac{dy}{dx}$?

In shorthand: $\frac{d}{dx}$ of $\frac{dy}{dx}$ or $\frac{d(dy)}{dx(dx)}$ or $\frac{d^2y}{dx^2}$

N.B. $\frac{d}{dx}$ of $\frac{dy}{dx}$ is pronounced 'dee-by-dee-x of dee-y by dee-x'

$\frac{d(dy)}{dx(dx)}$ is pronounced the same way

$\frac{d^2y}{dx^2}$ is pronounced 'dee-two-y by dee-x-squared'.

Thus for a *maximum* point $\frac{d^2y}{dx^2}$ is *negative* at that point and for a *minimum* point $\frac{d^2y}{dx^2}$ is *positive* at that point.

What happens if $\frac{d^2y}{dx^2}$ is neither positive nor negative, i.e. $\frac{d^2y}{dx^2}$ is zero?

In this case $\dfrac{dy}{dx}$ does not change from positive or negative or vice versa. Hence the graph *can* be as shown in Figure 15.7.

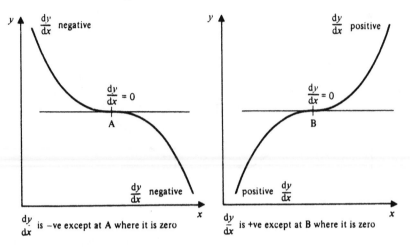

Figure 15.7

We call this a *point of inflection*.

We must, however, be *very careful* how we interpret a value $\dfrac{d^2 y}{dx^2} = 0$. *It does not always* imply a point of inflection!

Consider a very simple case, that of $y = x^4$ whose graph is sketched below:

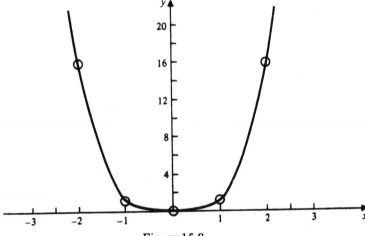

Figure 15.8

$$\frac{dy}{dx} = 4x^3 \qquad (1)$$

$$\therefore \qquad \frac{d^2y}{dx^2} = 12x^2 \qquad (2)$$

As usual, $\frac{dy}{dx} = 0$ at a turning point. Therefore $4x^3 = 0 \Rightarrow x = 0$. If we now substitute this into (2), then $\frac{d^2y}{dx^2} = 0$ at the turning point, which (from the sketch) is clearly a *minimum* turning point.

A full study of points of inflection is beyond the scope of this chapter, so to summarise what we have found:

If, at a certain point,

$$\frac{dy}{dx} = 0 \text{ and } \frac{d^2y}{dx^2} > 0, \text{ the point is a } local\ minimum$$

$$\frac{dy}{dx} = 0 \text{ and } \frac{d^2y}{dx^2} < 0, \text{ the point is a } local\ maximum$$

$$\frac{dy}{dx} = 0 \text{ and } \frac{d^2y}{dx^2} = 0, \text{ the point } could \text{ be max, min or inflection}$$

However, it is true to say that, if we *know* a point is a point of inflection, then $\frac{d^2y}{dx^2} = 0$, but $\frac{d^2y}{dx^2} = 0 \nRightarrow$ that the point is a point of inflection.

Example 15.3   If $y = x^2 + x - 6$ find the point on the curve at which the gradient is zero. Sketch the curve.

Solution

$$y = x^2 + x - 6 \qquad (1)$$

$$\therefore \qquad \frac{dy}{dx} = 2x + 1 \qquad (2)$$

The gradient of the curve is zero when $\frac{dy}{dx} = 0$, i.e. $2x + 1 = 0$.

$$\therefore \qquad x = -\tfrac{1}{2}$$

When $x = -\frac{1}{2}$:

$$y \;=\; (-\tfrac{1}{2})^2 + (-\tfrac{1}{2}) - 6 \qquad \text{from (1)}$$
$$= \tfrac{1}{4} - \tfrac{1}{2} - 6$$
$$= -6.25$$

Therefore the point $(-\frac{1}{2}, -6\frac{1}{4})$ is the point of zero gradient.

To test whether it is a maximum or minimum point, we find the sign of $\dfrac{d^2 y}{dx^2}$ at the point where $x = -\frac{1}{2}$.

From (2) $$\dfrac{d^2 y}{dx^2} = 2$$

This is positive (whatever the value of x). Therefore $(-\frac{1}{2}, -6\frac{1}{4})$ is a minimum point on the graph. To sketch the curve, first find:

     (i)   where the curve cuts the x axis;

     (ii)   where the curve cuts the y axis.

(i)   This is where $y = 0$

i.e. $$0 = x^2 + x - 6$$
$$0 = (x + 3)(x - 2)$$
$\therefore$ $x = -3$ or $+2$

(ii)   This is where $x = 0$

i.e. $$y = 0^2 + 0 - 6$$
$\therefore$ $y = -6$

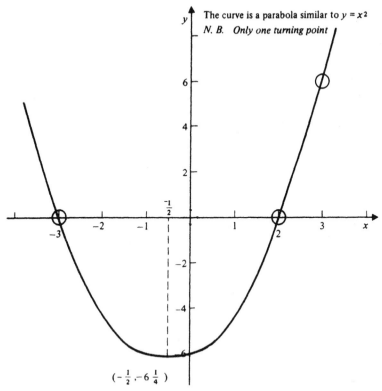

The curve is a parabola similar to $y = x^2$

N. B. *Only one turning point*

$(-\frac{1}{2}, -6\frac{1}{4})$

Figure 15.9

Example 15.4 Find the maximum and minimum values of $y = x^3 - 6x^2 + 9x + 6$ and sketch the curve.

Solution

$$y = x^3 - 6x^2 + 9x + 6 \qquad (1)$$

$$\therefore \quad \frac{dy}{dx} = 3x^2 - 12x + 9 \qquad (2)$$

For a turning point, $\dfrac{dy}{dx} = 0$

$$\therefore \qquad 3x^2 - 12x + 9 = 0$$

Divide by 3:

$$x^2 - 4x + 3 = 0$$

$$\therefore \qquad (x-1)(x-3) = 0$$

$$\therefore \qquad x = 1 \ \text{ or } \ x = 3$$

Substituting in (1):

when $x = 1$, $\quad y = 1 - 6 + 9 + 6 = 10$

when $x = 3$, $\quad y = 27 - 54 + 27 + 6 = 6$

$\therefore$ turning points occur at $(1,10)$ and $(3,6)$

To determine which of these is maximum and which is minimum we must look at the sign of $\dfrac{d^2y}{dx^2}$ at these points.

From (2): $\qquad\qquad\qquad\qquad\qquad \dfrac{d^2y}{dx^2} = 6x - 12$

When $x = 1$ $\quad \dfrac{d^2y}{dx^2}$ is negative $(= -6)$ $\therefore x = 1$, $y = 10$ is a maximum point

When $x = 3$ $\quad \dfrac{d^2y}{dx^2}$ is positive $(= +6)$ $\therefore x = 3$, $y = 6$ is a minimum point

Hence the maximum value of the function is 10, and the minimum value of the function is 6.

From (1), when $x = 0$, $y = 6$.

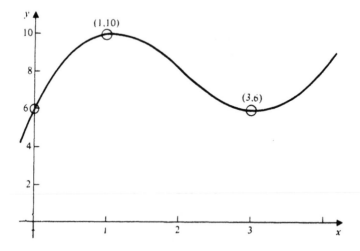

Figure 15.10

Exercise 15.2

In the following questions, find the maximum and/or minimum values of the given functions and the corresponding values of x. (You must discriminate between the maximum and minimum values by considering the sign of $\dfrac{d^2y}{dx^2}$ at the turning points.)

1. $y = 4x - x^2$ 2. $y = x^2 + x + 1$ 3. $y = x^2 - 3x + 1$

4. $y = 3x - x^3$ 5. $y = x^2(3 - x)$ 6. $y = (x + 3)(7 - x)$

7. The margin of profit in £ obtained by the sale of a certain article is given by

$$M = 120 - \frac{N}{10} \qquad \text{where } N \text{ is the number sold}$$

If I is the total income obtained by selling N articles, show that

$$I = 120N - \frac{N^2}{10}$$

and find the number of articles which must be sold to give the greatest income. (You must give a convincing proof that it is the greatest and not the least income.)

Find also the value of the greatest income.

8. Find the maximum value of the bending moment M of a beam at a point distance x from one end if M is given by

$$M = \tfrac{1}{2}w(lx - x^2)$$

where l is the length of the beam and w is the weight per unit length.

9. The power developed by a certain machine is given by

$$W = Ei - Ri^2$$

where i is the current and E and R are constants. Find the maximum value of W as i varies.

10. Find the maximum value of the sum of a positive number and its reciprocal.

11. The cost of a journey works out at £C per hour, given by

$$C = 16 + \frac{v^2}{100}$$

where v is the speed in km/h. Find the speed that gives the least cost for the whole journey.

(Hint: let the length of journey be d km and the time taken be t hours.)

12.   (i)   An open metal box with a square base side x and height y is to be made to hold 4 cubic metres. Write down two formulae giving the volume (which is 4) and surface area of the box in terms of x and y.

      (ii)   Eliminate y from these equations and show that $A = x^2 + \dfrac{16}{x}$.

      (iii)   Hence find the least area of sheet metal which can be used.

HOW TO DIFFERENTIATE MORE DIFFICULT EXPRESSIONS

The Chain Rule

Here we are concerned with such expressions as $(x + 1)^2$ or $(2x - 1)^3$ or $(5 - 3x)^2$, etc.

Take the first of these.

Let
$$y = (x + 1)^2$$
$$= x^2 + 2x + 1 \qquad (\text{using } (a + b)^2 = a^2 + 2ab + b^2)$$
$$\frac{dy}{dx} = 2x + 2$$

But we could have used another method as follows:

$$y = (\text{an expression})^2$$

Since the derivative of x^2 is $2x$ (and x can be anything!) then the derivative of (anything)^2 should be 2.(anything), i.e. the derivative of $(x + 1)^2$ should be $2(x + 1)$, i.e. $2x + 2$. This agrees with the correct answer above.

Let us try the second example.

Let
$$y = (2x - 1)^3 = (2x)^3 - 3(2x)^2.1 + 3.2x.1^2 - 1^3$$
$$\therefore \qquad\qquad y = 8x^3 - 12x^2 + 6x - 1$$
$$\frac{dy}{dx} = 24x^2 - 24x + 6 \qquad\qquad (1)$$

Using our other method:

$$y = (2x-1)^3 = (\text{something})^3 \quad \text{where } (2x-1) = (\text{something})$$
$$\frac{dy}{dx} = 3(\text{something})^2$$
$$= 3(2x-1)^2$$
$$= 3(4x^2 - 4x + 1)$$
$$= 12x^2 - 12x + 3 \qquad\qquad (2)$$

which does not check with (1). Why not?

Because we have not been differentiating with respect to x, but with respect to the (something), so that we have not quite finished.

Now
$$\frac{d}{dx}(2x-1) = 2 \qquad\qquad (3)$$

and if we multiply (2) by this we obtain

$$\frac{dy}{dx} = 2(12x^2 - 12x + 3)$$
$$= 24x^2 - 24x + 6 \qquad \text{which checks with (1)}$$

This suggests that we must differentiate the (something) w.r.t. x and multiply the answer obtained so far by this derivative.

We note that the first example gave us the correct answer because the derivative of the expression inside the brackets happened to be 1.

Now let us try the third example, by working it both ways.

First Method

Let
$$y = (5-3x)^2$$
$$= 25 - 30x + 9x^2$$
$$\frac{dy}{dx} = -30 + 18x$$

Second Method

Let
$$y = (5-3x)^2$$
$$\frac{dy}{dx} = 2(5-3x) \times (\text{the derivative of what is inside the brackets})$$
$$= 2(5-3x) \times (-3)$$
$$= -6(5-3x)$$

$$= -30 + 18x$$

The above method is known as the *chain rule* or the *function of a function rule* and expressed in symbols, states:

$$\boxed{\frac{dy}{dx} = \frac{dy}{dz} \cdot \frac{dz}{dx}}$$

where z represents the contents of brackets or similar function.

Example 15.5   Find the differential coefficient of $\sqrt{(3x+1)}$.

Solution   Let $y = \sqrt{(3x+1)} = (3x+1)^{\frac{1}{2}}$

$$\frac{dy}{dx} = \tfrac{1}{2}(3x+1)^{-\frac{1}{2}} \times \text{ (derivative of } 3x+1)$$

$$= \tfrac{1}{2}(3x+1)^{-\frac{1}{2}}.3$$

$$\frac{dy}{dx} = \frac{3}{2.\sqrt{3x+1}}$$

This would be difficult to do by any other method.

Exercise 15.3

Differentiate the following by using the chain rule:

1.	$(3x-2)^3$	2.	$(5-2x)^4$	3.	$(3-x)^5$
4.	$(4+3x)^6$	5.	$(x^3-2x+1)^2$	6.	$(3x^2+5x-1)^3$
7.	$\sqrt{4x-1}$	8.	$\sqrt{1-2x}$	9.	$\sqrt{ax+b}$
10.	$\sqrt[3]{1-ax}$	11.	$\dfrac{1}{1-x}$	12.	$\dfrac{1}{2+3x}$
13.	$\dfrac{1}{3x+4}$	14.	$\dfrac{1}{3x^4-7}$	15.	$\dfrac{1}{x^3-a^3}$

Differentiation of Products

We need to know how to differentiate expressions such as:

$$(2x+1)^2(1-5x) \quad \text{or} \quad (2x+1)\sqrt{(x+5)}$$

The rule for this is quite simple. We will state it without proof:

If u and v are both functions of x which are multiplied together to form a product uv, then:

$$\frac{d(uv)}{dx} = v\frac{du}{dx} + u\frac{dv}{dx}$$

In other words, to differentiate a product of two functions, multiply the second by the derivative of the first and add the first multiplied by the derivative of the second.

Example 15.6   Find the derivative of $(3x + 4)(2x + 1)$.

Solution   Here $u = 3x + 4$ and $\dfrac{du}{dx} = 3$

$$v = 2x + 1 \text{ and } \frac{dv}{dx} = 2$$

∴ the derivative of $(3x + 4)(2x + 1) = \dfrac{d}{dx}[(3x+4)(2x+1)]$

$$= (2x + 1).3 + (3x + 4).2 \quad \left[= v\frac{du}{dx} + u\frac{dv}{dx}\right]$$

$$= 6x + 3 + 6x + 8$$

$$= 12x + 11$$

Check by multiplying out the brackets and differentiating term by term.

Example 15.7   Find $\dfrac{d}{dx}(2x+1)(x+5)^3$.

Solution   Here $u = (2x + 1) \quad v = (x + 5)^3$

$$\frac{du}{dx} = 2 \qquad \frac{dv}{dx} = 3(x+5)^2.1$$

Using $\dfrac{d}{dx}(uv) = v\dfrac{du}{dx} + u\dfrac{dv}{dx}$:

$$\frac{d}{dx}(uv) = (2x+1).3(x+5)^2 + (x+5)^3.2$$

$$= (x+5)^2[3(2x+1)+2(x+5)]$$

$$= (x+5)^2(6x+3+2x+10)$$

$$= (x+5)^2(8x+13)$$

Differentiation of a Quotient

How can we differentiate an expression such as $\dfrac{d}{dx}\left[\dfrac{(x+1)}{(x+2)}\right]$?

We use a similar method to that for the differentiation of a product. It is a little more involved and is expressed as follows:

$$\frac{d}{dx}\left(\frac{u}{v}\right) = \frac{v\dfrac{du}{dx} - u\dfrac{du}{dx}}{v^2}$$

where $\dfrac{du}{dx}$ = derivative of the numerator and $\dfrac{dv}{dx}$ = derivative of denominator.

Example 15.8　　　Find the differential coefficient of $\dfrac{u}{v} = \dfrac{2x+1}{3x+2}$.

Solution　　　　　　$u = 2x + 1,\quad \dfrac{du}{dx} = 2$

　　　　　　　　　　$v = 3x + 2,\quad \dfrac{dv}{dx} = 3$

Then　　　　　$\dfrac{d}{dx}\left(\dfrac{u}{v}\right) = \dfrac{v\dfrac{du}{dx} - u\dfrac{du}{dx}}{v^2}$

　　　　　　　　　$= \dfrac{(3x+2).2 - (2x+1).3}{(3x+2)^2}$

　　　　　　　　　$= \dfrac{6x+4-6x-3}{(3x+2)^2}$

　　　　　　　　　$= \dfrac{1}{(3x+2)^2}$

Example 15.9　　　Differentiate $\dfrac{(x^3+1)^2}{1-2x^2}$.

Solution　　　$\dfrac{u}{v} = \dfrac{(x^3+1)^2}{1-2x^2}$

$u = (x^3+1)^2,\quad \dfrac{du}{dx} = 2(x^3+1).3x^2 \qquad v = (1-2x^2),\quad \dfrac{dv}{dx} = -4x$

$$\frac{d}{dx}\left(\frac{u}{v}\right) = \frac{v\dfrac{du}{dx} - u\dfrac{dv}{dx}}{v^2}$$

$$\frac{d}{dx}\left(\frac{u}{v}\right) = \frac{(1-2x^2).2(x^3+1).3x^2 - (x^3+1)^2.(-4x)}{(1-2x^2)^2}$$

$$= \frac{2x(x^3+1)[3x(1-2x^2)+2(x^3+1)]}{(1-2x^2)^2}$$

$$= \frac{2x(x^3+1)(3x-6x^3+2x^3+2)}{(1-2x^2)^2}$$

$$= \frac{2x(x^3+1)(3x-4x^3+2)}{(1-2x^2)^2}$$

Now we can differentiate sums, differences, products and quotients of quite complicated functions involving powers of x.

As we introduce new functions later in the book, we will expand further on this topic.

To summarise the basic rules:

1. $\quad \dfrac{d}{dx}(x^n) = nx^{n-1}$

2. $\quad \dfrac{d}{dx}(uv) = v\dfrac{du}{dx} + u\dfrac{dv}{dx}$ $\qquad$ where u, v are functions of x

3. $\quad \dfrac{d}{dx}\left(\dfrac{u}{v}\right) = \dfrac{v\dfrac{du}{dx} - u\dfrac{dv}{dx}}{v^2}$

4. $\quad \dfrac{d}{dx}$ (something)^n = n(something)$^{n-1}$ × (the derivative of the something)

Exercise 15.4

Differentiate the following:

1. $x^2 + 5x + 1$ $\qquad$ 2. $(x^2+5x+1)^3$ $\qquad$ 3. $(x+3)^{1/2}$ $\qquad$ 4. $(2x+3)^{1/2}$

5. $(3x+2)(5x-1)$ $\qquad$ 6. $\sqrt{1-t^2}$ $\qquad\qquad$ 7. $\sqrt[3]{(x+1)}$ $\qquad$ 8. $\sqrt[3]{(2x^2+1)}$

9. $(x^2+x+1)(x^2-x-1)$ $\qquad$ 10. $3x(x+1)^5$ $\qquad\qquad$ 11. $2x^2.(4x+1)^3$

12. $(2x^2+7)(x+1)^2$ $\qquad\qquad$ 13. $\sqrt{x}(x^2-3x+2)$ $\qquad$ 14. $\dfrac{x+1}{x-1}$

15. $\dfrac{2x^2+4}{3x-2}$ $\qquad$ 16. $\dfrac{a+x}{a-x}$ $\qquad$ 17. $\dfrac{(2t^{1/2}+1)}{2t^{1/2}}$ $\qquad$ 18. $\dfrac{(x^2+x-1)}{(x^2-x+1)}$

19. $\dfrac{2t}{\sqrt{1-t^2}}$ $\qquad$ 20. $\dfrac{t^2}{(1+t)^2}$ $\qquad$ 21. $\dfrac{1}{\sqrt{ax+b}}$

16

Integration

Having learnt to differentiate successfully, it is a logical step to reverse the process. This reversing process is known as *integration*.

When we differentiate a given function we are finding the gradient of the tangent at any point on the graph of the function. When we integrate, we are asking the question 'If we know the gradient of the tangent at every point on a curve, can we find the equation of the curve?'

Graphically, we are asking the following:

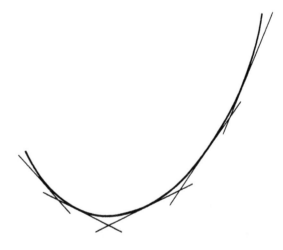

Figure 16.1

Can we fit a curve into a given set of tangents? It seems logical to see a curve being traced out as the tangent rolls round into different positions (Figure 16.1) and this is what we will try to do.

Thus, *differentiating* x^3 gives us $3x^2$ so that the reverse, i.e. *integrating* $3x^2$ with respect to x, gives us x^3.

Hence, we see that, for example,

$$\text{if } y = x^3 \text{ then } \frac{dy}{dx} = 3x^2$$

should mean that
$$\text{if } \frac{dy}{dx} = 3x^2 \text{ then } y = x^3$$

But is this the only answer? Suppose we differentiate $x^3 + 4$ or $x^3 + 10$ or $x^3 - 7$ or $x^3 \pm$ any constant. All these functions give $\frac{dy}{dx} = 3x^2$. In each case, the constant disappears when we differentiate. Therefore when reversing the process, we must put the constant back.

It is important to understand that *we do not know the value of this constant without being given additional information.* We have to include it because it might have been there before the differentiation was performed. Of course, it *might* have been zero but we cannot assume that.

Hence, the complete answer is

$$\text{if } \frac{dy}{dx} = 3x^2 \text{ then } y = x^3 + c$$

where c is an unknown constant, called the **constant of integration**.
We use a special symbol to denote integration. This 'integral sign', $\int$, is the old-fashioned elongated letter 's'. We shall see later that another way of viewing integration is as the limiting value of a special sum and this is what the sign stands for.

The above result is usually written

$$\int 3x^2 dx = x^3 + c$$

and this is read as 'the integral of $3x^2$ with respect to $x = x^3 + $ a constant'.

The following table shows how we can deduce several other results for different powers of x.

y	$\dfrac{dy}{dx}$		Conversely:	$f(x)$	$\int f(x)dx$
x^4	$4x^3$			$4x^3$	$x^4 + c$
x^3	$3x^2$			$3x^2$	$x^3 + c$
x^2	$2x$			$2x$	$x^2 + c$
x	1			1	$x + c$
Constant	0			0	c
x^{-1} or $\dfrac{1}{x}$	$-x^{-2}$ or $-\dfrac{1}{x^2}$			$\dfrac{1}{x^2}$	$-\dfrac{1}{x} + c$
x^{-2} or $\dfrac{1}{x^2}$	$-2x^{-3}$ or $-\dfrac{2}{x^3}$			$\dfrac{2}{x^3}$	$-\dfrac{1}{x^2} + c$

But if we divide both columns of the second table by the various coefficients we obtain the following:

$f(x)$	$\int f(x)dx$
x^3	$x^4/4 + c$ (dividing by 4)
x^2	$x^3/3 + c$ (dividing by 3)
x	$x^2/2 + c$ (dividing by 2)
1	$x^1/1 + c$ (dividing by 1)
0	c
$\dfrac{1}{x^2}$	$-\dfrac{1}{x} + c$ (dividing by -1)
$\dfrac{1}{x^3}$	$-\dfrac{1}{2x^2} + c$ (dividing by -2)

We can now see that all these results follow the following pattern:

$$\int x^n dx = \frac{x^{n+1}}{n+1} + c \qquad (n \neq -1)$$

i.e. to integrate a power of x increase the power by 1 and divide by the same number.

N.B. We cannot yet integrate $\dfrac{1}{x}$ (i.e. x^{-1}) since if $n = -1$, the above formula would give us a zero denominator and thus cannot be applied. We will see how to integrate $\dfrac{1}{x}$ in a later chapter.

Example 16.1   Integrate the following:

(a) x^5 (b) x^7 (c) $\sqrt[3]{x}$ (d) $3x^2 - 4x^2 + 5x - 3$

Solution   (a) $\displaystyle\int x^5 dx = \frac{x^{5+1}}{5+1} + c = \frac{x^6}{6} + c$

(b) $\displaystyle\int x^7 dx = \frac{x^8}{8} + c$

(c) $\displaystyle\int \sqrt[3]{x}\,dx = \int x^{\frac{1}{3}} dx = \frac{x^{\frac{1}{3}+1}}{\left(\frac{1}{3}+1\right)} + c = \frac{3x^{\frac{4}{3}}}{4} + c$

(d) Since we can differentiate term by term, we can do the same when integrating:

$$\int (3x^2 - 4x + 3)dx$$
$$= 3\times\frac{x^3}{3} - 4\times\frac{x^2}{2} + 3x + c$$
$$= x^3 - 2x^2 + 3x + c$$

N.B.1 It is important to remember to include the 'dx' each time you integrate and to remember to include the *constant of integration* each time.

N.B.2 Whenever possible, answers should be checked carefully by differentiating them. You should get back to what you were given.

Exercise 16.1

Integrate the following functions of x:

1.   (a) x^6 (b) x^{10} (c) x^9 (d) x^2 (e) x

   (f)   1   (g)   0   (h) $x^{1.4}$ (i) $x^{-1.4}$ (j) $x^{-0.5}$

2.   (a) $\dfrac{1}{x^3}$ (b) $\dfrac{1}{x^4}$ (c) $\dfrac{1}{x^5}$ (d) $\dfrac{1}{x^6}$

3.   (a) $x^{\frac{1}{2}}$ (b) $\dfrac{1}{\sqrt{x}}$ (c) $\dfrac{5}{x^2}$ (d) $3x^{\frac{2}{3}}$

4.   (a) $x^2 + x$ (b) $x^3 - a^3$ (c) $ax^2 + bx + c$

    (where a, b, c are constants)

5.   (a) $3x^2 + 2x - 1$ (b) $5 - x^3 + 2x^2 - x$ (c) $5x^{\frac{1}{2}} + 2x^{\frac{3}{2}}$

INTEGRATION AS THE LIMIT OF A SUM

We have introduced integration as a process that is the *converse of differentiation*. Given the gradient of a curve at every point x, we can find the equations of a family of curves that each have that gradient.

We can also view integration as a sum of all products $y.\delta x$ as δx approaches zero. Thus if y is some function of x, we can express an integral in the form

$$\int y\,dx = \underset{\delta x \to 0}{Lim} \sum y\delta x$$

We read this as: 'the limit as δx approaches zero of sigma y delta x'. The symbol $\sum$ is used here for adding up a large number of small pieces. We have mentioned previously that the sign $\int$ is a long S, meaning 'find the sum of all such things as'. $\int dx$ means 'add up all the little bits of x', which obviously gives a whole x (Figure 16.2).

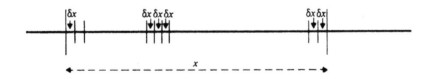

Figure 16.2

AREAS: DEFINITE INTEGRALS

The question now, of course, is why should we view an integral in such a way? A good example of this way of viewing an integral is its application to the evaluation of the area between a curve and the x axis (Figure 16.3).

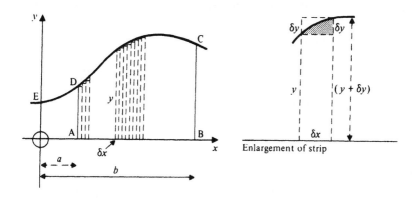

Figure 16.3(a)   Figure 16.3(b)

To find an area such as ABCD we divide the area into narrow strips of width δx and height y, a few of which are shown in Figure 16.3(a). The area required is the sum of the areas of these strips. Each strip is approximately a rectangle plus a small area just under the curve. (See shaded area in Figure 16.3(b).)

If we add the areas of the rectangles, we obtain an area approximately equal to the area ABCD, but with a serrated top instead of a smooth curve. Now imagine the strips becoming narrower until they become merely the thickness of a fine line as in Figure 16.4, in which only a few have been drawn.

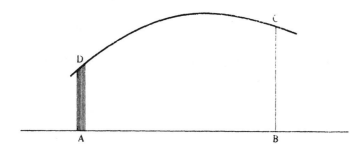

Figure 16.4

We see that if enough of these lines are drawn they will fill the whole area, and as δx becomes even smaller, and the number of strips becomes correspondingly larger, the serrated edge becomes indistinguishable from the curve itself. In this way we see that the area required is the sum of the areas of a very large number of very narrow rectangles of area $y\delta x$ and we write:

the area of ABCD = the limit of the sum of all areas $y\delta x$ as δx approaches zero

or, in mathematical shorthand:

$$\text{the area of ABCD} = \underset{\delta x \to 0}{Lim} \sum y\delta x \text{ between } x = a \text{ and } x = b$$

i.e. $$\text{Area ABCD} = \int y\delta x \text{ between the limits } x = a \text{ and } x = b$$

or $$\text{Area ABCD} = \int_a^b ydx$$

When we write an integral in this form, between definite limits, we call it a definite integral. If we are not using such limits, we call it an indefinite integral.

(Note that there is an unfortunate duplication of the word 'limit' here. The numbers that define the starting and finishing x values would perhaps be better named as boundaries but we are unfortunately stuck with the tradition of using the same word to mean two different things!)

How do we use this idea? Take the example of the area ABCD illustrated in Figure 16.3. When we need to evaluate the area ABCD, the instruction, $\int_a^b ydx$, means 'find the integral of ydx between the lower limit a and the upper limit b'.

In other words, we do not require the area OADE between the origin, O, and A.

But area ABCD = area OBCE – area OADE. This is saying that the area required is the *difference* between the *indefinite* integral evaluated for the upper limit and the indefinite integral evaluated for the lower limit. Some examples will illustrate the method.

Example 16.2   Find the area between the curve $y = x^2 + 3$, the x axis, and the limits $x = 1$ and $x = 4$.

Solution

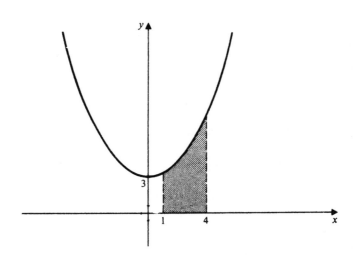

Figure 16.5

The curve is a parabola, symmetric about the y axis and cutting the y axis at $y = 3$.

$$A = \int (x^2 + 3)dx \text{ between the limits } x = 1 \text{ and } x = 4$$

i.e.
$$A = \int_1^4 (x^3 + 3)dx$$

$$= \left[\frac{x^4}{4} + 3x + \text{constant} \right]_1^4 \qquad \text{(note \textbf{square} brackets as a convention)}$$

This means that we substitute first $x = 4$, then $x = 1$ and subtract the second from the first as follows:

$$A = \left[\frac{4^4}{4} + 3.4 + \text{constant} \right] - \left[\frac{1^4}{4} + (3.1) + \text{constant} \right]$$

$$= 64 + 12 + C \qquad - \tfrac{1}{4} - 3 - C$$

(Note that the constant of integration will *always* cancel out for a definite integral, and it is therefore *not* necessary to put it in. For an indefinite integral, we must still include a constant of integration.)

$\therefore$ $\qquad\qquad\qquad A = 72.75$

Example 16.3　　　Find the areas of the two loops enclosed between the curve $y = (x - 1)(x + 2)(x - 3)$ and the x axis.

Solution　　　We note that this is a cubic curve having two turning points.

To find where it cuts the x axis put $y = 0$,

i.e. $\qquad\qquad\qquad 0 = (x - 1)(x + 2)(x - 3)$

$\qquad\qquad\qquad x = +1 \text{ or } -2 \text{ or } +3$

To find where it cuts the y axis put $x = 0$,

$$y = (-1)(+2)(-3) = +6$$

A sketch of the curve is therefore as in Figure 16.6, giving the shaded areas required, ABC and CDE.

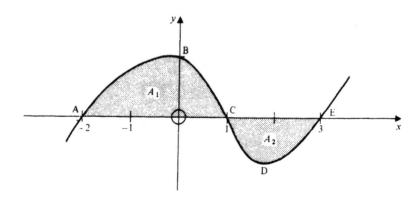

Figure 16.6

So that area ABC $(A_1) = \int y dx$ between limits $x = -2$ to $+1$

But $\qquad\qquad y = (x - 1)(x + 2)(x - 3)$

$\qquad\qquad\qquad = (x - 1)(x^2 - x - 6)$

$\qquad\qquad\qquad = x^3 - x^2 - 6x - x^2 + x + 6$

i.e. $y = x^3 - 2x^2 - 5x + 6$

$$A_1 = \int_{-2}^{+1}(x^3 - 2x^2 - 5x + 6)dx$$

$$= \left[\frac{x^4}{4} - \frac{2x^3}{3} - \frac{5x^2}{2} + 6x\right]_{-2}^{+1}$$

$$= \left[\frac{1}{4} - \frac{2}{3} - \frac{5}{2} + 6\right] - \left[\frac{(-2)^4}{4} - 2.\frac{(-2)^3}{3} - \frac{5(-2)^2}{2} - 12\right]$$

$$= \frac{37}{12} - \frac{16}{3} + 18$$

$$= \frac{63}{4}$$

Area CDE $(A_2) = \int ydx$ between limits $x = 1$ to 3

$$A_2 = \left[\frac{x^4}{4} - \frac{2.3^3}{3} - \frac{5x^2}{2} + 6x\right]_1^3$$

$$= \left[\frac{3^4}{4} - \frac{2.3^3}{3} - \frac{5.3^2}{2} + 6.3\right] - \left[\frac{1^4}{4} - \frac{2.1^3}{3} - \frac{5.1^3}{2} + 6.1\right]$$

$$= \left[\frac{81}{4} - 18 - \frac{45}{2} + 18\right] - \left[\frac{1}{4} - \frac{2}{3} - \frac{5}{2} + 6\right]$$

$$= \frac{81}{4} - \frac{45}{2} - \frac{1}{4} + \frac{2}{3} + \frac{5}{2} - 6$$

$$= \frac{80}{4} - \frac{40}{2} + \frac{2}{3} - 6$$

$$= -\frac{16}{3}$$

Note that areas *below* the x axis always come out negative since the 'heights' of all the small strips are negative. We always quote such areas as positive by taking the modulus of the answer.

Exercise 16.2

1.   Find $\int_{-2}^{+3} ydx$ for the function in the previous example. Did you expect this result?

Evaluate the following (in each case make a sketch of the curve):

2.　　The area between the curve $y = x^2 - x - 2$ and the x axis.

3.　　The area of the trapezium bounded by the line $y = 2x + 3$ and the ordinates at $x = 1$ and $x = 3$.

4.　　The area between the curve $y = x^2$, the x axis, and the ordinates at $x = -1$ and $x = +1$.

5.　　The areas of the loops formed by the curve $y = x(x + 1)(x - 2)$ and the x axis.

SOLIDS OF REVOLUTION

We can use integration to solve problems other than finding areas. Imagine a curve being rotated about the x axis. A surface is swept out which encloses a volume. It is possible to find the volume of this enclosed space by integration as follows:

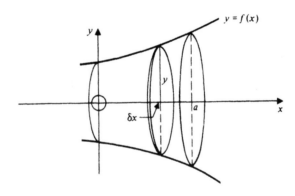

Figure 16.7

Consider any curve, $y = f(x)$. Let this be rotated about the x axis between the limits $x = 0$ and $x = a$. The solid generated is shown in Figure 16.7.

Now consider a small circular slice of thickness δx and radius y, taken at right angles to the x axis, anywhere within the solid. The volume of this slice $\cong \pi y^2 \delta x$, since the slice is approximately a cylinder, radius y and height δx.

∴   this small volume, $\delta V \cong \pi y^2 \delta x$

Summing these small volumes, we get, as $\delta x \to 0$,

$$\int dV = \pi \int y^2 dx \quad \text{from } x = 0 \text{ to } x = a$$

∴

$$\boxed{= \pi \int y^2 dx}$$

Let us use this fact in a few examples.

Example 16.4   Prove that the volume of any right cone is $\frac{1}{3} \times$ base area × height.

Solution   A right cone is a solid of circular cross-section, with its axis at right angles to the circular base as shown in Figure 16.8.

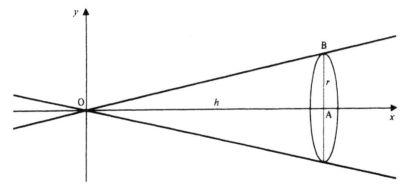

Figure 16.8

To prove $V = \frac{1}{3}\pi r^2 h$, where OA $= h$, AB $= r$.

Turning the cone on its side, we can generate the cone by rotating the correct line about the x axis. We must know the equation of the line. We find this as follows (Figure 16.9):

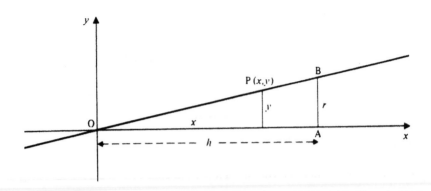

Figure 16.9

Let P be any point on the line with coordinates (x,y). Then the gradient of the line is $\dfrac{r}{h}$ and it passes through the origin.

$$\therefore \qquad\qquad y = \frac{r}{h}.x \qquad\qquad \text{(this is the equation of OB)}$$

Using this in our formula for volume we get

$$V = \pi \int_0^h \frac{r^2}{h^2}.x^2\,dx \qquad \text{(substituting for } y\text{)}$$

But $\dfrac{r^2}{h^2}$ is a constant and so can be taken outside the integral sign.

Hence
$$V = \frac{\pi r^2}{h^2}\left[\frac{x^3}{3}\right]_0^h$$

$$= \frac{\pi r^2}{h^2}\left[\frac{h^3}{3} - 0\right]$$

so
$$V = \frac{\pi r^2 h}{3} = \frac{1}{3}(\pi r^2)h$$

i.e.   volume of cone $= \dfrac{1}{3}$ area of base $\times$ height

Example 16.5   Find the volume swept out by the parabola $y = 5x^2 + 3$ when it is rotated round the x axis between $x = 1$ and $x = 2$.

Solution   When $x = 0, y = 3$.

Therefore the curve is roughly as shown in Figure 16.10.

We have to find the volume of solid ABCD.

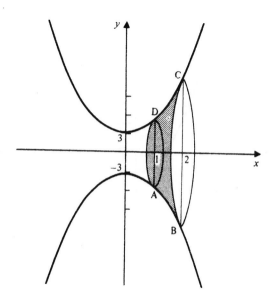

Figure 16.10

$$V = \pi \int y^2 dx \text{ from } x = 1 \text{ to } x = 2$$

$$= \pi \int_1^2 (5x^2 + 3)^2 dx$$

$$= \pi \int_1^2 (25x^4 + 30x^2 + 9) dx$$

$$= \pi \left[\frac{25x^5}{5} + \frac{30x^3}{3} + 9x \right]_1^2$$

$$= \pi[5x^5 + 10x^3 + 9x]_1^2$$
$$= \pi[258 - 24]$$
$$= 234\pi \text{ cubic units}$$

Example 16.6   Find the bowl-shaped volume generated when the curve in the previous example is rotated around the y axis (Figure 16.11) between $y = 3$ and $y = 5$.

Solution

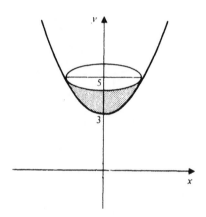

Figure 16.11

By a similar argument to that used for rotation round the x axis, it follows that when rotating round the y axis the volume $= \pi \int x^2 dx$ from $y = 3$ to $y = 5$. But

$$y = 5x^2 + 3$$

$\therefore$ $$y - 3 = 5x^2$$

so
$$\frac{y-3}{5} = x^2$$

and
$$V = \frac{\pi}{5}\int_3^5 (y-3)\,dy$$

(Note again that a constant factor can be taken outside the integral sign.)

$$\therefore \qquad V = \frac{\pi}{5}\left[\frac{y^2}{2} - 3y\right]_3^5$$

$$= \frac{\pi}{5}\left[\frac{25}{2} - 15\right] - \frac{\pi}{5}\left[\frac{9}{2} - 9\right]$$

$$= \frac{\pi}{5}[-2\tfrac{1}{2} + 4\tfrac{1}{2}]$$

$$V = \frac{2\pi}{5} \text{ cubic units}$$

Exercise 16.3

Find the volumes generated when the areas under the given curves, between the given values of x, are rotated round the x axis. In each case sketch the curve (or straight line) first. Leave answers in terms of π.

1. $\qquad y = x$ from $x = 2$ to $x = 4$.

2. $\qquad 2y = x + 3;$ $x = 0$ to $x = 8$.

3. $\qquad y = x^2;$ $x = 0$ to $x = 5$.

4. $\qquad y = x(1 - x);$ $x = 0$ to $x = 1$.

17

Introduction to Trigonometry

THE SINE, COSINE AND TANGENT OF AN ANGLE

We can explain the meaning of the word trigonometry very easily if we split the word in two. Since a polygon is a many-sided figure, a 'trigon' is a three-sided figure. The other part of the word, i.e. 'metry', is associated with measurement. For example, geometry literally means Earth measurement. Hence trigonometry simply means triangle measurement and trigonometry was originally the study of triangles.

We begin by looking at right-angled triangles but will soon want to extend our basic ideas into something much broader.

Look at these two triangles

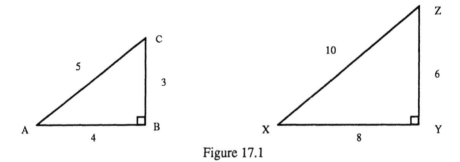

Figure 17.1

We call them *similar*, since they are exactly the same shape, but have different sizes. If the sizes were also the same, they would be *congruent*. All angles in the first triangle correspond to those in the second, i.e.

$$\angle A = \angle X, \ \angle B = \angle Y = 90° \ \text{and} \ \angle C = \angle Z$$

but of course the sides are all different in length.

However, if we find the ratio of any two sides of the first triangle and the ratio of the corresponding sides of the second, then we end up with the same number, i.e.

$$\frac{YZ}{XZ} = \frac{6}{10} = \frac{3}{5} = \frac{BC}{AC} \qquad \text{Similarly}: \quad \frac{AB}{AC} = \frac{XY}{XZ} \quad \text{and} \quad \frac{BC}{AB} = \frac{YZ}{XY}$$

These three ratios will always be equal for any two *similar* triangles and are called the *major trigonometrical ratios*. We find it useful to associate them with the actual angles of the triangle; so, concentrating on the second triangle, we will give the ratios special names. For the moment, choose angle X as being important. (We could equally well have chosen angle Z; it does not matter.) The ratios are known as the *sine, cosine* and *tangent* of the angle X. These are usually shortened to *sin, cos* and *tan*. (**N.B.** *Do not use capital letters!*)

Definitions for acute angles:

$$\sin X = \frac{YZ}{XZ} \qquad \cos X = \frac{XY}{XZ} \qquad \text{and} \qquad \tan X = \frac{YZ}{XY}$$

The ratios are difficult to remember in this form; so looking again at the second triangle and remembering that the longest side of a right-angled triangle is known as the *hypotenuse*, our definitions say that:

$$\sin X = \frac{\text{length of side } \textbf{opposite to } X}{\text{length of } \textbf{hypotenuse}}$$

$$\cos X = \frac{\text{length of side } \textbf{adjacent to } X}{\text{length of } \textbf{hypotenuse}}$$

$$\tan X = \frac{\text{length of side } \textbf{opposite to } X}{\text{length of side } \textbf{adjacent to } X}$$

(There are two sides adjacent to X but one of them is the hypotenuse!) If we had used the other triangle, then we would find that $\sin A$ would be the same as $\sin X$ because the angle A is the same as the angle X.

Notice that the side opposite to X is the side adjacent to Z so a little thought tells us that $\sin X$ is the same as $\cos Z$ and vice versa. In fact it is always true that:

If two angles add up to $90°$, the sine of one of them is the cosine of the other.

or: $$\sin X = \cos Z = \cos(90 - X) \quad \text{and} \quad \cos X = \sin Z = \sin(90 - X)$$

We can also see that

$$\tan X = \frac{\sin X}{\cos X}$$

since:

$$\tan X = \frac{YZ}{XY} = \left(\frac{YZ}{XY}\right)\cdot\left(\frac{XZ}{XZ}\right) = \left(\frac{YZ}{XZ}\right)\cdot\left(\frac{XZ}{XY}\right) = (\sin X)\cdot\left(\frac{1}{\cos X}\right) = \frac{\sin X}{\cos X}$$

USING A CALCULATOR

It is very easy to find the sin, cos or tan of an angle using a calculator, as the following examples show. At present, we are working with degrees but soon we will meet another way of measuring an angle, known as the *radian*. All you need to remember before working out trigonometric ratios, is to make sure your calculator is in *degree mode* or *radian mode*. The instruction leaflet that came with your calculator will tell you how to do this. (You may also have a third angle mode marked '*grad*'. Take care *never* to use this!)

Example 17.1   Find the sin, cos and tan of (a) 35° (b) 77°.

Solution

(a)   Put calculator in degree mode

      enter angle 35

      press **sin** button   answer should be 0.574 to 3 d.p.

      enter angle 35

      press **cos** button   answer should be 0.819 to 3 d.p.

      enter angle 35

      press **tan** button   answer should be 0.700 to 3 d.p.

(b)   This time we will use the memory buttons, to save re-entering the angle.

      enter angle 77

press **memory in** button (or equivalent)

press **sin** button   answer should be 0.974 to 3 d.p.

press **memory return** (or equivalent)

press **cos** button   answer should be 0.225

press **memory return**

press **tan** button   answer should be 4.331

Exercise 17.1

1.   Use your calculator to find the following trigonometrical ratios:

(a)   sin 30°   (b)   cos 57°   (c)   tan 49°

(d)   sin 87.2°   (e)   cos 15.9°   (f)   tan 48.54°

2.   Verify the statement $\sin X = \cos(90 - X)$ for angles less than 90° by calculating:

(a)   cos 30°   and   sin 60°

(b)   cos 73°   and   sin 17°

(c)   sin 53.4°   and   cos 36.6°

3.   Verify the statement $\tan X = \dfrac{\sin X}{\cos X}$ for the following angles:

(a)   30°   (b)   27°   (c)   74.9°   (d)   11.452°

ANGLES OF ANY MAGNITUDE

In practice, we have to deal with angles greater than 90°. For such angles, our previous definitions make no sense. We cannot have an angle greater than 90° in a right-angled triangle. A much wider set of definitions is necessary, which includes our previous definitions as a special case.

We begin by remembering that an angle measures a *rotation*. The following *quadrant diagram* (Figure 17.2) helps us to picture a rotation of any magnitude. The line OP, which is fixed at O, is allowed to rotate in the plane of the page. We call OP a *rotating vector*. Think of looking downwards on a revolving door, hinged at O.

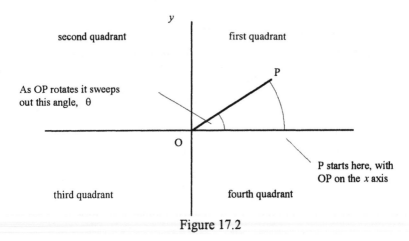

Figure 17.2

OP can be of any length and *always starts by lying along the positive x axis*. As OP rotates, it sweeps out an angle, θ *(theta)*. We will immediately introduce a new convention: if OP rotates in an anticlockwise sense, it gives a positive angle, and a clockwise rotation gives a negative angle. Negative angles are no stranger than negative numbers, they simply mean 'clockwise'.

Here are some angles, together with their quadrant diagrams (not to scale).

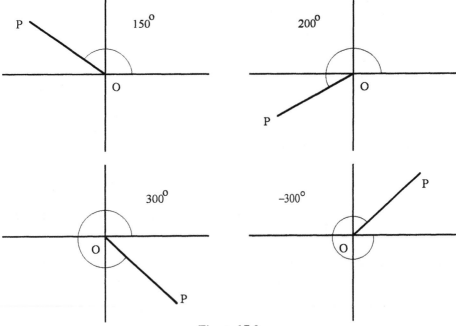

Figure 17.3

EQUIVALENT ANGLES

Two completely different rotations can give us similar diagrams. In the following, OP ends up in an identical position, even though the angles represented are very different.

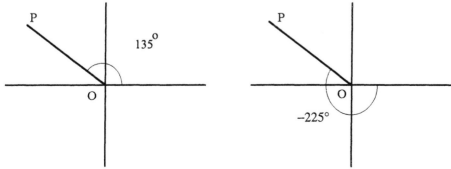

Figure 17.4

Such angles are said to be equivalent. In both the above, the rotating vector ends up making an acute angle of 45° with the x axis.

Exercise 17.2

1. Draw quadrant diagrams for the following angles:

    (a)   50°   (b)   −50°   (c)   130°
    (d)   −130°   (e)   230°   (f)   −230°
    (g)   310°   (h)   −310°   (i)   420°
    (j)   −600°   (k)   810°   (l)   −900°

2. In each case above, find the acute angle that your rotating vector makes with the x axis and hence state which angles are equivalent.

TRIGONOMETRIC RATIOS FOR ANGLES OF ANY MAGNITUDE

Now give our rotating vector, OP, a length of 1 unit (it could be 1 centimetre, 1 inch, 1 mile, it does not matter) and consider OP lying in the first quadrant as shown below:

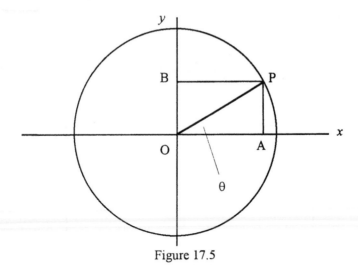

Figure 17.5

Draw AP and BP parallel to the axes as shown. OA is known as the projection of OP on the x axis and OB is the projection on the y axis. Let θ be the angle AOP.

From our previous definitions of sin and cos:

$$\sin\theta = \frac{AP}{OP} = \frac{OB}{OP} = OB, \text{ since } OP = 1, \text{ and } \cos\theta = \frac{OA}{OP} = OA$$

so we may simply define:

$$\sin\theta = \text{projection of OP on the } y \text{ axis}$$
$$\cos\theta = \text{projection of OP on the } x \text{ axis}$$
$$\tan\theta = \frac{\sin\theta}{\cos\theta}$$

These definitions can now be used for any angle. Remember that the projections are OA and OB (*not* AO and BO) and can therefore be positive or negative depending on which way they lie on the axes. OP is a *length* of 1 unit and is therefore always positive.

This has some consequences on the *signs* of our trigonometric ratios, as the following diagrams show.

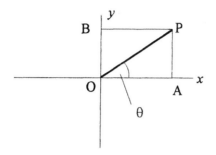

Angles in 1st. quadrant

OA is positive, OB is positive.

Hence

sin θ is positive

cos θ is positive

tan θ is positive

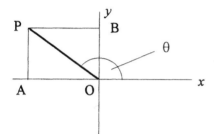

Angles in 2nd. quadrant

OA is negative, OB is positive.

Hence

sin θ is positive

cos θ is negative

tan θ is negative

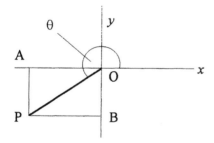

Angles in 3rd. quadrant

OA is negative, OB is negative.

Hence

sin θ is negative

cos θ is negative

tan θ is positive

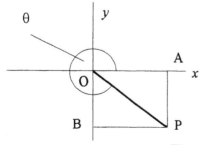

Angles in 4th. quadrant

OA is positive, OB is negative.

Hence

sin θ is negative

cos θ is positive

tan θ is negative

Figure 17.6

This can be summarised as follows:

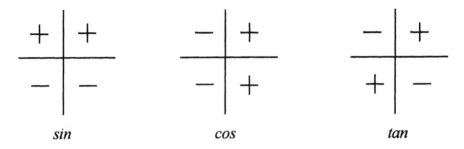

$$sin \qquad\qquad cos \qquad\qquad tan$$

Figure 17.7

TRIGONOMETRIC RATIOS WHEN OP LIES ON THE AXES

As an example, consider $0°$. In this case, the projection on the y axis is zero and the projection on the x axis is OP itself.

$$\sin 0° = \frac{0}{OP} = 0 \qquad \cos 0° = \frac{OP}{OP} = 1 \qquad \tan 0° = \frac{\sin 0°}{\cos 0°} = \frac{0}{1} = 0$$

Also $$\sin 90° = \frac{OP}{OP} = 1 \qquad \cos 90° = \frac{0}{OP} = 0 \qquad \tan 90° = \frac{\sin 90°}{\cos 90°} = \frac{1}{0} = \infty$$

(See the earlier discussion on division by zero!)

$$\therefore \qquad \begin{array}{lll} \sin 0° = 0 & \cos 0° = 1 & \tan 0° = 0 \\ \sin 90° = 1 & \cos 90° = 0 & \tan 90° = \infty \end{array}$$

The rest are left as an exercise and summarised here:

$$\begin{array}{lll} \sin 180° = 0 & \cos 180° = -1 & \tan 180° = 0 \\ \sin 270° = -1 & \cos 270° = 0 & \tan 270° = -\infty \\ \sin 360° = 0 & \cos 360° = 1 & \tan 360° = 0 \text{ which are the same as for } 0° \end{array}$$

As usual, this information is easier to see in a diagram; so we will superimpose it on Figure 17.7 to give:

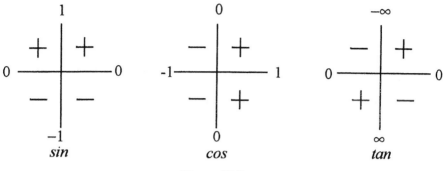

Figure 17.8

Note that if you try to find tan 90° or tan 270° on your calculator, it will signal an error. You need to be aware that, in this case, it is not your fault!

THE ASSOCIATED ACUTE ANGLE

Unless OP ends up lying along one of the axes, it will always make an acute angle with the *x axis*. For example:

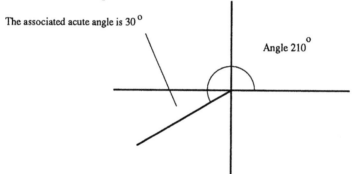

Figure 17.9

In each case, the sin, cos and tan of θ are related to the sin, cos and tan of this acute angle which we will call α (alpha). For example, in the above diagram, $\alpha = 30°$.

$$\sin 210° = -0.5 \qquad\qquad \sin 30° = 0.5$$
$$\cos 210° = -0.866 \qquad\qquad \cos 30° = 0.866$$
$$\tan 210° = 0.577 \qquad\qquad \tan 30° = 0.577$$

In other words, the trigonometric ratios for any angle are *numerically* equal to the corresponding ratios for the associated acute angle. Their sign, of course, depends upon which quadrant OP lies in.

USE OF CALCULATORS

Before the advent of cheap electronic calculators, we relied on books of tables to find trigonometric ratios. These tables only gave angles between $0°$ and $90°$. Hence for any other angle, we relied on first finding the associated acute angle and then working out the ratios for our angle. This is unnecessary today as you simply use your calculator.

Example 17.2   Find the sin, cos and tan of (a) $300°$ (b) $-495°$.

Solution

(a)   Put calculator in degree mode

       enter 300

       press **sin** button   answer should be -0.866 to 3 d.p.

       enter 300

       press **cos** button   answer should be 0.5 exactly

       enter 300

       press **tan** button   answer should be -1.732 to 3 d.p.

(b)

       enter 495

       press +/– button

       press **sin** button   answer should be -0.707 to 3 d.p.

       enter 495

       press +/– button

       press **cos** button   answer should be -0.707 to 3 d.p.

       enter 495

       press +/– button

       press **tan** button   answer should be 1

As an alternative, you could use your memory buttons so that the angle is only entered once.

The diagrams below should make it clear what is going on.

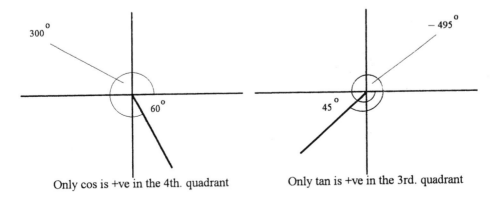

Only cos is +ve in the 4th. quadrant   Only tan is +ve in the 3rd. quadrant

Figure 17.10

$\sin 300° = -\sin 60° = -0.866$ $\sin(-495°) = -\sin 45° = -0.707$

$\cos 300° = +\cos 60° = 0.5$ $\cos(-495°) = -\cos 45° = -0.707$

$\tan 300° = -\tan 60° = -1.732$ $\tan(-495°) = \tan 45° = 0.866$

Exercise 17.3

Use your calculator to find the following trigonometric ratios. In each case, draw quadrant diagrams and use them to find the associated acute angle.

1. $\tan 230°$ 2. $\sin -115°$ 3. $\cos 743°$
4. $\sin -544°$ 5. $\tan 419.64°$ 6. $\cos -67.3°$

Confirm the trigonometrical relationships between the angles and their associated acute angles. For example, in question 1 $\tan 230° = 1.192$. The associated acute angle is 50° and $\tan 50° = 1.192$.

GRAPHS OF THE TRIGONOMETRIC RATIOS

$y = \sin x$

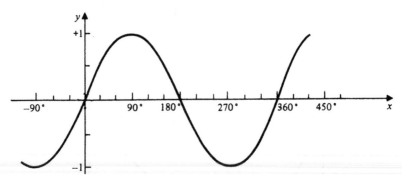

Figure 17.11

Note the following important properties of this graph:

(i)   The graph repeats every 360°. It is said to be periodic, with period 360°.

(ii)   The whole graph lies in the region $|\sin x| \le 1$.

$y = \cos x$

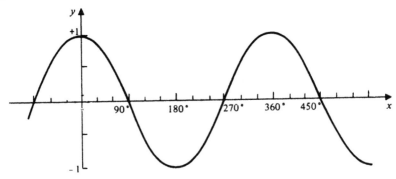

Figure 17.12

Note the following important properties of this graph:

(i)   The graph has an identical shape to $y = \sin x$. It also has period 360°.

(ii)   It is *out of phase* with $y = \sin x$, being shifted 90° to the left.

(iii)   The whole graph lies in $|\cos x| \le 1$.

$y = \tan x$

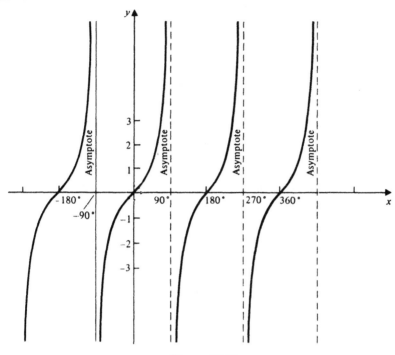

Figure 17.13

Note the following important properties of this graph:

(i)   It is periodic, with period 180°.

(ii)   The graph becomes infinite at $x = \pm 90°, \pm 270°$, etc.

TO FIND ALL THE ANGLES WITH A GIVEN SIN, COS OR TAN

We now need to consider what happens when we know a number to be the sin (say) of a given angle and wish to find the angle.

Example 17.3   Given the number 0.7549. Find the angle which has this number as a sine.

Solution   We use the **INV** or **2ND FUNCTION** button to help us:

Enter the number 0.7549

press **INV** (or equivalent)

press **sin**

the answer should be 49° to the nearest degree

No matter what number we enter (between -1 and $+1$, of course), the calculator will always give an answer between $-90°$ and $+90°$. However, 49° is not the only angle whose sin is 0.7549.

Use your calculator to show that 131°, $-229°$ and 409° all have the same sin.

We need to be able to find *all* angles with a given sin, cos or tan. Fortunately, because of the repeating or *periodic* nature of the trigonometric ratios, we can develop formulae to help us. This can be done either by using quadrant diagrams or from the graphs. Consider the position of angles with sine equal to 0.7549 on the graph of $y = \sin x$:

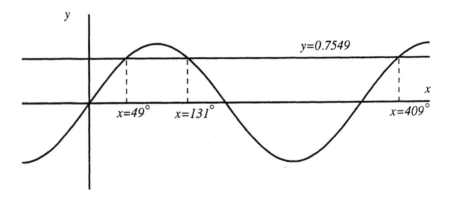

Figure 17.14

It is easy to see why these angles have the same sine.

Example 17.4   Find all the angles whose tan is 0.42.

Solution   Using a calculator, the first or *principal angle* is 22.8° to the nearest tenth of a degree (shown by the position of OP in Figure 17.15).

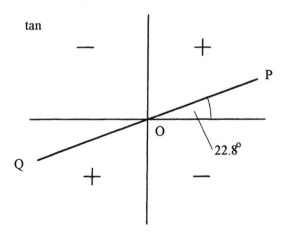

Figure 17.15

Think of OP continuing to rotate anticlockwise. Since the tan is positive, the angles we want must be in either the first or third quadrant. The first such angle that has an associated acute angle of 22.8° is half a full turn (180°) extra rotation (OQ in the diagram). Then the next time OP is back in the original position, it will have rotated through an additional 180°, i.e. a total of one full turn of 360°. If we keep on adding 180° we will always end up in position OP or OQ.

Similarly, OP could rotate clockwise from its original position, by multiples of 180°, and end up either in position OP or OQ. Hence the possible angles are:

$\quad$ ...22.8° − 360°, $\qquad$ 22.8° − 180°, 22.8°, $\qquad$ rotating clockwise

and $\quad$ 22.8° + 180°, 22.8° + 360°,... etc. $\qquad$ rotating anticlockwise.

Hence we have shown that

if $\qquad\qquad\qquad$ $\tan \theta = 0.42 = \tan 22.8°$

then $\qquad\qquad\qquad$ $\theta = 180n + 22.8°$ $\quad$ where $n = ...,-3,-2,-1,0,1,2,3,...$

We can see this on the graph of $y = \tan x$:

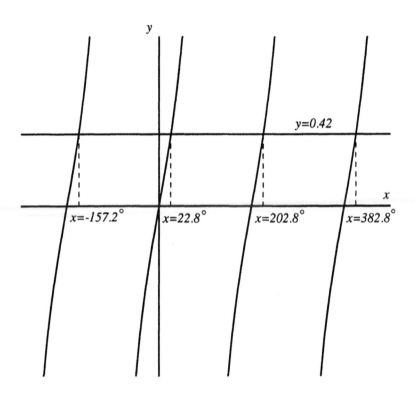

Figure 17.16

Giving, in general:

$$\text{If} \quad \tan \theta = \tan \alpha$$
$$\text{then} \quad \theta = 180n + \alpha$$

where α is found from your calculator and $n = ..., -3, -2, -1, 0, 1, 2, 3, ...$ α will be a positive acute angle if you enter a positive tan and will be a negative acute angle if you enter a negative tan.

Example 17.5　　　Find all the angles whose cos is 0.5342.

Solution　　　The calculator gives us 57.7° to 1 d.p. but a glance at the quadrant diagram below shows us that it could equally well have given −57.7°.

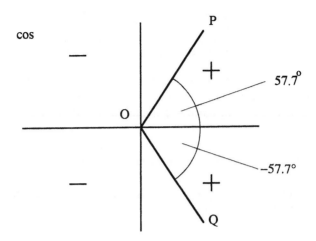

Figure 17.17

Hence the first two solutions are ±57.7° and it is easily seen that we get back to these two positions by adding or subtracting 360°, 720° , etc.

So, if $\qquad\qquad \cos \theta = 0.5342 = \cos (\pm 57.7°)$

then $\qquad\qquad\qquad \theta = 360n \pm 57.7°$

where $n = ..., -3, -2, -1, 0, 1, 2, 3, ...$ as before.

Graphically:

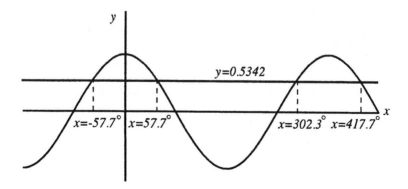

Figure 17.18

And the general solution is:

$$\text{If} \qquad \cos \theta = \cos \alpha$$
$$\text{then} \qquad \theta = 360n \pm \alpha$$

where, again, α is found from your calculator and n is again a positive or negative whole number (or zero). Notice, by the way, that your calculator will give an acute angle answer if you enter a positive cos and an *obtuse* angle if you enter a negative cos.

Example 17.6   Find all the angles whose sin is 0.5.

Solution   The calculator gives $30°$. The following quadrant diagram shows that the obtuse angle $150°$ has the same sine.

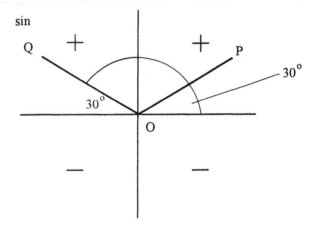

Figure 17.19

So the first two angles are $30°$ and $(180° - 30°)$

the next two positive angles are $(360° + 30°)$ and $(540° - 30°)$

the next two are $(720° + 30°)$ and $(900° - 30°)$ and similarly for clockwise rotations.

Notice that if the multiple of $180°$ is even ($360°$, $720°$, etc.), we add the $30°$ but if the multiple is odd ($180°$, $540°$, etc.), we subtract the $30°$.

We may therefore write:

if $\qquad\qquad\qquad\qquad \sin \theta = 0.5 = \sin 30°$

then $\qquad\qquad\qquad\qquad \theta = 180n + 30° \quad$ if n is even and $\qquad\qquad\qquad\qquad \theta = 180n - 30°$ if n is odd

This can again be confirmed by reference to the graph and we leave that as an exercise.

Hence:

$$
\boxed{
\begin{array}{ll}
\text{If} & \sin\theta = \sin\alpha \\
\text{then} & \theta = 180n + \alpha \ \text{ if } \ n \ \text{ is even} \\
& \theta = 180n - \alpha \ \text{ if } \ n \ \text{ is odd}
\end{array}
}
$$

The following examples and exercise should make things clear.

Example 17.7   Find all the angles whose tan is -1.4.

Solution

$\qquad\qquad$ $\tan\theta = -1.4 = \tan(-54.5°)$ to the nearest tenth of a degree

so $\qquad\qquad$ $\theta = 180n + (-54.5°)$

i.e. $\qquad\qquad$ $\theta = 180n - 54.5°$

Example 17.8   Find all angles between $-360°$ and $360°$ satisfying $\sin\theta = 0.7219$.

Solution

$\qquad\qquad$ $\sin\theta = 0.7219 = \sin 46.2°$

i.e. $\qquad\qquad$ $\theta = 180n + 46.2°$ $\qquad$ if n is even

$\qquad\qquad\qquad$ $\theta = 180n - 46.2°$ $\qquad$ if n is odd

We now substitute various values of n, *remembering to add when n is even and subtract when n is odd.*

$\qquad$ $n = -2$ gives $\theta = 180.(-2) + 46.2° = -360 + 46.2° = -313.8°$

$\qquad$ $n = -1$ gives $\theta = 180.(-1) - 46.2° = -180 - 46.2° = -226.2°$

$\qquad$ $n = 0$ gives $\theta = 46.2°$

$\qquad$ $n = 1$ gives $\theta = 180(1) - 46.2° = 133.8°$

$\qquad$ $n = 2$ gives $\theta = 180(2) + 46.2°$ which is outside the range

Hence, there are four angles in the given range: $-313.8°$, $-226.2°$, $46.2°$ and $133.8°$.

Example 17.9   Find all angles between $0°$ and $360°$ such that $\cos 2\theta = -0.69$.

Solution   This time we will generate a formula for 2θ but we need to find θ:

$$\cos 2\theta = -0.69 = \cos 133.6°$$
$$2\theta = 360n \pm 133.6°$$
$$\theta = 180n \pm 66.8° \quad (\text{dividing by 2})$$

$n = 0$ gives $\theta = \pm 66.8°$, one of which is outside the range

$n = 1$ gives $\theta = 180 \pm 66.8°$, leading to the two angles $113.2°$ and $246.8°$

$n = 2$ gives $\theta = 360 \pm 66.8°$ leading to just one angle, $293.2°$, in given range

Exercise 17.4

1.   Find a formula for all the angles whose sine is 0.2701. Hence find the angles between $0°$ and $720°$ whose sine is 0.2701.

2.   Repeat question 1 for angles whose cosine is 0.2701.

3.   Repeat question 1 for angles whose tangent is 0.2701.

4.   Solve the following equations, to the nearest tenth of a degree, for angles between $-360°$ and $360°$.

(a) $\sin \theta = -0.3542$ (b) $\tan \theta = -7.3901$
(c) $\cos \theta = 3 \cos 70°$

5.   Find the values of θ in the range $0° \leq \theta \leq 360°$ such that:

(a) $\cos 2\theta = 0.7051$ (b) $\sin 2\theta = 0.3062$
(c) $\tan 3\theta = 1.5$

(**N.B.** Remember, for example, $\tan 3\theta = 1.5$ does NOT mean $\tan \theta = 0.5$!)

6.   Using the fact that $\tan \theta = \dfrac{\sin \theta}{\cos \theta}$, solve $4 \sin \theta = 3 \cos \theta$.

THE MINOR RATIOS

At this stage, it is useful to introduce three more trigonometric ratios. There is no need to panic, since they are not really new ratios, just the *reciprocals* of the ones we have already met. They are called *secant, cosecant* and *cotangent* and for any angle A, they are usually abbreviated to *sec A, cosec A* and *cot A* (note again, *no capitals!*). Define, for any angle A:

$$\sec A = \frac{1}{\cos A} \qquad \operatorname{cosec} A = \frac{1}{\sin A} \qquad \cot A = \frac{1}{\tan A}$$

The Ancient Greeks studied trigonometry and used the ratios sine, secant and tangent. For them, the ratios had geometrical significance. (The Greeks were great geometers.) When you see the graph of secant, you will see why it has fallen out of favour and been replaced by cosine.

GRAPHS

$y = \sec x$

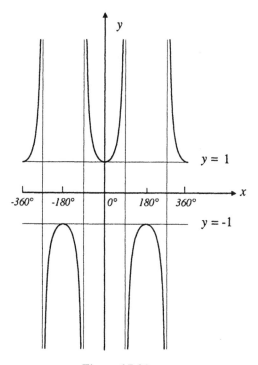

Figure 17.20

$y = \operatorname{cosec} x$

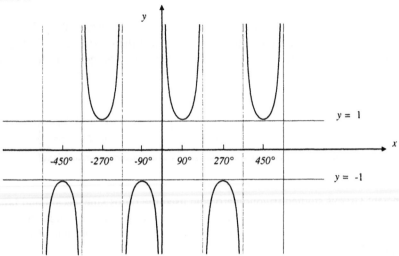

Figure 17.21

$y = \cot x$

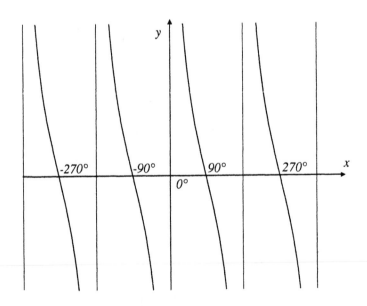

Figure 17.22

Notes on the graphs:

1. All the graphs are *discontinuous*. They cannot be drawn without taking the pencil from the paper!

2. The graphs of sec x and cosec x do not exist in the region $-1 \leq y \leq 1$.

3. They are periodic, having same period as the corresponding major ratio.

USE OF CALCULATOR WITH THE MINOR RATIOS

With most calculators, we cannot calculate the minor ratios directly. We simply calculate the corresponding major ratio and then find its reciprocal.

A warning about $sin^{-1}x$

Many of you will have calculators which have 'sin^{-1}' written above the sin button. This is *not* for finding the cosec of an angle!

Normally:

$$x^{-1} \text{ means } \frac{1}{x} \text{ but } sin^{-1}x \text{ does not mean } \frac{1}{sin\, x}.$$

$sin^{-1}x$ means *the inverse of* sin x, i.e '*the angle whose sin is x*'. Hence if you follow the procedure:

enter 60

press **INV**   (or equivalent)

press **sin**

you will generate an error because you have tried to find the angle whose sin is 60. This just doesn't exist!

You would, however, get away with no *apparent* error starting with 0.9:

enter 0.9

press **INV**

press **sin**   gives answer 64.16

However, you will have found the angle whose sin is 0.9, not cosec 0.9°.

The correct procedure for finding cosec 0.9° is as follows:

Make sure your calculator is in degree mode

enter 0.9

press **sin**   gives 0.0157

press **1/x**   gives 63.66, which is cosec 0.9°

Secant and cotangent are evaluated in a similar way, as the following examples show.

Example 17.11   Find sec 60°.

Solution   enter 60

press **cos**   gives 0.5   this is cos 60°

press **1/x**   gives answer 2   this is sec 60°

Example 17.12   Find cot (−711).

Solution   enter 711

press **+/−**   to make the angle negative

press **tan**   gives 0.1584   this is tan (−711°)

press **1/x**   gives 6.3138   this is cot (−711°)

Example 17.13   Find sec 90°.

Solution   enter 90

press **cos**   gives zero

press **1/x**   gives an error!

This is similar to trying to find tan 90°. The true value is ∞. You have not made a mistake, but you need to be aware that most calculators cannot handle this procedure.

Example 17.14   Find the acute angle whose cosec is 3.7.

Solution   Suppose the angle is A; then

$$\operatorname{cosec} A = 3.7$$

$$\sin A = \frac{1}{3.7} = 0.2703$$

$$A = 15.68°$$

The calculator procedure is

enter 3.7		this is cosec A
press 1/x	giving 0.2703	this is sin A
press **INV**		
press **sin**	giving the required angle to be 15.68°	

Example 17.15   Find the angle whose cot is 0.6.

enter 0.6		this is cot A
press 1/x	giving 1.6667	this is tan A
press **INV**		
press **tan**	giving 59° to the nearest degree	

Example 17.16   Find all the angles whose sec is 2.

Solution

$$\sec A = 2$$

$$\cos A = 0.5$$

$$A = 60° \quad \text{is the principal angle}$$

$$A = 360n \pm 60° \quad \text{gives all the required angles}$$

Exercise 17.5

1. Use your calculator to find

    (a) sec 67°   (b) cosec 125°   (c) cot 290°

    (d) sec (−219°)   (e) cosec (−17°)   (f) cot (−91°)

2. Find a formula for all the angles whose sec is 5.

    Hence find angles in the range $0° \le \theta \le 360°$ whose sec is 5.

3. Repeat question 2 for angles whose cosec is 150.

4. Repeat question 2 for angles whose cot is 0.7654.

5. Solve the following equations in the range $-360° \le \theta \le 360°$:

    (a) $\sec \theta = 1.2$ (b) $\csc \theta = -5.9$

    (c) $\cot 2\theta = 7$ (d) $\sec 2\theta = 16$

THE THEOREM OF PYTHAGORAS

In the triangle ABC shown, we use the convention of labelling the sides so that the side opposite angle A is called 'a' etc.

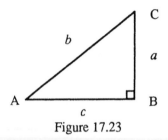

Figure 17.23

Pythagoras' Theorem states that for any such *right-angled* triangle:

$$b^2 = a^2 + c^2$$

The converse is also true, i.e. if the sides of a triangle obey the above formula, then the triangle has a right angle at B.

We can use the theorem to work out the values of the trigonometric ratios for 30°, 45° and 60°. It is useful to know these values as they often lead to quicker, more elegant solutions than using a calculator.

sin, cos and tan of 45°

Consider a right-angled triangle whose shorter sides are both of length 1 unit, as shown. Then by Pythagoras' Theorem, the hypotenuse will have length √2.

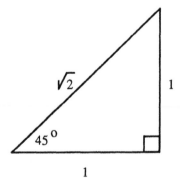

Figure 17.24

Hence, we can see that

$$\sin 45° = \frac{1}{\sqrt{2}} = \frac{\sqrt{2}}{2}, \quad \cos 45° = \frac{1}{\sqrt{2}} = \frac{\sqrt{2}}{2} \quad \text{and} \quad \tan 45° = 1$$

sin, cos and tan of 30° and 60°

This time we use a more indirect method. Look at the following equilateral triangle. All the angles are 60°. Let all the sides have length 2 units.

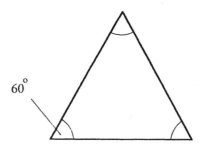

Figure 17.25

Now draw a line from the vertex at the top to bisect the base and throw away the right hand side to give the right-angled triangle below. The two known sides are now of lengths 1 unit and 2 units. Pythagoras' Theorem then gives the third length as √3.

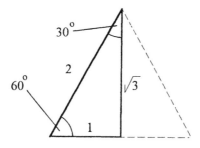

Figure 17.26

Therefore

$$\sin 60° = \frac{\sqrt{3}}{2}, \quad \cos 60° = \frac{1}{2} \quad \text{and} \quad \tan 60° = \frac{\sqrt{3}}{1} = \sqrt{3}$$

$$\cos 30° = \frac{\sqrt{3}}{2}, \quad \sin 60° = \frac{1}{2} \quad \text{and} \quad \tan 30° = \frac{1}{\sqrt{3}} = \frac{\sqrt{3}}{3}$$

Pythagorean Triples

In most cases, if two sides of a right angle have known *integer* lengths, the third side will not be of integer length. For example, if the two short sides have lengths 1 and 2, then

$$\text{hypotenuse}^2 = 1^2 + 2^2 = 1 + 4 = 5$$

and so the hypotenuse would have length $\sqrt{5}$ which is irrational.

There are, however, special cases where all three sides are integers. The four most commonly encountered sets of sides are:

$$3, 4, 5 \qquad 5, 12, 13 \qquad 7, 24, 25 \qquad \text{and} \qquad 8, 15, 17$$

We will confirm the first one and leave the rest to you.

$$3^2 + 4^2 = 9 + 16 = 25 = 5^2$$

Note that multiples of these sets, such as 6, 8, 10 and 15, 36, 39 also work.

Pythagorean Identities

The following important identities are true for all angles. We will give a proof for acute angles only. You can verify their validity for other angles in the exercise that follows.

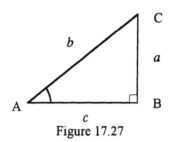
Figure 17.27

From Figure 17.27:

$$c^2 + a^2 = b^2$$

$$\frac{c^2}{b^2} + \frac{a^2}{b^2} = 1 \qquad\qquad \text{dividing by } b^2$$

i.e. $$\left(\frac{c}{b}\right)^2 + \left(\frac{a}{b}\right)^2 = 1$$

But $$\cos A = \frac{c}{b} \quad \text{and} \quad \sin A = \frac{a}{b}$$

Hence $$(\cos A)^2 + (\sin A)^2 = 1$$

or $$\mathbf{\cos^2 A + \sin^2 A = 1}$$ at it is usually written

Similarly, $$1 + \left(\frac{a}{c}\right)^2 = \left(\frac{b}{c}\right)^2$$ dividing by c^2

i.e. $$\mathbf{1 + \tan^2 A = \sec^2 A}$$

and finally $$\mathbf{1 + \cot^2 A = \operatorname{cosec}^2 A}$$

These formulae are true for all angles and are hence identities, which we summarise here:

$$\boxed{\begin{array}{l} \cos^2\theta + \sin^2\theta \equiv 1 \\ 1 + \tan^2\theta \equiv \sec^2\theta \qquad \text{for any angle } \theta \\ 1 + \cot^2\theta \equiv \operatorname{cosec}^2\theta \end{array}}$$

These identities are very useful. We can use them to solve more complicated trigonometrical equations than those we have met so far.

Example 17.17   Solve $1 + \sin\theta = 2\cos^2\theta$ for angles lying between $0°$ and $360°$.

Solution

$$1 + \sin\theta = 2\cos^2\theta$$

$$1 + \sin\theta = 2(1 - \sin^2\theta) \qquad \text{substituting for } \cos^2\theta$$

$$1 + \sin\theta = 2 - 2\sin^2\theta$$

$$2\sin^2\theta + \sin\theta - 1 = 0 \qquad \text{which is a quadratic equation}$$

$$(2\sin\theta - 1)(\sin\theta + 1) = 0$$

Hence $\sin\theta = \frac{1}{2} = \sin 30°$ giving $\theta = 180n + (-1)^n 30°$

or $\sin\theta = -1 = \sin(-90°)$ giving $\theta = 180n + (-1)^n(-90°)$

or $\sin \theta = -1 = \sin(-90°)$ giving $\theta = 180n + (-1)^n(-90°)$

Example 17.18   Given that $\sin \theta = \frac{3}{4}$, find the possible values of $\cos \theta$ and $\tan \theta$, *without using a calculator.*

Solution $\cos^2\theta + \sin^2\theta = 1$

and since $\sin \theta = \frac{3}{4}$ $\cos^2\theta + 9/16 = 1$

Hence $\cos^2\theta = 7/16$

Therefore $\cos \theta = \pm \text{ root } 7/16 = $ etc.

and so, $\tan \theta = \sin \theta/\cos \theta = \frac{3}{4}$

Exercise 17.6

1.   Use calculators to verify that $\cos^2 \theta + \sin^2 v = 1$ for various non-acute angles.

2.   Repeat for the other two Pythagorean identities.

3.   Solve the following equations, in the range $-360° \le \theta \le 360°$:

   (a) $4 \cos^2 \theta + 5 \sin \theta = 3$

   (b) $2 \sin^2 \theta - \cos \theta = 1$

   (c) $\sec^2\theta + \tan^2\theta = 2$

   (d) $\sec \theta = 1 - 2 \tan^2 \theta$

   (e) $\cot^2 \theta = \text{cosec } \theta$

   (f) $\tan \theta + \cot \theta = 6$

4.   Find the general solution of the following equations:

   (a) $2 \cos^2\theta = 1 + 4 \sin \theta$

   (b) $4 \sin^2\theta - 5 \cos \theta + 2 = 0$

   (c) $\tan \theta = 2 \sec^2 \theta - 1$

   (d) $4 \sec^2\theta = 5 + 3 \tan \theta$

5.   If $\sin A = \frac{3}{5}$ and A is obtuse, find $\cos A$ and $\tan A$ without using calculators.

6.   If $\tan A = \frac{7}{24}$ and A lies in the third quadrant, find all the trigonometrical ratios for angle A, without using a calculator.

7.   If $\sec A = \frac{17}{15}$ and A lies in the fourth quadrant, find all the trigonometrical ratios for angle A, without using a calculator.

18

Calculus of Trigonometry

RADIANS: A NEW WAY OF MEASURING ANGLES

The idea of dividing a full turn into 360 smaller rotations called degrees is really rather arbitrary. It would be better if we could find a more 'mathematical' way of measuring rotation.

Suppose you wish to cut a circular flower bed of radius 2 m into an existing lawn. Hammer a post into the ground at the centre C of your circle and loop one end of a rope around the post. Tie a sharp stick to the other end of the rope, so that the rope has length 2 m. If you now scribe out a complete circle, the rope will rotate through one full turn or 360° (Figure 18.1). However, if you stop when you have marked out an *arc* of length of 2 m, the length of this arc will be the same as the radius. What angle has the rope turned through?

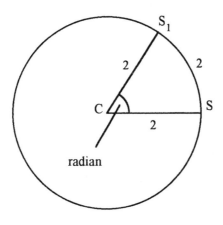

Figure 18.1

As we have already seen, the length of the circle from Start to Stop1 (S to S_1) is called an arc. The shape bounded by this arc and the two positions of the rope (it looks like a 'cheese triangle') is known as a *sector*. If all three sides of the sector were straight, then we would have an equilateral triangle and the angle at C would be 60° but the curve of the arc reduces the angle slightly so it will be slightly less than 60°.

We call this angle 1 *radian*.

The angle is the same no matter what the radius of the circle is; as long as we mark out an arc whose length is the same as the radius, the angle through which the radius turns will be 1 radian.

It is now natural to ask: 'How many radians are there in one full turn?' We can find out by considering the formula for the circumference of a circle.

There is a relationship between the circumference of a circle and its radius, given by the formula

$$C = 2\pi r$$

where π is the irrational number 3.14159... introduced in Chapter 1. Hence by substituting this value into the formula we see that, for any circle,

$$C = 6.28r \qquad \text{approximately}$$

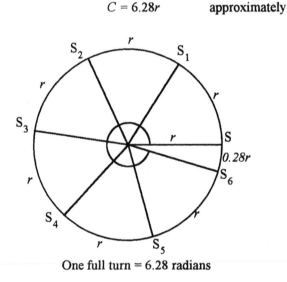

One full turn = 6.28 radians

Figure 18.2

Hence, as Figure 18.2 shows, we can mark out six radius lengths ($6r$) from our starting point, S, and turn through 6 radians without ending up back at S. To complete the circle requires another 0.28 radius lengths ($0.28r$) and so to complete one full turn requires turning through a further 0.28 radians.

Hence   one full turn $= 360° = 2\pi$ radians

Obviously, there are not a whole number of radians in a full turn but this is a minor inconvenience compared with the extra benefits of using radians rather than degrees.

Sometimes, we will need to be able to convert from degrees to radians and vice versa. Modern calculators will have some way of converting from one unit to the other and since they are all slightly different, we will leave it to you to find out how your personal calculator does it. However, we still need to know how to convert without using a calculator so the following conversion factors are useful:

Since $360° = 2\pi$ radians

then $180° = \pi$ radians

giving

$$1 \text{ degree} = \frac{\pi}{180} \text{ radians}$$

and

$$1 \text{ radian} = \frac{180}{\pi} \text{ degrees}$$

Hence we see that 1 radian $= 57.3°$ approximately and therefore just less than $60°$ as we stated earlier.

From the above conversions, we can work out the values of certain important angles, when expressed in radians.

Example 18.1   Convert these angles to radians. Leave answers in terms of π:

(a) $30°$ (b) $45°$ (c) $60°$ (d) $90°$

Solution

(a) $30° = 30 \times \dfrac{\pi}{180} = \dfrac{\pi}{6}$ radians

(b) $45° = 45 \times \dfrac{\pi}{180} = \dfrac{\pi}{4}$ radians

(c) $60° = 2 \times 30° = 2 \times \dfrac{\pi}{6} = \dfrac{\pi}{3}$ radians

(d) $90° = 2 \times 45° = 2 \times \dfrac{\pi}{4} = \dfrac{\pi}{2}$ radians

Note that these angles are normally referred to as $\dfrac{\pi}{6}$, $\dfrac{\pi}{4}$, $\dfrac{\pi}{3}$ and $\dfrac{\pi}{2}$, etc., the word 'radians' being understood. They are read as 'pi by three', 'pi by four', etc.

Exercise 18.1

1. Use your calculator to convert the following from degrees to radians:

    (a)   10°   (b)   13°   (c)   60°

    (d)   115°   (e)   409°   (f)   −712°

2. Use your calculators to convert the following from radians to degrees:

    (a)   0.3   (b)   0.5   (c)   2.1

    (d)   4.7   (e)   6.28   (f)   −1.1

3. Complete the following conversion table. Leave all radian results in terms of π:

Degrees	30	45	60	90	120		150	180
Radians	$\frac{\pi}{6}$	$\frac{\pi}{4}$	$\frac{\pi}{3}$	$\frac{\pi}{2}$		$\frac{3\pi}{4}$		π
Degrees	210		240	270		315		360
Radians		$\frac{5\pi}{4}$			$\frac{5\pi}{3}$		$\frac{11\pi}{6}$	2π

4. Convert the following angles to radians, *leaving answers in terms of* π:

    (a)   5°   (b)   10°   (c)   12°   (d)   15°   (e)   22.5°

    (f)   24°   (g)   36°   (h)   72°   (i)   450°   (j)   720°

THE ADDITION FORMULAE

The distributive law of multiplication states that, if a, b and c are real numbers, then

$$a(b + c) = ab + ac$$

It would be rather convenient if our trigonometric ratios behaved the same way but unfortunately they don't! Hence, for example:

$$\sin(A + B) \neq \sin A + \sin B$$

We can verify this by example. Suppose $A = 30°$ and $B = 60°$.

Then $\quad\quad\quad sin(A + B) = sin(30° + 60°) = sin\ 90° = 1$

But $\quad sin\ A + sin\ B = sin\ 30° + sin\ 60° = \dfrac{1}{2} + \dfrac{\sqrt{3}}{2} = \dfrac{2.732...}{2} = 1.367$ to 3 d.p.

Hence $\quad\quad\quad\quad sin(\ 30° + 60°) \neq sin\ 30° + sin\ 60°$

You will have an opportunity to verify this result for further angles in the next exercise.

The correct formula for calculating the sine of sum is as follows:

$$sin(A + B) \equiv sin\ A.cos\ B + cos\ A.sin\ B$$

We state this without proof but will verify it, again using 30° and 60°. You will be able to verify it in the next exercise, for different angles. Note that this is an identity and true for all angles, in either degrees or radians.

Hence, if $A = 30°$ and $B = 60°$, then

$$sin(A + B) = sin(30° + 60°) = sin\ 90° = 1 \text{ as before}$$

And $\quad\quad sin\ A.cos\ B + cos\ A.sin\ B = sin\ 30°.cos\ 60° + cos\ 30°.sin\ 60°$

$$= \left(\frac{1}{2} \times \frac{1}{2}\right) + \left(\frac{\sqrt{3}}{2} \times \frac{\sqrt{3}}{2}\right) = \frac{1}{4} + \frac{3}{4} = 1$$

Hence $\quad\quad sin(30°+60°) = sin\ 30°.cos\ 60° + cos\ 30°.sin\ 60°$

We can verify three similar formulae, for $sin(A - B)$, $cos\ (A + B)$ and $cos\ (A - B)$. All four are collected together here for quick reference:

$$\boxed{\begin{aligned} sin(A + B) &\equiv sin\ A\cos B + cos\ A\sin B \\ sin(A - B) &\equiv sin\ A\cos B - cos\ A\sin B \\ cos(A + B) &\equiv cos\ A\cos B - sin\ A\sin B \\ cos(A - B) &\equiv cos\ A\cos B + sin\ A\sin B \end{aligned}}$$

Note how the minus signs are reversed in the cosine identities.

These identities are used extensively in many applications but one of our first tasks will be to use them so that we can differentiate $y = sin\ x$ and $y = cos\ x$.

Exercise 18.2

1. Verify the identity $\sin(A + B) \equiv \sin A.\cos B + \cos A.\sin B$ for the following pairs of angles:

   (a) $A = 45°, B = 120°$ (b) $A = 13°, B = 217°$

   (c) $A = 0.521$ radians, $B = 7.904$ radians

2. Repeat question 1 for the identity $\sin(A - B) \equiv \sin A.\cos B - \cos A.\sin B$.

3. Repeat question 1 for the identity $\cos(A + B) \equiv \cos A.\cos B - \sin A.\sin B$

4. Repeat question 1 for the identity $\cos(A - B) \equiv \cos A.\cos B + \sin A.\sin B$

5. Use angles of your own to verify the truth of the following inequalities:

   $\sin(A + B) \neq \sin A + \sin B$ $\sin(A - B) \neq \sin A - \sin B$

   $\cos(A + B) \neq \cos A + \cos B$ $\cos(A - B) \neq \cos A - \cos B$

SMALL ANGLES

The following table gives values of the sines and cosines of small angles from $6°$ reducing to $0.1°$. The values are given to 6 decimal places.

angle θ in degrees	angle θ in radians	$\sin \theta$	$\cos \theta$
6	0.104720	0.104528	0.994522
5	0.087266	0.087156	0.996195
4	0.069813	0.069756	0.997564
3	0.052360	0.052336	0.998630
2	0.034907	0.034899	0.999391
1	0.017453	0.017452	0.999848
0.5	0.008727	0.008727	0.999962
0.1	0.001745	0.001745	0.999998

We notice that as θ is made smaller and smaller, there is no obvious relationship between the angle in degrees and its sine, but the angle measured in radians is almost the same as its sine. The smaller we make the angle (in radians!), the closer these two numbers become.

Notice also that as θ becomes smaller, its cosine approaches closer and closer to the value 1.

We recognise that we are talking about limiting values here; so we may state that *if* θ *is measured in radians*:

$$\text{as } \theta \to 0, \text{ then } \sin \theta \to \theta \text{ and } \cos \theta \to 1$$

Therefore if θ is very small indeed *and measured in radians*, $\sin \theta$ can be replaced by θ. In addition, $\cos \theta$ can be replaced by 1.

These facts will help us to differentiate $\sin x$ and $\cos x$.

DIFFERENTIATION OF TRIGONOMETRIC FUNCTIONS

To differentiate $\sin x$

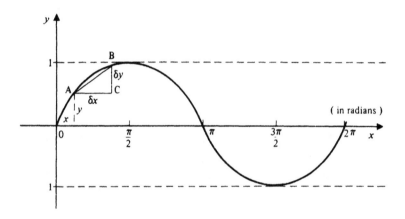

Figure 18.3

Let $y = \sin x$ *where x is measured in radians*.

Using the usual method for deducing the gradient at a point:

let A be the point (x, y) and B be the point $(x + \delta x, y + \delta y)$

In the small triangle ABC, $AC = \delta x$ and $BC = \delta y$.

But, since $y = \sin x$, then $y + \delta y = \sin(x + \delta x)$ and so, subtracting one from the other,

$$\delta y = \sin(x + \delta x) - \sin x$$

Hence, by the second addition formula:

$$\delta y = \sin x . \cos \delta x - \cos x . \sin \delta x - \sin x$$

But the angle δx is very small indeed and measured in radians, so we may use

$$\sin \delta x = \delta x \text{ and } \cos \delta x = 1 \text{ approximately}$$

Therefore $\quad\quad\quad\quad \delta y = \sin x - \delta x . \cos x - \sin x = \delta x . \cos x$

This gives, for gradient of chord AB

$$\frac{\delta y}{\delta x} = \frac{\delta x . \cos x}{\delta x}$$

i.e. $\quad\quad\quad\quad\quad\quad \dfrac{\delta y}{\delta x} = \cos x$

As $\delta x \to 0$, the LHS $\to \dfrac{dy}{dx}$ by definition and the RHS does not change since it does not depend on δx.

Therefore, $\quad$ $\boxed{\text{if } y = \sin x, \text{ then } \dfrac{dy}{dx} = \cos x}$ $\quad$ i.e. $\quad$ $\boxed{\dfrac{d}{dx}(\sin x) = \cos x}$

To differentiate cos x

This time, let $y = \cos x$ *where x is measured in radians.*

Let A be the point (x,y) and B be the point $(x + \delta x, y + \delta y)$.

Since $y = \cos x$, then $y + \delta y = \cos(x + \delta x)$ so, by subtraction,

$$\delta y = \cos(x + \delta x) - \cos x$$

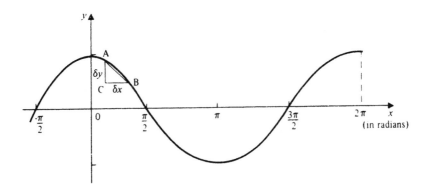

Figure 18.4

Hence, by the fourth addition formula:

$$\delta y = \cos x.\cos \delta x - \sin x.\sin \delta x - \cos x$$

Again, the angle δx is very small indeed and measured in radians, so we may use

$$\sin \delta x = \delta x \text{ and } \cos \delta x = 1$$

Therefore $\delta y = \cos x - \delta x.\sin x - \cos x = -\sin x$

This time, chord AB has gradient

$$\frac{\delta y}{\delta x} = \frac{-\delta x.\sin x}{\delta x}$$

i.e. $$\frac{\delta y}{\delta x} = -\sin x$$

As $\delta x \to 0$, the LHS $\to \dfrac{dy}{dx}$ by definition and the RHS does not change, giving:

$$\boxed{\text{if } y = \cos x, \text{ then } \frac{dy}{dx} = -\sin x}$$ i.e. $$\boxed{\frac{d}{dx}(\cos x) = -\sin x}$$

We can now use these basic results to deduce further results, using our knowledge of product, quotient and chain rules, as the following examples show.

Example 18.2   Differentiate   (a) $\sin 3x$ (b) $\cos 5x$ (c) $\sin nx$

Solution   (a)   Let $y = \sin 3x$. Then, by the chain rule

$$\frac{dy}{dx} = (\cos 3x) \times (\text{the derivative of } 3x)$$

$$= 3\cos 3x$$

(b)   Let $y = \cos 5x$

$$\frac{dy}{dx} = (-\sin 5x) \times (\textit{the derivative of } 5x)$$

$$= -5\sin 5x$$

(c)   Again using the chain rule

$$\frac{d}{dx}(\sin nx) = (\cos nx) \times (\text{the derivative of } nx)$$

$$= n\cos nx$$

Example 18.3   Differentiate   (a) $3x\cos x$ (b) $\sin x\cos x$ (c) $\sin^2 x$

Solution   (a)   If $y = 3x\cos x$, then by the product rule

$$\frac{dy}{dx} = 3(x.\frac{d}{dx}\cos x + \cos x.\frac{d}{dx}x)$$

$$= 3(x.-\sin x + \cos x.1)$$

$$= 3(\cos x - x\sin x)$$

(b)   Again using the product rule

$$\frac{d}{dx}(\sin x\cos x) = \sin x\frac{d}{dx}\cos x + \cos x\frac{d}{dx}(\sin x)$$

$$= \sin x.(-\sin x) + \cos x.(\cos x)$$

$$= \cos^2 x - \sin^2 x$$

(c)   By the chain rule

$$\frac{d}{dx}(\sin^2 x) = 2\sin x \times (\text{the derivative of } \sin x)$$

$$= 2\sin x\cos x$$

Example 18.4   Differentiate   (a) $\dfrac{x^2}{\cos x}$ (b) $\tan x$ (c) $\sec x$

Solution   (a)   By the quotient rule

$$\frac{d}{dx}\left(\frac{x^2}{\cos x}\right) = \frac{\cos x . \frac{d}{dx}(x^2) - x^2 . \frac{d}{dx}(\cos x)}{(\cos x)^2}$$

$$= \frac{2x\cos x - x^2 .(-\sin x)}{\cos^2 x}$$

$$= \frac{x(2\cos x + x\sin x)}{\cos^2 x}$$

(b) $$\frac{d}{dx}(\tan x) = \frac{d}{dx}\left(\frac{\sin x}{\cos x}\right)$$

We can now use the quotient rule:

$$\frac{d}{dx}(\tan x) = \frac{\cos x . \frac{d}{dx}(\sin x) - \sin x . \frac{d}{dx}(\cos x)}{(\cos x)^2}$$

$$= \frac{\cos^2 x + \sin^2 x}{\cos^2 x} = \frac{1}{\cos^2 x}$$

Hence

$$\frac{d}{dx}(\tan x) = \sec^2 x$$

(c)   By similar use of the quotient rule:

$$\frac{d}{dx}(\sec x) = \frac{d}{dx}\left(\frac{1}{\cos x}\right) = \frac{\cos x .(0) - 1.(-\sin x)}{\cos^2 x}$$

$$= \frac{\sin x}{\cos^2 x} = \frac{1}{\cos x} . \frac{\sin x}{\cos x}$$

Therefore

$$\frac{d}{dx}(\sec x) = \sec x \tan x$$

Exercise 18.3   Differentiate the following w.r.t. x:

1. $3\sin x - 4\cos x$ 2. $\sin 8x$ 3. $\dfrac{1}{6}\cos 3x$

4. $\sin(2x+5)$ 5. $\cos(\frac{\pi}{6}-x)$ 6. $3\sin(3x-2)$

7. $6\sin\frac{2}{3}x$ 8. $\sin 2\pi x$ 9. $\cos^2 x$

10. $4\sin\dfrac{x}{2}$ 11. $\sqrt{\sin x}$ 12. $\sin x - \frac{1}{6}\sin 3x$

13. $\sqrt{x}\,\sin x$ 14. $\sin 2x\cos x$ 15. $2x^2\cos 3x$

16. $\cos ax \cos bx$ 17. $x^2\sin x$ 18. $\dfrac{\sin x}{x}$

19. $\dfrac{1+\sin x}{1-\sin x}$ 20. $\cot x$ 21 $\operatorname{cosec} x$

Derivatives of all the trigonometric functions

From the previous examples and the last two parts of Exercise 18.3 we now know how to differentiate all six trigonometric functions. These are summarised below.

$$\frac{d}{dx}(\sin x) = \cos x \qquad \frac{d}{dx}(\cos x) = -\sin x \qquad \frac{d}{dx}(\tan x) = \sec^2 x$$

$$\frac{d}{dx}(\sec x) = \sec x \tan x \qquad \frac{d}{dx}(\operatorname{cosec} x) = -\operatorname{cosec} x \cot x \qquad \frac{d}{dx}(\cot x) = -\operatorname{cosec}^2 x$$

19

The Law of Natural Growth

Most of us are aware that some quantities not only increase but that the rate at which they increase is also increasing. We are continually reminded, for instance, that world population is increasing at an increasing rate or, more impressively, that the speed of travel since the invention of the steam locomotive has grown from about 13 mph to 18,000 mph with space travel (Figure 19.1).

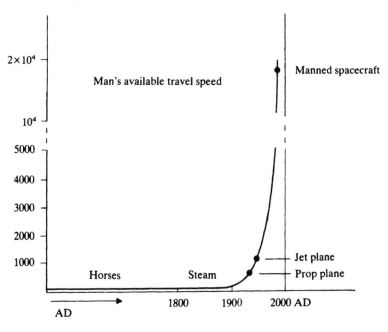

Figure 19.1

The graph clearly shows that after about 1900, not only is the graph increasing but so is its gradient! The growth of explosive power available shows a similar pattern, as does the number of books published per year.

All the above quantities grow with time. We will try to find a law that governs such quantities that increase at an increasing rate.

COMPOUND INTEREST

Let us consider a quantity with which we are all only too familiar – money. Is it true that money 'grows'? In our society money accumulates interest, not just simple interest, but what is known as compound interest, i.e. not only does the original sum of money gain interest, but the interest itself accumulates interest, and the interest on the interest accumulates more interest, and so on. Thus in one sense, we see that money does indeed grow.

Suppose the bank rate is 8%. If we invest £100 at 8% per annum *simple interest* then at the end of 1 year we would earn £8. Simple interest means that the bank now *does not include the* £8 and at the end of the second year *it only pays interest on the original investment*. The money therefore grows at just £8 p.a. and the graph of the growth of our money would be linear as shown below, with a gradient of £8 p.a.

Simple interest on investment of £100 at 8% per annum

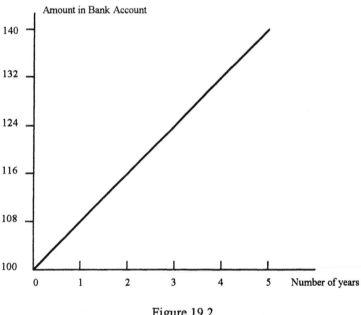

Figure 19.2

On the other hand, if the interest is left to accumulate in the bank account, a different picture arises. We started with £100 at the beginning and gained £8 during the first year making a total of £108.

The whole of this £108 then accumulates interest at 8%, so that the interest during the second year is a bit more than £8, in fact £8.64, and during the third year this extra bit also accumulates interest and so on. This type of growth is due to what is known as *compound interest*. The graph showing the growth is given in Figure 19.3.

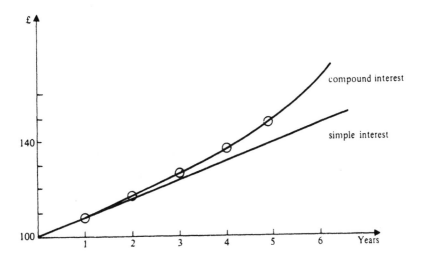

Figure 19.3

The compound interest graph diverges more and more from the simple interest graph as time goes on. At the beginning of the first year we have £100. At the end of the first year, or the beginning of the second year, we have £100 plus £8 interest. We can write this as $£100\left(1+\dfrac{8}{100}\right)$, showing that our original £100 has been multiplied by a factor of $\left(1+\dfrac{8}{100}\right)$.

If we had started with £50 it would have increased to £50 plus £4 interest. This can be written as $£50\left(1+\dfrac{4}{50}\right)=£50\left(1+\dfrac{8}{100}\right)$, and our £50 investment has been multiplied by the same factor.

This will happen whatever amount we start with and whatever year we are dealing with. The multiplying factor is the same.

Hence we may deduce:

If interest is calculated yearly

There will be one calculation at the end of each year and

$$£100 \text{ increases to } £100\left(1+\frac{8}{100}\right)$$

If interest is calculated half-yearly

Suppose interest is calculated at the end of every 6 months instead of at the end of every year. Our 8% per year would now be split into two 4% interests every half year, so that during the second half year, interest is gained on the previous half year's 4% interest.

There will be two calculations each year and at the end of each 6 month period, the amount invested is multiplied by a factor $\left(1+\frac{4}{100}\right)$ which is the same as $\left(1+\frac{8}{200}\right)$.

Hence after 6 months our £100 investment becomes $£100\left(1+\frac{8}{200}\right)$

At the end of the year, this becomes $£100\left(1+\frac{8}{200}\right)\times\left(1+\frac{8}{200}\right) = £100\left(1+\frac{8}{200}\right)^2$

There will be **two** calculations each year and

$$£100 \text{ increases to } £100\left(1+\frac{8}{200}\right)^2$$

If interest is calculated monthly

If interest is reckoned monthly, i.e. at $\frac{8}{12}\%$ interest each month there will be twelve calculations each year and at the end of each month, the amount invested is multiplied by a factor $\left(1+\frac{8}{1200}\right)$.

Hence after 1 month our £100 investment becomes $\qquad £100\left(1+\frac{8}{1200}\right)$

After 2 months, this becomes $\qquad £100\left(1+\frac{8}{1200}\right)^2$

After 3 months, this becomes $\qquad £100\left(1+\frac{8}{1200}\right)^3$

At the end of the year this becomes $\qquad\qquad\qquad £100\left(1+\dfrac{8}{1200}\right)^{12}$

There will be **twelve** calculations each year and

$$£100 \text{ increases to } £100\left(1+\frac{8}{1200}\right)^{12}$$

If interest is calculated daily

By now, we can spot a pattern and deduce that if we were to calculate interest daily, there will be **365** calculations each (non-leap) year and

$$£100 \text{ increases to } £100\left(1+\frac{8}{36,500}\right)^{365}$$

If interest is calculated n times per year

We are now justified in saying that if £100 is invested at an interest of 8% per annum, reckoned n times per year,

the amount at the end of the year is $£100\left(1+\dfrac{8}{100n}\right)^{n}$

We can further deduce that if the interest is $x\%$ annum reckoned n times per year, we have an interest at the end of 1 year of $£\left(1+\dfrac{x}{100n}\right)^{n}$ on every £1 invested.

An Investigation of the Expression $\left(1+\dfrac{a}{n}\right)^{n}$

Our above study of compound interest has led us to an expression of the type $\left(1+\dfrac{a}{n}\right)^{n}$ where a is a constant $\left(=\dfrac{x}{100}, \text{here}\right)$. This type of expression crops up in many branches of knowledge and it is now interesting to consider what happens if the growth is continuous. In other words, the growth takes place steadily instead of in distinct steps, as in our compound interest example.

We can begin our study by considering what would happen above if the interest were calculated hourly, or every minute or every second or every millionth of a second. In each case, the time between calculations is becoming less and less and the number of calculations (n) is becoming larger and larger.

What happens when $n \to \infty$?

Let us look at the simplest possible case with $x = 1$, i.e. let us try to find $\underset{n\to\infty}{Lim}\left(1+\dfrac{1}{n}\right)^n$.

We must be very careful how we approach this problem. It is so easy to let our intuition lead us into a trap!

Some people might say that as $n \to \infty$ then $\dfrac{1}{n}$ becomes very, very small and can be ignored. Thus we only need to consider $\underset{n\to\infty}{Lim}(1)^n$ which is 1, since 1 raised to any power is 1. This is the *wrong* conclusion!

Others might argue that no matter how small $\dfrac{1}{n}$ gets, it still makes $\left(1+\dfrac{1}{n}\right)$ bigger than 1 and any number greater than 1, raised to a very large power, is very large indeed. Thus $\underset{n\to\infty}{Lim}\left(1+\dfrac{1}{n}\right)^n$ would seem to be ∞. This is also wrong.

In both the above wrong conclusions, we have concentrated on either $\dfrac{1}{n}$ or the power n individually. To reach the correct conclusion we must consider them both together! A calculator can help us here! Make a table similar to the one below and fill in the gaps. You should be able to use some very large values of n before your calculator gives up! However, do not just start with 1,000,000, say. Try all the values suggested and note what happens.

n	$\left(1+\dfrac{1}{n}\right)^n$	Value to 10 d.p.
1	$(1+1)^1$	2.0000000000
2	$(1+\frac{1}{2})^2$	2.2500000000
3	$(1+\frac{1}{3})^3$	2.3703703703
10		
100		
1000		
1,000,000		
1,000,000,000		

It looks as though $\left(1+\dfrac{1}{n}\right)^n$ is approaching a definite value as n gets bigger, which is neither 1 nor ∞. This value, to 5 d.p., is 2.71828 and we give it the special symbol e. Like π, e is an irrational number, although this is difficult to prove.

This investigation leads us to a definition:

$$e = \lim_{n \to \infty}\left(1+\frac{1}{n}\right)^n$$

A similar investigation would reveal that $\lim_{n \to \infty}\left(1+\frac{x}{n}\right)^n$ gives us e raised to the power x and hence we define:

$$e^x = \lim_{n \to \infty}\left(1+\frac{1}{n}\right)^n$$

In this form it looks rather formidable, but let us see what we can do with it. Using the standard binomial expansion

$$(1+x)^n = 1+nx+\frac{n(n-1)}{2!}x^2 + \cdots$$

Replacing x by $\left(\dfrac{x}{n}\right)$ we get

$$\left(1+\frac{x}{n}\right)^n = 1+n.\left(\frac{x}{n}\right)+\frac{n(n-1)}{2!}\left(\frac{x}{n}\right)^2 + \frac{n(n-1)(n-2)}{3!}\left(\frac{x}{n}\right)^3 + \cdots$$

Rearranging the ns in the denominator:

$$\left(1+\frac{x}{n}\right)^n = 1+\frac{n}{n}x+\frac{\dfrac{n}{n}\left(\dfrac{n-1}{n}\right)}{2!}x^2 + \frac{\dfrac{n}{n}\left(\dfrac{n-1}{n}\right)\left(\dfrac{n-2}{n}\right)}{3!}x^3 + \cdots$$

$$= 1+x+\frac{1\left(1-\dfrac{1}{n}\right)}{2!}x^2 + \frac{1\left(1-\dfrac{1}{n}\right)\left(1-\dfrac{2}{n}\right)}{3!}x^3 + \cdots$$

Now let $n \to \infty$. Then $\dfrac{1}{n} \to 0$ and

$$\lim\left(1+\frac{x}{n}\right)^n = 1+x+\frac{1.1}{2!}x^2 +\frac{1.1.1}{3!}x^3 + \cdots$$

$$e^x = 1+x+\frac{x^2}{2!}+\frac{x^3}{3!}+\frac{x^4}{4!}+\cdots$$

This is an infinite series, i.e. a series containing an infinite number of terms (it has no ending). There is no need to worry about this; we do not need to perform an infinite number of calculations, since the terms become very small very quickly. For example, if $x = 2$, the tenth term of the series is approximately 0.0002821 and so will only affect the fourth decimal place.

We can now deduce the value of e by putting $x = 1$ in the above formula.

$$e^x = 1+1+\frac{1}{2!}+\frac{1}{3!}+\frac{1}{4!}+\cdots$$

The first seven terms give $e = 2.5+\frac{1}{6}+\frac{1}{24}+\frac{1}{120}+\frac{1}{720}+\cdots = 2.71806$ to 5 d.p.

By adding more terms, the correct value to 5 d.p. is $e = 2.71828$, confirming our calculator investigation.

This constant e is very important, as it has some very special characteristics.

DIFFERENTIATION AND INTEGRATION OF THE FUNCTION $y = e^x$

The above infinite series definition actually defines for us a new function $y = e^x$. This is known as the *exponential function*. Can we find the gradient of this function?

By making an assumption, we can avoid differentiating from first principles (i.e. using δx and δy). We assume that the series can be differentiated term by term. Proof of the validity of this assumption is beyond the scope of this book.

$$e^x = 1+x+\frac{x^2}{2!}+\frac{x^3}{3!}+\frac{x^4}{4!}+\cdots$$

Differentiating term by term:

$$\frac{d}{dx}(e^x) = 0+1+\frac{2x}{2.1}+\frac{3x^2}{3.2.1}+\frac{4x^3}{4.3.2.1}+\cdots$$

$$= 1+x+\frac{x^2}{2.1}+\frac{x^3}{3.2.1}+\cdots$$

Therefore

$$\frac{d}{dx}(e^x) = e^x$$

This is a remarkable result! We have found a function, $y = e^x$, which has the gradient at every point on its graph equal to the value of the function at that point. There is no other function that has this property.

We can immediately write down the inverse of this result:

$$\int e^x dx = e^x + c$$

We can now use the chain rule to differentiate other powers of e.

Example 19.1   Differentiate the following:

(a) e^{2x} (b) e^{ax} (c) e^{-x} (d) $e^{(3x+2)}$ (e) e^{-x^2}

Solution

(a)   Let $y = e^{2x}$

$\dfrac{dy}{dx} = e^{2x} \times$ (derivative of $2x$)   (by function of a function rule)

$= 2e^{2x}$

(b)   Let $y = e^{ax}$ then $\dfrac{dy}{dx} = a.e^{ax}$

(c)   Let $y = e^{-x}$

$\dfrac{dy}{dx} = e^{-x} \times$ (derivative of $-x$)

$= -e^{-x}$

(d)   Let $y = e^{3x+2}$

$\dfrac{dy}{dx} = e^{(3x+2)} \times$ [derivative of $(3x+2)$]

$= 3e^{(3x+2)}$

(e)   Let $y = e^{-x^2}$

$$\frac{dy}{dx} = e^{-x^2} \times \text{(derivative of } -x^2)$$

$$= -2xe^{-x^2}$$

The above examples lead to the following general results when e is raised to a *linear* power $ax+b$:

$$\boxed{\frac{d}{dx}(e^{ax+b}) = ae^{ax+b}}$$ and $$\boxed{\int e^{ax+b}\,dx = \frac{1}{a}e^{ax+b} + c}$$

Exercise 19.1

1.   Use the chain rule to differentiate:

     (a) e^{3x} (b) e^{4x-1} (c) $3e^{2x}$ (d) $\frac{1}{4}e^{4x+2}$

     (e) $e^{-\frac{x}{2}}$ (f) e^{2x^2} (g) $e^x + e^{-x}$ (h) $e^{\sin x}$

2.   Use the product rule to differentiate:

     (a) $3x^2e^x$ (b) xe^{ax} (c) $e^x\sin x$ (d) $e^{5x}\tan x$

3.   Use the quotient rule to differentiate:

     (a) $\dfrac{e^x - e^{-x}}{e^x + e^{-x}}$ (b) $\dfrac{x^2 + 2x}{e^x}$

4.   Integrate the following:

     (a) e^{-x} (b) e^{3x} (c) $e^{(4x-1)}$ (d) $3e^{2x}$

     (e) $\frac{1}{4}e^{(4x+2)}$ (f) $e^{-\frac{x}{2}}$ (g) $e^x + e^{-x}$ (h) xe^{2x^2}

(Hint: in part (e) see Example 19.1(e).)

Natural or Napierian Logarithms

These are logarithms to base e.
Let $y = e^x$;   then $\log_e y = x$ by definition

However, we use the symbol $\ln x$ instead of $\log_e x$ (read 'ell enn') meaning log natural,

and thus:

$$\boxed{\text{If } y = e^x \text{ then } x = \ln y.}$$

DIFFERENTIATION OF $y = \ln x$

We will now differentiate the function $y = \ln x$. The easiest way to perform this differentiation is not to use first principles but to use the fact that $y = \ln x \Rightarrow x = e^y$.

Hence,   if $y = \ln x$ then $x = e^y$

Therefore $\dfrac{dx}{dy} = e^y = x$

giving $\dfrac{dy}{dx} = \dfrac{1}{x}$

and so, we may write:

$$\boxed{y = \ln x \implies \frac{dy}{dx} = \frac{1}{x}} \qquad \text{or} \qquad \boxed{\frac{d}{dx}(\ln x) = \frac{1}{x}}$$

This result now fills the missing link in our integration results from our earlier chapter.

Remember that we have a formula to integrate x^n for any value of n except -1. But since $\dfrac{1}{x} = x^{-1}$ we can now write

$$\int x^{-1} dx = \ln x + c$$

which is more usually written as:

$$\boxed{\int \frac{dx}{x} = \ln x + c}$$

GRAPHS OF $e^x, e^{-x}, \ln x$

The graph of $y = e^x$ is a steadily increasing function. If we interchange x and y in the equation $y = e^x$ we get

$$x = e^y \qquad \text{or} \qquad \ln x = y$$

It can be shown that the graphs of $y = e^x$ and $y = \ln x$ (i.e. $x = e^y$) are reflections of each other in the line $y = x$ (Figure 19.4).

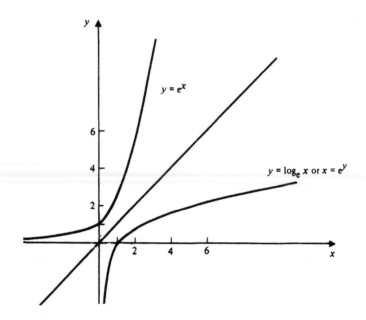

Figure 19.4

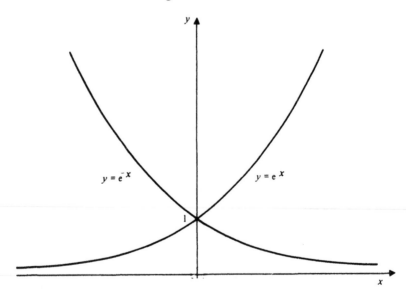

Figure 19.5

Properties of e^x

The following properties of e^x are worth noting. The graph of e^x helps to memorise them.

- $e^\infty = \infty$ $e^0 = 1$ $e^{-\infty} = 0$

- $e^x > 1$ for all positive values of x *and* $e^x < 1$ for all negative values of x.

- The graph lies above the x axis.

- The equation $y = e^x$ for a given value of y has only one solution. This solution is $x = \ln y$.

Properties of ln x

- $\ln (+\infty) = +\infty$ $\ln 1 = 0$ $\ln 0 = -\infty$

- $\ln x$ is positive for $x > 1$, $\ln x$ is negative for $0 < x < 1$.

- $\ln x$ has no real value if x is negative.

- $\ln x$ increases steadily from $-\infty$ to $+\infty$ as x increases from 0 to ∞.

- $e^{\ln x} = x$ by definition.

THE EXPONENTIAL FUNCTION AND THE LAW OF NATURAL GROWTH

A quantity obeys the law of natural growth (or decay) if the rate of increase (or decrease) of the quantity is always proportional to the quantity itself.

This law is very important as it occurs frequently in mechanics, physics, engineering sciences and biology.

Let the quantity be y, which is a function of x (say).

Then the rate of increase of y w.r.t. x is $\dfrac{dy}{dx}$.

If this is proportional to y itself, we have

$$\frac{dy}{dx} = ky \qquad \text{where } k \text{ is a constant}$$

This is known as a *differential equation* which we have to solve. It is impossible to integrate the right hand side directly.

This would give $y = \int ky\,dx + c$ but since we are trying to find y, we don't know what it is in terms of x *and so it cannot be substituted in.*

However, $\dfrac{dy}{dx} = ky$ can be inverted to read $\dfrac{dx}{dy} = \dfrac{1}{ky}$.

Now we can integrate with respect to y:

$$x = \int \frac{1}{ky}\,dy + c = \frac{1}{k}\ln y + c$$

Suppose that y is some value (say C) when $x = 0$.

Substituting:

$$0 = \frac{1}{k}\log C + c$$

$$\therefore \qquad c = -\frac{1}{k}\log C$$

Hence

$$x = \frac{1}{k}(\log y - \log C)$$

$$= \frac{1}{k}\log \frac{y}{C}$$

$$\therefore \qquad kx = \log \frac{y}{C}$$

$$\therefore \qquad e^{kx} = \frac{y}{C} \text{ (by definition of logs)}$$

$$\text{and} \qquad y = Ce^{kx}$$

Thus, the solution of the equation $\dfrac{dy}{dx} = ky$ is $y = Ce^{kx}$ where C is a constant depending on initial conditions.

Therefore, the law of natural growth describes a law based on the exponential function. There are many examples in the real world of quantities that can be modelled by the law and some of them are considered briefly here.

Expansion of a metal rod

The law states that the rate of increase of the length of a metal rod with respect to temperature is proportional to the length of the rod at that temperature.

Let l = length of rod at temperature $T°$. Then $\dfrac{dl}{dT} = kl$ where k is a constant depending on the type of metal used.

From preceding theory:

$$l = l_0 e^{kT} \qquad \text{where } l_0 = \text{original length when } T = 0$$

This increase is so small that in practice the terms of the exponential series after the first two terms are usually ignored giving

$$l = l_0(1 + kT)$$

Newton's law of cooling

This states that the rate of cooling of a hot body is proportional to the temperature difference between the body and its surroundings. Since the temperature is falling the gradient of the temperature–time graph is negative. This is a law of natural *decay*, rather than growth:

i.e. $\qquad \dfrac{dT}{dt} = -kT \qquad$ where T is the temperature difference at time t

$\therefore \qquad T = T_0 e^{-kt} \qquad$ where T_0 is the temperature difference when $t = 0$

Decay of current in an inductive circuit

The current i amperes in a circuit of resistance R ohms and inductance L henrys satisfies the equation

$$L\frac{di}{dt} + Ri = E$$

where E volts is the EMF applied to the circuit.

If the EMF is cut off, i.e. $E = 0$, we get

$$L\frac{di}{dt} + Ri = 0$$

$$\therefore \quad \frac{di}{dt} = -\frac{R}{L} \cdot i$$

$$\therefore \quad i = i_0 e^{-\frac{R}{L} \cdot t}$$

where i_0 = current at the time when the EMF was cut off. Again, this is a law of decay.

The population explosion

If it is assumed that the rate of increase of the population is proportional to the population at any given time, then we have

$$\frac{dp}{dt} = kp \qquad \text{where } k \text{ is a known constant and } p \text{ is the population at time } t$$

k will depend on how many more births there are than deaths in any one year. We will assume a model where there are 1005 births for every 1000 deaths; then

$$\frac{dP}{dt} = 0.005P$$

This is of the same form as our standard form

$$\frac{dy}{dx} = ky \qquad \text{whose solution is } y = Ce^{kx}$$

Suppose the population was roughly 53.5 million in 1970, when $t = 0$. The solution of our equation is

$$P = Ce^{0.05t}$$

But when $t = 0$, $P = 53.5$ million.

Substituting:

$$\therefore \qquad\qquad 53.5 \text{ million} = Ce^0$$

But $e^0 = 1$ so $C = 53.5$ million

$$\therefore \qquad\qquad P = 53.5\, e^{0.005t} \text{ in millions}$$

To find the population in the year 2000, when $t = 30$

$$P = 53\tfrac{1}{2}.5e^{0.005 \times 30}$$

$$= 53.5e^{0.15}$$

$$= 53.5 \times 1.1618 \quad \text{from calculator}$$

$$\cong 62.2 \text{ million}$$

Exercise 19.2

1. If a population of flies under certain conditions satisfies the law that

    $$\frac{dP}{dt} = 0.35P$$

    where P is the population at time t days:

    (i) estimate the probable population at the end of 10 days if initially it is 200;

    (ii) find the number of days it will take to double the colony correct to 1 place of decimals.

2. The rate at which a radio-active substance decays is proportional to the number of atoms N present at any time t. If the constant of proportionality is λ (the decay constant) and initially there are N_0 atoms present, express

    (i) N as a function of t;

    (ii) find also the time taken for half of the initial amount to decay, i.e. the half-life of the substance.

20

Some Applications of Logarithms

USE OF LOGARITHMS TO LINEARISE GRAPHS

In Chapter 13, we found that it was possible to create straight line graphs from data that did not normally produce a straight line. We now look at some further ideas, involving the use of logarithms. There are two main types of laws that can be handled this way

(i) $y = ax^n$ where a and n are unknown constants

To linearise this law, we need to get n out of the index. To do this, we can take logs of both sides. We can use either base 10 or base e. Consider base 10:

$$\log y = \log(ax^n)$$
$$= \log a + \log x^n \qquad \text{(logs rule 1)}$$
$$= \log a + n \log x \qquad \text{(logs rule 3)}$$

Rearrange: $\log y = n \, . \, \log x + \log a$

Compare with $Y = m \, . \, X + C$

If we plot $\log y$ against $\log x$ we should get a linear graph whose gradient tells us n and whose intercept on the '$\log y$' axis is $\log a$.

If we had used natural logs we would have merely ended up with

$$\ln y = n \ln x + \ln a$$

and compared with $Y = mX + C$

Plotting $\ln y$ against $\ln x$ will give a straight line gradient n and intercept $\ln a$.

Example 20.1   We suspect that the following data represents a law of the form
$y = ax^n$ where a and n are positive integers. Find a and n.

x	1	2	3	4	5
y	2	16	54	128	250

Solution   (This one is quite easy to see: the y values are $2(1)^3$, $2(2)^3$ and $2(3)^3$ etc.
Hence we can anticipate the correct answers to be $a = 2$ and $b = 3$.) Take logs to base
10 and form a new table. The graph is shown in Figure 20.1.

$\log x$	0	0.30	0.48	0.60	0.69
$\log y$	0.30	1.20	1.73	2.11	2.40

Note that when dealing with logarithms, although, as in Chapter 13, it is not worth
quoting to more than 2 d.p., it is very important to be as accurate as possible with the
graph. A very small error in calculating log a from the intercept can lead to quite a
substantial error in a itself.

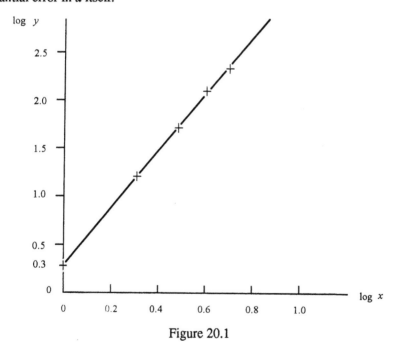

Figure 20.1

From the graph

$$\log y = n\log x + \log a$$

Hence $\qquad n = \text{gradient} = \dfrac{2.4-0.3}{0.69-0} = \dfrac{2.1}{0.69} = 3.032...$

i.e. $n = 3$ to nearest integer and

$$\log a = \text{intercept on vertical axis} = 0.30$$

$\Rightarrow a = 10^{0.3} = 1.995 = 2$ to nearest integer.

Therefore, the data obeys a law of the form $y = 2x^3$.

(ii)   **The exponential law $y = ae^{bx}$ where a and b are unknown constants**

Remember, the idea is to get rid of the power. Note that it is even more important now since the variable x is actually in the power. Clearly, because of the presence of e, it is best to take natural logs:

$$\ln y = \ln ae^{bx}$$
$$= \ln a + \ln e^{bx} \qquad \text{(rule 1)}$$
$$= \ln a + bx \ln e \qquad \text{(rule 3)}$$

i.e. $\qquad\qquad\qquad \ln y = bx + \ln a$

Compare with $\qquad\qquad Y = mx + c$

This time we plot $\ln y$ against x. This gives a linear graph gradient b, intercept $\ln a$.

Example 20.2   If the following data represents a law of the form $N = ae^{bt}$, find a and b.

t	2	3	4	5	6
N	2.575	1.562	0.947	0.575	0.349

Solution

If $N = ae^{bt}$ then $\qquad\qquad \ln N = bt + \ln a$

This time we only need to take logs of N values, so the new table is

t	2	3	4	5	6
ln N	0.946	0.446	−0.054	−0.553	−1.053

which gives the graph

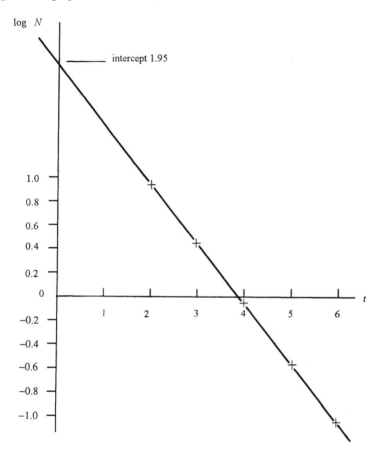

Figure 20.2

This time c = gradient = − 0.5 and 'y intercept' = ln a = 1.95 $\Rightarrow a = e^{1.95} = 7.4$.

Hence
$$N = 7.4e^{-t/2}$$

THE USE OF LOGARITHMIC GRAPH PAPER

The work involved in the above two cases is reduced if logarithmic graph paper is used, i.e. special graph paper which converts either both or one variable directly to logs without the use of calculators. The two types are:

- Semi–logarithmic graph paper, often called 'log–linear' graph paper.

- Logarithmic graph paper, usually called 'log–log' graph paper.

The logarithmic scale

The logarithms to base 10 of the integers between 1 and 10 are as follows:

Number	1	2	3	4	5	6	7	8	9	10
log	0	0.301	0.477	0.602	0.699	0.778	0.845	0.903	0.954	1

For numbers between 10 and 100:

Number	10	20	30	40	50	60	70	80	90	100
log	1	1.301	1.477	1.602	1.699	1.778	1.845	1.903	1.954	2

Lengths are marked off along an axis corresponding to the logarithms (see Figure 20.3). This gives the logarithms of numbers without further use of tables. The intervals are further subdivided to give intermediate values.

The unit used is called one 'log cycle' and may be extended to several cycles to represent the logarithms of numbers between 10 and 100, 100 and 1000, and so on. Printed log paper is available with as many as five log cycles for plotting a range of values from 1 to 10^5, or 0.1 to 10^4, or any similar range.

Semi–logarithmic graph paper is usually printed with a logarithmic scale vertically, and a uniform (i.e. linear) scale horizontally, but of course these may be reversed if necessary by rotating the paper.

Vertical Logarithmic Scale

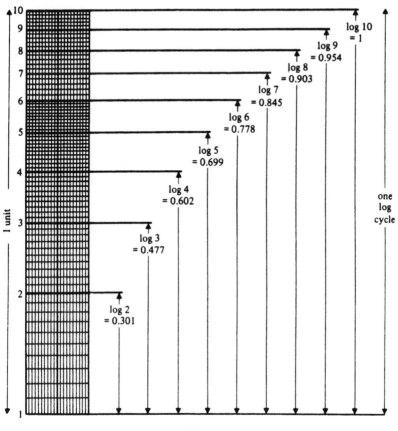

Figure 20.3

Logarithmic graph paper is printed with logarithmic scales on both axes. The use of each of these types is illustrated by the following examples.

Example 20.3

In an experiment on a certain woollen material the values of one of its properties, y, were observed for various values of another property x, as follows:

x	4	8	12	16	20	24	28	32	36	40	44	48	52
y	1.45	1.8	2.4	3.15	3.9	5.1	6.6	8.4	11.0	14.0	17.5	23.0	29.9

It is thought that these results may obey a relationship of the form $y = ae^{bx}$ where a and b are constants. Verify this, and find the probable values of a and b.

Solution   If $y = ae^{bx}$

then $\log y = \log a + bx \log e$

Let $Y = \log y$, $A = \log a$, $B = b \log e$

Then $Y = A + Bx$

Hence if Y is plotted against x, the graph should be a straight line.

Thus we could plot Y (i.e. $\log y$) vertically against x horizontally, as before, or we can use a semi–log graph paper.

N.B.1   The range of values of y is 1.45 to 29.9. This means we need two log cycles 1 to 10 and 10 to 10^2.

N.B.2   The logarithms of the y values are automatically given by plotting the values of y on the vertical logarithmic scale (see Figure 20.4).

The plotted points appear to be in a straight line, allowing for experimental errors, thus verifying that $y = ae^{bx}$ is the form which fits the given values of x and y.

To find the equation of the line we evaluate A and B in the previous equation.

To find a and b, we use

$$\log a = A \qquad \text{and} \qquad b = B/\log e$$

To find A and B, and hence a and b, we proceed as follows:

P and Q are two points on the line $Y = A + Bx$

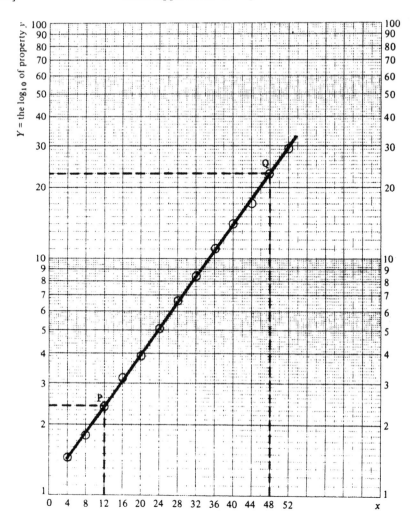

Figure 20.4

At P, when $x = 12$, $Y (= \log y) = \log 2.4$.

$\therefore$ $$\log 2.4 = A + 12B$$

At Q, when $x = 48$, $Y = \log 23.0$

$\therefore$ $$\log 23.0 = A + 48B$$

Subtract:

$$\log 23.0 - \log 2.4 = 36B$$

$$\therefore \quad B = \frac{\log 23.0 - \log 2.4}{36} \cong 0.0273$$

But
$$b = \frac{B}{\log e} \cong \frac{0.0273}{0.4343} = 0.063$$

Hence,
$$A \cong \log 23.0 - 48B = 0.0530$$
$$\therefore \quad a \cong 10^A = 10^{0.0530} \cong 1.13$$

The law of the curve is $\quad y \cong 1.13e^{0.063x}$

Example 20.4   In an experiment similar to that in the example above, another property of the material, z, was observed for the same values of x as before.

x	4	8	12	16	20	24	28	32	36	40	44	48	52
z	3.6	6.4	9.2	11.9	14.1	17.0	18.5	21.0	23.0	25.5	27.2	29.7	31.9

Verify that these results obey the law $z = kx^n$ and find the probable values of k and n.

Solution   If $z = kx^n$

then $\log z = \log k + n \log x$

Let $Z = \log z \quad K = \log k \quad X = \log x$

Then $Z = K + nX$

Hence if $Z (= \log z)$ is plotted against $X (= \log x)$, a straight line should be obtained.

Here we need logarithmic scales on both axes and can use 'log–log' graph paper.

N.B.1   The range of values of x is the same as before and needs two log cycles. The range of values of Z is 3.6 to 31.9 and also needs two log cycles.

N.B.2   The logarithms of both x and z values are shown automatically by using 'log–log' paper (see Figure 20.5).

The points lie approximately on the line drawn, thus verifying that the law connecting x and z is of the form $z = kx^n$.

To find n and K and hence k, we proceed as follows:

P and Q are two points on the line $Z = K + nX$

At P, $x = 4$, $z = 3.6$. Hence $X = \log 4$, $Z = \log 3.6$.

Therefore $\log 3.6 = K + n \log 4$

At Q, similarly, $X = \log 52 \quad Z = \log 32$

$\therefore$ $\log 32 = K + n \log 52$

Subtracting:

$\log 32 - \log 3.6 = n(\log 52 - \log 4)$

$\therefore$ $\log (32/3.6) = n \log(52/4)$

$\therefore$ $n = \log(8.889) \div \log 13 = 0.85$

Substituting back:

$\log 32 = K + 0.85 \log 52$

$\therefore$ $K = \log 32 - 0.85 \log 52 \cong 0.0472$

$\therefore$ $k = 10^{0.0472} \cong 1.12$

Hence the law connecting x and z is given by

$$z = 1.12x^{0.85}$$

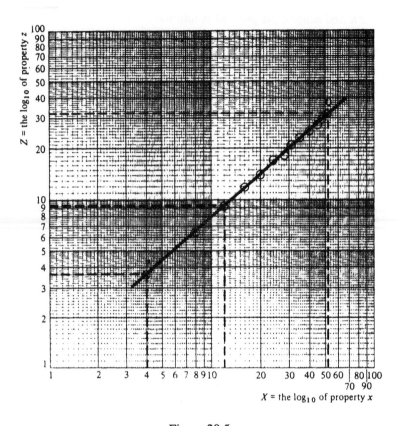

Figure 20.5

Exercise 20.1

Carry out the following exercises using ordinary graph paper and logarithmic graph paper.

1.   The following table gives pairs of values of two quantities p and q, which probably follow the law $q = ap^k$.

p	0.92	1.81	3.7	5.54	8.65	12.5
q	2.5	4.2	7.6	11	15	21

Plot a graph to determine whether this is so, and if so what are the most likely values of a and k?

2.   Verify that a law of the form $y = Ce^{kx}$ holds for the following pairs of values:

x	0	1.0	1.5	2.5	4.0
y	4.2	0.937	0.4427	0.0987	0.0105

Find values for C and k.

3.   The tension in a rope which is just on the point of slipping on a pulley is given in terms of the length l turns in contact with the pulley by the following table:

l	¼	¾	1	1¼	2	2½
T	3.17	7.90	12.68	19.9	80.3	201.5

Verify the law $T = T_0\, e^{2\pi\mu l}$ and find T_0 and μ.

DIFFERENTIATION OF LOGARITHMIC FUNCTIONS

In Chapter 19, we developed the results:

$$\boxed{\frac{d}{dx}(\ln x) = \frac{1}{x}} \qquad \text{and} \qquad \boxed{\int \frac{dx}{x} = \ln x + c}$$

These basic results are easily extended by use of the chain rule.

Example 20.5   Differentiate (a) $\ln(x + 5)$ (b) $\ln(3x - 7)$

Solution   Using the same reasoning as in earlier chapters, when we differentiate $\ln(\text{'something'})$ we get $\dfrac{1}{\text{something}}$ and to complete the differentiation, we must multiply by the derivative of the 'something'. Hence

(a) $\dfrac{d}{dx}[\ln(x+5)] = \dfrac{1}{x+5} \times$ the derivative of $(x+5)$

$$= \frac{1}{x+5}$$

(b) $\qquad \dfrac{d}{dx}[\ln(3x-7)] = \dfrac{1}{3x-7} \times$ the derivative of $(3x-7)$

$$= \dfrac{3}{3x-7}$$

We can reverse the above results to deduce that

$$\int \dfrac{dx}{x+5} = \ln(x+5)+c \text{ and } \int \dfrac{dx}{3x-7} = \tfrac{1}{3}\ln(3x-7)+c$$

Similar results will hold true for logarithms of all linear functions and so we may deduce the two general results:

$$\dfrac{d}{dx}[\ln(ax+b)] = \dfrac{1}{ax+b} \qquad \text{and} \qquad \int \dfrac{dx}{ax+b} = \dfrac{1}{a}\ln(ax+b)+c$$

Exercise 20.2

1. Differentiate the following:

     (a) $\ln(x+6)$ (b) $\ln(x-11)$ (c) $\ln(5-x)$

     (d) $\ln(5x+1)$ (e) $\ln(4x-3)$ (f) $\ln(3-7x)$

2. Integrate the following:

     (a) $\int \dfrac{dx}{x+3}$ (b) $\int \dfrac{dx}{x-7}$ (c) $\int \dfrac{dx}{3-x}$

     (d) $\int \dfrac{dx}{3x+4}$ (e) $\int \dfrac{dx}{3x-4}$ (f) $\int \dfrac{dx}{11-8x}$

Example 20.6   Differentiate:

     (a) $\ln(2-x^2)$ (b) $\ln\dfrac{(3+x)}{(3-x)}$ (c) $\ln(3x^2-2x+1)$

Solution

(a)   Let $y = \ln(2-x^2)$

then $\dfrac{dy}{dx} = \dfrac{1}{2-x^2} \times$ the derivative of $(2-x^2)$

$$= \dfrac{-2x}{2-x^2} = \dfrac{2x}{x^2-2}$$

(b) Let $y = \ln\dfrac{3+x}{3-x} = \ln(3+x) - \ln(3-x)$

$\therefore \quad \dfrac{dy}{dx} = \dfrac{1}{(3+x)}.1 - \dfrac{1}{(3-x)}.(-1)$

$$= \dfrac{1}{3+x} + \dfrac{1}{3-x}$$

$$= \dfrac{3-x+3+x}{(3+x)(3-x)} = \dfrac{6}{(9-x^2)}$$

(c) Let $y = \ln(3x^2 - 2x + 1)$

$\dfrac{dy}{dx} = \dfrac{1}{3x^2 - 2x + 1} \times$ the derivative of $(3x^2 - 2x + 1)$

$$= \dfrac{6x-2}{3x^2 - 2x + 1}$$

Note that the last part of this example gives us a clue as to how to perform certain special integrals.

These integrals are of the type $\displaystyle\int \dfrac{\text{derivative of denominator}}{\text{denominator}}.dx$

For example, reversing the process in (c) above:

$$\int \dfrac{6x-2}{3x^2 - 2x + 1} dx \qquad \text{(numerator is derivative of denominator)}$$

$$= \ln(3x^2 - 2x + 1) + c \quad \text{(hence answer} = \text{log of denominator)}$$

This will always happen, giving the general rule:

$$\int \dfrac{\text{derivative of denominator}}{\text{denominator}}.dx = \ln(\text{denominator})$$

Example 20.7   Integrate:

(a) $\dfrac{3x^2 - 8x + 7}{x^3 - 4x^2 + 7x - 15}$ (b) $\dfrac{3x + 2}{3x^2 + 4x - 11}$ (c) $\cot x$

Solution

(a)   Since the numerator is the derivative of the denominator, we may immediately write

$$\int \frac{3x^2 - 8x + 7}{x^3 - 4x^2 + 7x - 15}\,dx = \ln(x^3 - 4x^2 + 7x - 15) + c$$

(b)   This time the numerator is only half the derivative of the denominator but we can get around this by writing

$$\int \frac{3x + 2}{3x^2 + 4x - 11}\,dx = \tfrac{1}{2}\int \frac{6x + 4}{3x^2 + 4x - 11}\,dx = \tfrac{1}{2}\ln(3x^2 + 4x - 11) + c$$

(c)   This is a clever one! Since $\cot x = \dfrac{1}{\tan x} = \dfrac{\cos x}{\sin x}$ then

$$\int \cot x\,dx = \int \frac{\cos x}{\sin x}\,dx = \ln \sin x$$

because $\cos x$ is the derivative of $\sin x$.

Exercise 20.3

1.   Differentiate the following:
   (a) $\ln(3x)$ (b) $\ln(4x - 1)$ (c) $2\ln(x^2)$ (d) $\ln(\sin x)$
   (e) $\ln(\cos x)$ (f) $\ln\dfrac{a + x}{a - x}$ (g) $\ln(e^x + e^{-x})$ (h) $\ln(\sqrt{\sin x})$

2.   Use the product rule to differentiate:
   (a) $x^3 \ln x$ (b) $x \ln 2x$ (c) $\sin ax \ln bx$ (d) $\ln ax.\ln bx$

3.   Integrate the following:
   (a) $\int \dfrac{2x\,dx}{1 + x^2}$ (b) $\int \dfrac{6x^2}{2 + 2x^3}\,dx$ (c) $\int \dfrac{2x - 3}{x^2 - 3x}\,dx$ (d) $\int \dfrac{\sin x}{\cos x}\,dx$

4.   Evaluate the following definite integrals
   (a) $\int_1^2 \dfrac{1}{1 + x}\,dx$ (b) $-\int_2^3 \dfrac{3}{x - 1}\,dx$ (c) $\int_0^{1.5} \dfrac{x}{1 + x^2}\,dx$ (d) $\int_{1.5}^{2.5} \dfrac{2}{1 + 3x}\,dx$

21

A First Look at Statistics

We shall not attempt a formal definition of what we mean by statistics. It is sufficient to say that we are bombarded by statistics in our everyday lives. Since the Domesday Book was written, and probably much earlier than that, humans have been fascinated with collecting and recording facts. We meet them today under the headings of batting averages, average earnings, inflation rates; through advertising such as 'nine out of ten people use our margarine'; and so on.

It is the statistician's job to make sense of these collected facts or *data* and to look into the future, predicting possible consequences and planning strategies to cope with them.

The difference between statistical data and data collected in, say, a physics laboratory, is the *lack of control*. In physics we can design experiments so that unwanted variables are eliminated. We can control very precisely how one variable depends on another, eliminating all other influences. As a consequence, we can expect our results to be accurate and our conclusions to be certain.

Data collected by statisticians is much harder to pin down. If we are manufacturing new cars, say, and want to predict how much investment to inject into the development of a particular new model and how much to spend on an advertising campaign, we can only base our ideas on past knowledge. There are so many variables involved that we could never predict *exactly* what to do. It is up to the statistician to find the *best* strategy. Then our 'guess' at what to do is based on intelligent choice rather than intuition.

Statistics therefore gives a guide, however rough, to what is happening in society. It can be a very useful tool and it can easily be abused, presenting facts in a distorted or twisted way, in order to confuse or mislead people.

A knowledge of the correct use of statistics is therefore very useful in modern life and so we will attempt to put statistics on a rational basis and look at some of the essential tools used by statisticians. We cannot do more than this in the present book.

ON THE AVERAGE

We are all familiar with batting averages, bowling averages, average rainfall, etc. We recognise this as a quick way of conveying an understanding of what a batsman or bowler or Nature has done, and it conveys the information much more clearly than a whole set of figures.

There are several different ways of measuring averages all of which have special uses. Consider the following eleven numbers:

$$2 \quad 3 \quad 5 \quad 4 \quad 5 \quad 6 \quad 5 \quad 4 \quad 8 \quad 1 \quad 2$$

The first thing we notice is that the number 5 is 'most popular'. It occurs three times. From this point of view, it gives us an average figure from the set. This type of average is called the *mode* of the set of figures. Now rearrange the numbers in ascending (or descending) order:

$$1 \quad 2 \quad 2 \quad 3 \quad 4 \quad 4 \quad 5 \quad 5 \quad 5 \quad 6 \quad 8$$

The number 'in the middle' is 4 and in that sense gives us an idea of 'average'. This kind of average is called the *median*. (How would we find the median of ten numbers?)

The most common average used in everyday life is the *arithmetic mean* (or simply, *the mean*). To find the mean of a set of numbers we simply find their sum and divide by the number of numbers, i.e. for the above set

$$\text{Mean} = \frac{2+3+5+4+5+6+5+4+8+1+2}{11}$$

$$= \frac{45}{11}$$

$$= 4.1 \quad \text{to 1 decimal place}$$

Notice that the mean is not necessarily a number in the given set. Notice also that we have given the answer to just 1 decimal place. The original data was in integer form so there is no point going further.

(We can now see, by the way, that an easy measurement for the median of a set of 10 numbers is to find the mean of the middle two.)

When looking at averages, we tend to look at 'middle' values of a set of figures, ignoring extremes. Thus averages are known as *measures of central tendency*. They all have their uses and misuses. In this book we will concentrate on calculations involving the mean. Firstly two examples of how the mean should be treated carefully. It can easily convey the wrong idea.

Example 21.1   The boss of a small business wants to attract you to apply for a vacant post in the office. There are five people in the office, including the boss. Their annual wages are

$$£4000 \quad £4000 \quad £5000 \quad £10,000 \quad £40,000$$

(guess which one is the boss!).

Write down the mode and the median and then calculate the mean.

What average salary should the boss quote to give a fair idea of what you might expect to earn?

What would you say to a boss who offered you the job at £4000 p.a. after quoting the above mean in the advert?

Example 21.2   An aeroplane flies in a straight line from A to B and from B to C. Its speed from A to B is 400 km/h and from B to C is 450 km/h. If $AB = 200$ km and $BC = 150$ km what is its average speed?

Solution

We could say

average speed = the average of the two speeds

$$= \frac{400 + 450}{2}$$

$$= 425 \text{ km / h}$$

But is it?

What do we mean by 'average speed'? When we make a long journey by car and compliment ourselves on 'averaging' 60 km/h in spite of 'hitting the rush hour' in a busy town, stopping for meals or refuelling, don't we mean that if we had been able to travel straight through without let or hindrance at the speed of 60 km/h we would have taken the same time for the journey as that actually taken? In other words:

$$\text{Average speed} = \frac{\text{total distance}}{\text{total time}}$$

In this case:

$$\text{time from } A \text{ to } B = \frac{200}{400} \text{ hours} = \frac{1}{2} \text{ hour}$$

$$\text{time from } B \text{ to } C = \frac{150}{450} \text{ hours} = \frac{1}{3} \text{ hour}$$

Therefore $$\text{total time} = \frac{5}{6} \text{ hour}$$

so $$\text{average speed} = \frac{\text{total distance}}{\text{total time}}$$

$$= 350 \div \frac{5}{6} = 350 \times \frac{6}{5}$$

$$= 420 \text{ km/h}$$

This is the correct answer. We fell into a trap earlier because speed is a combination of two quantities, distance and time. It is permissible to consider the mean of quantities such as pounds or runs or ages since these are simple units in their own right.

THE MEAN

Mathematically we define the arithmetic mean as

$$\bar{x} = \frac{x_1 + x_2 + x_3 + \ldots + x_n}{n}$$

where $\{x_1, x_2, x_3, \ldots x_n\}$ is a set of n numbers

or using our 'sigma' notation

$$\bar{x} = \frac{1}{n} \sum x_i \qquad \text{where } i \text{ takes the values 1 to } n$$

This is what is normally understood as the 'average'; we add up all the numbers in a set and divide by the number of numbers in the set.

Example 21.3   The following set gives the lifetime x (in years) of six washing machines. Calculate the mean lifetime.

$$x = \{14, \ 6, \ 11, \ 7.5, \ 4, \ 9.5\} \text{ years}$$

Solution

$n = 6$; therefore

$$\bar{x} = \frac{1}{n}\sum x = \frac{1}{6}(14 + 6 + 11 + 7.5 + 4 + 9.5)$$

$$= \frac{52}{6}$$

Therefore, the mean lifetime $= 8$ years, 8 months

CODING

The above method is quite laborious if the numbers involved are large. There are methods which help to simplify the calculations. Be aware that these methods are still useful even when using a calculator. Since they enable us to work with easier numbers, it is harder to make a mistake!

The idea is to manipulate the set of numbers to give a simpler set, by subtracting a convenient number from all members of the set. We can subtract the smallest number in the set as in Example 21.4(a) or make a rough guess at the mean as in 22.4(b) and subtract this number.

We find the mean of the simpler set and then reverse the process to find the mean of the original set. These methods are called *coding methods*.

The number we choose to subtract is called the *working mean* or *assumed mean*. In the following examples, we will call it α.

Example 21.4 Find the mean of the following numbers:

453, 475, 450, 461, 492, 450, 455, 485, 470, 481

Solution

To illustrate the method, we will perform the calculation two different ways:

(a) We notice that 450 is the least number. Let $\alpha = 450$ and subtract α from all the numbers in the list, arranging the work as follows. This gives a much simpler set.

	x	$(x - \alpha)$
$n = 10$	453	3
	475	25
$\alpha = 450$	450	0
	461	11
	492	42
	450	0
	455	5
	485	35
	470	20
	481	31
		$\sum(x-\alpha) = 172$

$$\text{mean of the simplified set} = \frac{\sum(x-\alpha)}{n} = \frac{172}{10} = 17.2$$

But we subtracted 450 from all numbers in the original set. We must now add it back:

$$\text{mean of the original set, } \bar{x} = 17.2 + 450$$

$$= 467.2$$

(b)   We notice the numbers range from 450 to 492, so we might take $\alpha = 470$ as a rough estimate of the middle of the set, and set out the work as follows:

This time let $u = (x - \alpha)$; then we can talk about $\bar{u}$:

	x	$u = (x - \alpha)$	
$n = 10$	453		−17
	475	+ 5	
$\alpha = 470$	450		−20
	461		− 9
	492	+ 22	
	450		−20
	455		−15
	485	+ 15	
	470	0	
	481	+ 11	
		$\sum u = -28$	

therefore $\quad\quad\quad\quad \bar{u} = \dfrac{\Sigma u}{n} = -\dfrac{28}{10} = -2.8$

Hence $\quad\quad\quad\quad \bar{x} = \bar{u} + 470$ adding back the assumed mean.

i.e. $\quad\quad\quad\quad \bar{x} = -2.8 + 478 = 467.2$ as before

The above is very easy to prove using simple algebra and the properties of Σ.

Let $u = x - \alpha$. Then obviously $x = u + \alpha$ but this does not necessarily mean that $\bar{x} = \bar{u} + \alpha$. However since $u = x - \alpha$,

$$\Sigma u = \Sigma(x - d) = \Sigma x - \Sigma \alpha$$
$$= \Sigma x - n\alpha \quad\quad \text{(since } \alpha \text{ is constant)}$$

so $\quad\quad\quad\quad \bar{u} = \dfrac{\Sigma u}{n} = \dfrac{\Sigma x - n\alpha}{n}$

$$= \dfrac{\Sigma x}{n} - \alpha$$

i.e. $\quad\quad\quad\quad \bar{u} = \bar{x} - \alpha$

so $\quad\quad\quad\quad \bar{x} = \bar{u} + \alpha$

Therefore, if we subtract α from all our xs to find $\bar{u}$, we then add it back to find $\bar{x}$.

Exercise 21.1

1. Calculate the mean of the following percentages (20 in all) gained by a set of students in an examination.

    45, 71, 64, 32, 37, 48, 25, 62, 54, 60,
    58, 29, 36, 42, 38, 43, 65, 72, 15, 46.

2. (a) Do the same for the following:

    45, 61, 64, 32, 37, 48, 25, 62, 54, 60,
    58, 29, 36, 42, 38, 43, 65, 62, 35, 46.

    (b) Can you account for these answers noting that the highest and lowest marks in the two cases (i.e. the ranges of marks) are quite different?

In the above examples, although the mean gives part of the picture, there could be many sets of numbers having the same mean. We therefore need another indicator to give us some idea of the *spread* of the values. We shall see later how this is done.

SAMPLING

Suppose the government is interested in finding out something about the incomes of families in the UK week by week, how they spend their incomes, how the incomes are grouped, what kind of houses they live in, how many children there are in each income group, etc. It would obviously be an impossible task to ask every household in the country every week. Therefore, it makes a limited survey, taking a *sample* of the *population*.

For example, in a certain survey conducted in 1972, the total number of households surveyed was 7017 and these were taken as a random sample. The results were as follows.

Average weekly expenditure in 1972 of households grouped according to gross income of household.

	* under £15	£15 and under £20	£20 and under £25	£25 and under £30	£30 and under £35	£35 and under £40	£40 and under £45	£45 and under £50	£50* and under £60	£60* and under £80	£80* and over
Total No. of H'holds	1047	423	476	513	553	630	565	530	833	880	567
Total No. of Persons	1446	892	1214	1417	1679	2010	1883	1756	2866	3068	2241
Av. No. Persons per H'hold	1.38	2.11	2.55	2.76	3.04	3.19	3.33	3.31	3.44	3.49	3.95

*Note the larger income ranges for the higher and lower income groups.

The average income per head of population in the various groups is also found. Expenditures are also taken under various headings such as food (split up into sections, e.g. bread, milk, meat, etc.), rent and rates, clothing, etc. This does not give the whole picture of what is happening to the incomes and expenditures in the UK but it gives some idea of the standard of living of the various groups in a much clearer way than if we just worked out (total income/total population), taken over the whole country.

We are only too aware these days of the phenomenon of 'opinion polls', whether they concern elections or how much marmalade people buy per week. These are 'sampling processes' and the samples are taken in a random way from the whole ''population'. This meaning of population is not confined to people. We can talk about the population of all the cars in this country or the population of trees in a park.

FREQUENCY DISTRIBUTIONS

We notice that when taking the results of a large sample from some population, the same numbers may occur several times. In Example 21.4 the number 450 occurred twice, i.e. with a *frequency* of two. When this occurs we arrange our work in tabular form as in the following example.

Example 21.5   When a die is thrown 100 times the scores 1 to 6 turn up with the following frequencies. Find the mean score.

Score (x)	1	2	3	4	5	6
Frequency (f)	16	20	9	14	21	20

Solution   Total throws, $n = \Sigma f = 100$

$$\begin{aligned} \text{Total score} &= 1 \times 16 + 2 \times 20 + 3 \times 9 + 4 \times 14 + 5 \times 21 + 6 \times 20 \\ &= \Sigma \text{ (frequency} \times \text{score)} \\ &= \Sigma f . x \end{aligned}$$

Hence $\text{mean score} = \dfrac{1}{n}\Sigma f . x$

i.e.

$$\bar{x} = \frac{\Sigma fx}{\Sigma f}$$

The solution can be tabulated as below. We take an assumed mean, $\alpha = 3$.

Score x	$u = (x - \alpha)$	f	fu
1	−2	16	−32
2	−1	20	−20
3	0	9	0
4	1	14	+14
5	2	21	+42
6	3	20	+60
		$\Sigma f = 100$	$\Sigma fu = 64$

$$n = \Sigma f = 100 \qquad \Sigma fu = 64$$

Therefore $\bar{u} = \frac{1}{n} \sum fu = 0.64$

Hence $\bar{x} = \bar{u} + \alpha$

$$= 3.64$$

The proof of why the above works is very similar to the one given earlier and follows the same principles.

DISCRETE AND CONTINUOUS MEASUREMENTS, GROUPING

In the above, the possible scores, 1 to 6, are *discrete values*, i.e. they are definite exact numbers. Other examples are the number of children in a family, the marks scored in an examination or the number of cars that roll off the production line in 1 day. These variables are called *discrete random variables*. They do not need to be integers. Think of buying shoes in half sizes!

We can think of other examples where the measurements are not discrete but vary *continuously*: the time taken by various workers to perform a specific task; the heights of children of school age. These variables are called *continuous random variables* and we may be able to measure them only approximately. The heights of children may be given to the nearest centimetre for convenience. Thus a child given as 145 cm may be anything between 144.5 cm and 145.5 cm, but would be classed as 145 cm. When we do this, we are *grouping* a set of values within a range into one group or class, e.g. the '145 cm class'. This is called *a grouped frequency distribution*, and the class from 144.5 cm to 145.5 cm is distinguished by 145 cm, which is the *midpoint of the range*.

Note that we must always make it perfectly clear what we intend to do if a measurement falls on exactly one or other of the boundaries, i.e. we could say 144.5 cm (exactly) falls into the 145 cm class, but 145.5 cm falls into the class above, i.e. the 146 cm class.

TALLY MARKS

When a sample has a large number of values, the work in calculating the mean may be simplified as in the following example by using tally marks. Here the results are grouped and the mean found by grouping may be slightly different from that of the original data (but only slightly). Therefore there is no point in quoting your mean to 10 decimal places, even if your calculator can do so.

Example 21.6   An efficiency expert takes a sample of 100 'breakdown' times (in minutes) of the machines in a factory. (Breakdown time is the length of time during which a machine is out of action either awaiting repair or being repaired or replaced.) The results are as follows. Find the mean breakdown time.

21	23	19	27	43	24	20	45	33	22
21	38	36	49	23	20	18	22	17	35
38	24	11	22	45	21	9	29	21	41
59	27	12	23	27	10	7	34	22	33
36	32	24	44	16	13	27	23	33	37
22	42	19	47	29	31	10	48	15	21
40	13	24	25	22	23	19	20	34	39
25	37	16	52	17	28	24	18	23	53
35	20	31	28	25	44	30	32	29	30
27	28	21	23	12	17	24	40	14	15

Solution   We organise the work as follows:

1.   Pick out the least and greatest readings; here 7 and 59.

2.   Decide the total range. Here we take 5 to 60, i.e. a range of 55.

3.   This suggests eleven classes of five. We choose 5–9, 10–14, 15–19, etc., inclusive.

4.   Corresponding to each value in the table we put a stroke in the appropriate range, and to help counting we make 'gates' of five (卌).

5.   Find the class midpoint, e.g. for numbers 5, 6, 7, 8, 9, the class midpoint is 7.

6.   Calculate the mean as before. (Here we use $\alpha = 22$ as an assumed mean.)

7.   We avoid any confusion about the class in which a value should be placed by taking the boundaries as shown.

Class Values	Class Boundaries	Class Midpoint x	Tally Marks	Frequency f
5–9	4.5–9.5	7	//	2
10–14	9.5–14.5	12	//// ///	8
15–19	14.5–19.5	17	//// //// //	12
20–24	19.5–24.5	22	//// //// //// //// //// ////	30
25–29	24.5–29.5	27	//// //// ////	14
30–34	29.5–34.5	32	//// ////	10
35–39	34.5–39.5	37	//// ////	9
40–44	39.5–44.5	42	//// //	7
45–49	44.5–49.5	47	////	5
50–54	49.5–54.5	52	//	2
55–59	54.3–59.5	57	/	1

To simplify this data, we will take an assumed mean of 22. This gives column[1] below. We see then that all figures in this column are divisible by 5, so we may divide by 5 to give column[2]. Therefore our coding is

$$u = \frac{(x-\alpha)}{5}$$

　　　　　　　　　　　[1]　　　　　　[2]

x	$(x-\alpha)$	$u = \dfrac{x-\alpha}{5}$	f	fu
7	−15	−3	2	−6
12	−10	−2	8	−16
17	−5	−1	12	−12
22	0	0	30	0
27	5	1	14	14
32	10	2	10	20
37	15	3	9	27
42	20	4	7	28
47	25	5	5	25
52	30	6	2	12
57	35	7	1	7
			$n = \Sigma f = 100$	$\Sigma fu = 99$

Therefore　　　　　　　$\bar{u} = \dfrac{\Sigma fu}{\Sigma f} = \dfrac{99}{100} = 0.99$

By similar reasoning to that given in our earlier proof:

$$\bar{u} = \frac{\bar{x} - \alpha}{5}$$

Hence $\qquad 5\bar{u} = \bar{x} - \alpha$

i.e. $\qquad \bar{x} = 5\bar{u} + \alpha \qquad$ (transposing)

So $\qquad \bar{x} = 5(0.99) + 22 \quad$ in this case

i.e. $\qquad$ mean breakdown time $= 4.95 + 22 = 26.95$ minutes

N.B.1 An easy way to remember the *order* of what to do when 'decoding':

When coding:	1)	subtract α	
	2)	divide by factor	*Last thing done*
			is first thing
When decoding:	1)	multiply by factor	*reversed*
	2)	add α	

N.B.2 If we had simply added together all the original values and divided by 100, we would find the mean to be 27.06. Hence using coding does give some inaccuracies in the result. But remember that *the original data is only given to the nearest minute* so there is no point quoting to 10 d.p.! Notice that 26.95 and 27.06 are both 27 minutes *to the nearest minute* and this is accurate enough for any efficiency expert!

Exercise 21.2

1.   (a)   Calculate the mean of the distribution

x	1	2	3	4	5	6
f	1	3	5	8	2	1

using $\bar{x} = \dfrac{1}{n}\Sigma fx, \; n = \Sigma f$.

       (b)   What difference would it make to $\bar{x}$ if the first and last frequencies were both increased by: (i)   5;   (ii)   10?

             Account for this.

2.　Using a working mean $\alpha = 5$, calculate the mean score of the following results:

Score x	0	1	2	3	4	5	6	7	8	9	10
Frequency f	7	3	1	12	16	37	14	4	3	2	1

3.　Check the result $\bar{x} = 26.95$ for Example 21.6 taking x as the class midpoint, frequency f as shown, and $\alpha = 32$.

GRAPHICAL REPRESENTATION OF FREQUENCY DISTRIBUTIONS

The Histogram

We represent each class by means of a rectangle whose *width* is the *class width* and whose *area* is the *frequency*.

Example 21.7

In a certain city there is a 'population' of milk delivery vans. The ages of these vans are as follows:

Age in years	0–1	1–2	2–3	3–4	4–5	5–6	6–7	classes
No. of vans	26	39	52	61	31	20	6	frequency

Solution　　The age is a continuous variable but we have grouped ages as shown. Plotting frequency against age we obtain the following histogram (Figure 21.1).

N.B.1 The *area* of each rectangle represents the frequency.

N.B.2 We can view the diagram as a series of piles of blocks, each block representing a van. The histogram shows how the vans are distributed according to their ages.

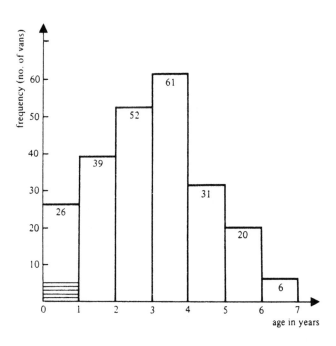

Figure 21.1

The Frequency Polygon

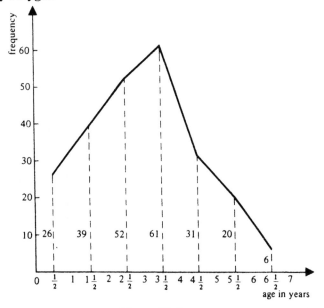

Figure 21.2

If we plot the frequency against the *class midpoint* we obtain the frequency polygon of the distribution, shown above. This corresponds to joining the midpoints of the tops of the rectangles in the histogram.

The Main Features of Histograms

A comparison of histograms is often useful in giving a clear picture when we wish to convey information quickly. Below we see the results of two examination papers set to the same group of 70 students.

Examination A

Marks	20–30	30–40	40–50	50–60	60–70	70–80
No. of students	3	4	20	27	12	4

Examination B

Marks	10–20	20–30	30–40	40–50	50–60	60–70
No. of students	2	7	15	24	16	6

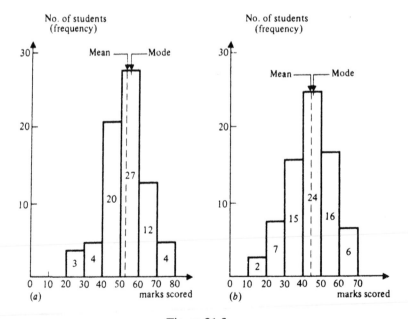

Figure 21.3

If the histograms (Figure 21.3) were drawn on transparent paper and then one were superimposed on the other, it would be obvious that in exam A there is a very definite shift to the right compared with exam B. In other words a higher standard of performance was achieved in exam A.

Mean of A marks = 52.6   Mean of B marks = 44.0

Mode of A marks = 55   Mode of B marks = 45.0

The *central values* of the *tallest blocks* are the *modes*. Note that the mean and the mode can be very different.

MEASURES OF SPREAD

Look at these two sets of six numbers:

1, 2, 3...18, 19, 20

1.....................9, 10, 10, 13.....................20

They both have the same mean (i.e. 10.5) but they are very different sets. In the first set, none of the numbers is anywhere near the mean and in the second they are nearly all bunched up and clustered close to the mean value.

Hence the mean alone is not enough to describe fully a set of figures. We need a second number which measures how the data is *spread* on either side of the mean.

There are several ways we can measure spread. The simplest way is to use the range of the data (i.e. largest value – smallest value). Note that this is not very useful here since both the above sets have the same range as well as the same mean!
Let us take a simple example. How can we measure the spread of the following numbers?

15   25   29   32   36   37   38   40   43   45

First calculate the mean. This is easily found to be 34.

It might now be an idea to find out how far each of the original numbers 'deviates' from the mean; so, subtract 34 from all these values:

x	deviations from the mean $(x-\bar{x})$	Note that:
15	-19	
25	-9	Numbers below the
29	-5	mean have a negative
32	-2	deviation.
36	2	
37	3	
38	4	Numbers above the
40	6	mean have a positive
43	9	deviation.
45	11	
	$\sum(x-\bar{x}) = -35 + 35 = 0$	

It would now be excellent if we could calculate the mean of all the deviations

But　　　　**mean of the deviations** $= \dfrac{\sum(x-\bar{x})}{10} = \dfrac{0}{10} = 0$

This calculation will *always* give zero for *any* data, because:

$$\frac{\sum(x-\bar{x})}{n} = \frac{\sum x}{n} - \frac{\sum \bar{x}}{n}$$

$$= \bar{x} - \frac{n\bar{x}}{n} \qquad \text{(since } \bar{x} \text{ is constant)}$$

$$= \bar{x} - \bar{x} = 0$$

We need to prevent this cancelling of positive and negative deviations. We can do this in two ways, both of which involve ensuring that the deviations are always positive.

The easiest (but not the most useful!) way of doing this is to *ignore the minus signs* by taking the *modulus* of the deviations: $|x-\bar{x}|$. The table now becomes

x	$|x - \bar{x}|$
15	19
25	9
29	5
32	2
36	2
37	3
38	4
40	6
43	9
45	11
	$\sum |x - \bar{x}| = 70$

Most data lies between 27 and 41 → (points to the x column)

deviations are now +ve and won't cancel! (points to the $|x-\bar{x}|$ column)

We can now find the mean of these deviations by dividing by 10.

$$\text{Mean deviation from the mean} = \frac{\sum |x - \bar{x}|}{n} = \frac{70}{10} = 7$$

This now gives an indication of the *average* spread and we can say that *most* of the data lies between $(34 - 7)$ and $(34 + 7)$, i.e. between 27 and 41.

The disadvantage of this method is that it is difficult to handle the modulus *algebraically*. A more complicated (but statistically better!) way of ensuring positive deviations is to square them and then find the mean of these 'squared deviations'. In this case we get

x	$x - \bar{x}$	$(x - \bar{x})^2$
15	-19	261
25	-9	81
29	-5	25
32	-2	4
36	2	4
37	3	9
38	4	16
40	6	36
43	9	81
45	11	121
		$\sum (x - \bar{x})^2 = 638$

We can now find the mean of these squared deviations:

$$\text{the mean of the squared deviations} = \frac{\Sigma(x-\bar{x})^2}{10} = 63.8$$

This number is known as the *variance* of the data. However, 63.8 is far too big a number for our purposes of measuring spread! We need to scale it down to the size of the original data. Remembering that we squared all the deviations, we merely need to *take the square root* to reverse the process. Hence, we have a new measure of spread called the *standard deviation* defined by

$$\text{standard deviation} = \sqrt{\frac{\Sigma(x-\bar{x})^2}{n}}$$

$$= \sqrt{63.8} \qquad \text{in this case}$$

$$= 8.9 \qquad \text{to 1 decimal place}$$

If we are talking about a sample, we use s to represent standard deviation. If we are talking about a whole population we use σ (this is small sigma just as Σ is 'capital' sigma).

To sum up:

VARIANCE AND STANDARD DEVIATION

Let $\bar{x}$ be the mean of n observations $x_1, x_2,..., x_n$. Then $(x_1-\bar{x})^2 + (x_2-\bar{x})^2 + ... + (x_n-\bar{x})^2$ is called the sum of the squares of the deviations from the mean. The average of these, i.e. $\frac{1}{n}[(x_1-\bar{x})^2 + (x_2-\bar{x})^2 + ... + (x_n-\bar{x})^2]$, is called the variance, s^2, of the n observations,

i.e. $$s^2 = \frac{1}{n}\Sigma(x-\bar{x})^2$$

The square root of this is called the standard deviation, s, from the mean,

i.e. $$s = \sqrt{\frac{\Sigma(x-\bar{x})^2}{n}}$$ where x takes values $x_1, x_2,..., x_n$

With frequency data, if the observations $x_1, x_2,..., x_n$ occur with frequency $f_1, f_2,..., f_n$ respectively, then the variance is given by

$$s^2 = \frac{f_1(x_1 - \bar{x})^2 + f_2(x_2 - \bar{x})^2 + ... + f_n(x_n - \bar{x})^2}{f_1 + f_2 + f_3 + ... + f_n}$$

i.e. $s^2 = \frac{1}{N} \sum f(x - \bar{x})^2$ where $N = \sum f$ and x takes the values $x_1, x_2,..., x_n$.

Similarly the standard deviation, s, is given by:

$$s = \sqrt{\frac{\sum f(x - \bar{x})^2}{N}} = \sqrt{\frac{\sum f(x - \bar{x})^2}{\sum f}}$$

This basic definition is not easy to use in practice and an alternative form can be proved to be:

$$s_x = \sqrt{\frac{\sum fx^2}{N} = \bar{x}^2}$$

In practice, we might be using coding with a working mean α, in which case, if $u = x - \alpha$, we calculate the standard deviation of the us, using

$$s_u = \sqrt{\frac{\sum fu^2}{N} - \bar{u}^2}$$

Then, from the properties of $\sum$, it can be shown that $s_x = s_u$.

An example will illustrate the method.

Example 21.8   A hundred children were examined in school and their heights noted to the nearest inch. Calculate the mean and standard deviation of the distribution. What is the mode of the distribution?

x = Height to nearest inch	60	61	62	63	64	65	66	67	68
f = Frequency	2	0	15	29	25	12	10	4	3

Solution   We arrange the calculation as follows:

Working mean $= \alpha = 64$

x	$u = (x-\alpha)$	f	fu	u^2	fu^2
60	-4	2	-8	16	32
61	-3	0	0	9	0
62	-2	15	-30	4	60
63	-1	29	-29	1	29
64	0	25	0	0	0
65	1	12	+12	1	12
66	2	10	+20	4	40
67	3	4	+12	9	36
68	4	3	+12	16	48
		$\Sigma f = 100$	$\Sigma fu = -11$		$\Sigma fu^2 = 257$

Therefore
$$\bar{u} = \frac{\Sigma fu}{\Sigma f} = -\frac{11}{100} = -0.11$$

This is measured from the working mean of 64

so   true mean $= 64 - 0.11$

i.e. $\bar{x} = 63.89$ inches

Also $\Sigma fu^2 = 257$ gives $\dfrac{\Sigma fu^2}{N} = 2.57$

Hence $s_x^2 = s_u^2 = \dfrac{\Sigma fu^2}{N} - \bar{u}^2$

$$= 2.57 - (-0.11)^2$$

$$= 2.25 - 0.0121$$

$$= 2.5579$$

Therefore $s_x = s_u = 1.6$ to 1 decimal place

From the table the height occurring most frequently is 63 inches.

Hence, the class midpoint is 63 inches, i.e. the mode = 63 inches.

Exercise 21.3

1. Calculate the mean and standard deviation of the following set of results:

x	30	31	32	33	34	35	36	37	38	39
f	4	8	23	35	62	44	18	4	1	1

2. The weights of 50 men to the nearest 10 lb are measured and grouped as follows:

x	115	125	135	145	155	165	175	185	195	205	215
f	2	3	3	6	9	8	7	8	1	1	2

where x = the class midpoint in lb.

Calculate the mean and standard deviation of the grouped frequency distribution.

22

Probability

CHANCE, PROBABILITY AND POSSIBILITY

(a) When we throw a die there is an equal chance of any of the six faces finishing uppermost. We say that there is a 'one in six' chance of scoring any of the numbers on the faces. What do we mean by a 'one in six' chance?

(b) When someone tells you that if you smoke 24 cigarettes a day you have a one in nine chance of contracting lung cancer, what are they implying?

In case (a) we *do not mean* that if we throw a die six times we shall throw one of each of a one, a two, a three, a four, a five and a six. We could score any combination of these.

In case (b) we *do not mean* that if nine people each smoke at least 24 cigarettes a day one of them will eventually contract lung cancer and the others will escape.

We do mean that:

(a) If we throw a die a great number of times (say 2400) then for about one-sixth of the throws, i.e. 400 times, we should throw a six or a five etc., i.e. about 400 of each. It could be 398 or 403 but it would be in the region of one-sixth of the total. We say, therefore, that the probability of throwing any particular number is 1/6, meaning 'in the long run'. (We note that the sum of the probabilities of all the outcomes when throwing a die is $1/6 + 1/6 + 1/6 + 1/6 + 1/6 + 1/6 = 1$.)

(b) If a large number of people all smoke at least 24 cigarettes a day about one-ninth of them will get lung cancer, and eight-ninths will escape. We say:

The probability of getting lung cancer $= 1/9$
The probability of not getting lung cancer $= 8/9$ } Total $= 1$

Mathematically we state:

The probability of an event E occurring is given by

$$p(E) = \lim_{n \to \infty} \left(\frac{n_E}{n} \right)$$

where n_E = the number of times E occurs in a sequence of n trials.

Note that what we are saying here is that when we make a succession of trials the fraction $\frac{n_E}{n}$ can be almost any fraction less than 1 for small trials but the more trials we make, the nearer it gets to some definite fraction which we call $p(E)$. This is represented in Figure 22.1.

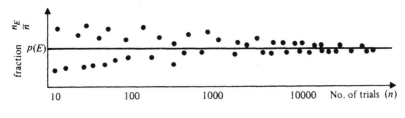

Figure 22.1

The dots represent the value of $\frac{n_E}{n}$ after n trials.

POSSIBILITY

'Possibility' is a different thing from 'probability'. It is possible that all the people in a certain country will commit suicide on the same day at the same time, but it isn't probable. It is possible that all the cars in Yorkshire could try to travel along the M1 on a Wednesday. But that isn't probable either. It is possible that all the air molecules which are bouncing about in a room could all finish up just under the ceiling, but it isn't probable.

Physicists have to deal with very large numbers of molecules, atoms, etc. They cannot deal with each molecule separately, not only because molecules are so small, but because each one has an effect on others by collisions, attractions, bouncing off the walls and so on. Calculations, therefore, when dealing with such a large number of individuals, would be impossible even with a computer. The scientist does not know which particular atom of a radio-active substance is going to decay next, but does know how many will *probably* decay within a given period of time. This has been walls and so on. Calculations, therefore, when dealing with such a large number of individuals, would be impossible even with a computer. The scientist does not know which particular atom of a radio-active substance is going to decay next, but does know how many will *probably* decay within a given period of time. This has been found by observation. Therefore, physicists and others have to deal with probabilities and work with these. Hence the necessity for trying to understand *the laws of probability*.

Causal laws (i.e. laws which say 'to every effect there is a precise cause'), have their uses in large-scale effects, but are severely limited at the atomic level. For example, if a few atoms are moving about in a given space, it is impossible to calculate what is going to happen to them, even if we send them off in given directions. Owing to the limited accuracy when dealing with such microscopically small objects, we don't know whether they will collide or not. If, however, we are dealing with a large number of atoms we could perhaps expect some of them to collide and in such a case predict the outcome. This is where probability is useful.

Probability theory is used extensively in work concerned with the reliability of aircraft, atomic power stations, chemical plant and other complex systems. To clarify the laws of probability we must discuss some elementary problems. Here is one:

Example 22.1 What are the probabilities that, when three coins are thrown, we shall turn up:

    (a) three heads;
    (b) two heads, one tail;
    (c)1 one head, two tails;
    (d) three tails?

Solution First let us find out how many different outcomes are possible.

When tossing one coin there are two possible outcomes, either a head or a tail, i.e. H or T.

We have to assume here these are *equiprobable*, i.e. there is a long-run expectancy that equal numbers of each would turn up

Therefore
$$\left. \begin{array}{l} p(H) = \frac{1}{2} \\ p(T) = \frac{1}{2} \end{array} \right\} \text{Total} = 1$$

(We read this as 'The probability of a head $= \frac{1}{2}$' etc.)

When tossing two coins, for each outcome of the first toss there are two outcomes for the second so there are four outcomes altogether, or 2^2 outcomes. These are, *HH, HT, TH, TT*, where, for example, *HH* means heads on the first coin and heads on the second.

When tossing three coins, for each outcome of the first two there are two outcomes for the third. Therefore, there are eight outcomes altogether, or 2^3 outcomes.

Let us write these down:

[1]	[2]	[3]	
H	*H*	*H*	Notice the quick way of writing these down
H	*H*	*T*	Column [1] contains groups of four
H	*T*	*H*	Column [2] contains groups of two
H	*T*	*T*	Column [3] contains groups of one, i.e. *H* and *T* alternately

We note that:

T	*H*	*H*	p (throwing $3H$) = 1 out of 8 = ⅛
T	*H*	*T*	p (throwing $2H$, $1T$) = 3 out of 8 = ⅜
T	*T*	*H*	p (throwing $1H$, $2T$) = 3 out of 8 = ⅜
T	*T*	*T*	p (throwing $3T$) = 1 out of 8 = ⅛

Total = 1

This last result reminds us of the binomial expansion:

$$(a+b)^n = a^n + na^{n-1}b + \frac{n(n-1)}{2!}a^{n-2}b^2 + \ldots + b^n$$

If n is 3, $(a+b)^3 = a^3 + 3a^2b + 3ab^2 + b^3$

so, if $a = ½$, $b = ½$

$$(½+½)^3 = (½)^3 + 3(½)^2(½) + 3(½)(½)^2 + (½)^3$$
$$= p(3H) + p(2H,1H) + p(1H,2T) + p(3T)$$

Notice that these terms are exactly the probabilities we have found for three heads; two heads, one tail; one head, two tails; and three tails.

This is a particular case of the general binomial probability distribution which we will return to later.

Note that the binomial distribution is only useful when two *mutually exclusive* outcomes are involved, e.g. right or wrong, black or white, heads or tails, win or lose, a success or a failure, a good product or a defective product – the two outcomes cannot both occur together.

Exercise 22.1

1. How many different outcomes are there when four coins are tossed? Is this the same as when the same coin is tossed four times?

2. Write down these outcomes using the pattern given in the last example and then say what the probabilities are of throwing:

   (a) four heads;
   (b) three heads, one tail;
   (c) two heads, two tails;
   (d) one head, three tails;
   (e) four tails.

3. Expand the binomial $(\frac{1}{2} + \frac{1}{2})^4$. Does this give the same terms as you have just obtained in question 2?

4. There are ten balls in a bag, nine white balls and one black ball. What is the probability of drawing a black ball from the bag? A white ball? Call these $p(B)$, $p(W)$ respectively. If a ball is taken at random from the bag and then replaced and a second one taken and replaced, what are the probabilities of getting:

   (a) two white balls in succession;
   (b) one white and one black;
   (c) the black ball twice?

5. If there are a very large number of articles in a batch of which it is known 10% are defective, what is the probability that if two articles are picked at random:

   (a) both will be good;
   (b) one will be good and one defective;
   (c) both will be defective?

   (The answers should be as in question 4 since the probabilities are not substantially affected when drawing from a large number.)

EQUIPROBABLE EVENTS

In the examples of tossing a coin or throwing a die we are making a very fundamental assumption that any outcome is equally likely. This is a big assumption, but we have to make it in the first instance. In other words, we are assuming that the tosses or throws are made in an identical manner and that the coin or die is equally balanced, i.e. is symmetrical.

If, on the other hand, these conditions are *not* satisfied we have a *weighted* probability.

When two results A and B are unpredictable and when all the known factors which could influence the result are *symmetrical* between A and B we say that the two results are ***equiprobable***.

It is only possible to assess this symmetry by means of past trials, i.e. by experience. If, for example, over a long period, the number of times A has occurred is twice that of B occurring, this would show conclusively that the factors influencing the outcome were not symmetrical. If the results were equiprobable, the more trials that we take, the nearer the results would be to a 50–50 division.

Choosing at Random

We can obtain equiprobable results by choosing at random provided there are equal numbers of the different kinds of objects.

Example 22.2   A man puts two tickets labelled 6, three tickets labelled 5, one ticket labelled 4, and one of each of 3, 2 and 1 into a bag, shakes them up and draws one. Would it be equally likely that a 5 or a 3 would be drawn?

Solution   Of course not. The fact that there are three 5s and only one 3 destroys symmetry.

THE LAWS OF PROBABILITY

Probability Law I

If the probability of an event occurring is p, the probability of not occurring is $(1 - p)$.

We have illustrated this by throwing a die.

The probability of getting any one of the numbers 1, 2, 3, 4, 5 or 6 is $\frac{1}{6}$.

The probability of not getting that number, i.e. the probability of throwing any of the other five numbers, is $\frac{5}{6}$.

Example 22.3   If a coin is tossed three times there are eight possible outcomes, i.e. 2^3.

The probability of getting a head each time $= p(3H) = \frac{1}{8}$

The probability of getting at least one tail $=$ the probability of not getting three heads

$$= 1 - \frac{1}{8}$$
$$= \frac{7}{8}$$

Exercise 22.2

1.   If I toss a coin ten times, what is the probability:

     (a)   of getting a head each time;
     (b)   of getting at least one head?

2.   If I throw three dice, what is the probability of throwing three sixes?

     What is the probability of throwing any other combination of 1, 2, 3, 4, 5 or 6?

Example 22.4   The probability of drawing an ace from an ordinary pack of 52 cards is 4 out of 52, i.e. $\frac{4}{52}$. Therefore, the probability of drawing any other type of card is $1 - \frac{4}{52}$ or $\frac{48}{52}$.

This checks since there are 48 cards in the pack which are not aces.

Probability Law II   Addition of probabilities

If A and B are mutually exclusive events, i.e. they cannot both occur simultaneously, then $P(A \text{ or } B) = p(A) + p(B)$.

Example 22.5   What is the probability of drawing a 'picture card' or an ace from a pack of cards?

Solution

The probability of drawing an ace $= \dfrac{4}{52}$ since there are four aces.

The probability of drawing a picture $= \dfrac{12}{52}$ since there are twelve pictures.

So, the probability of drawing an ace or a picture $= \dfrac{4}{52} + \dfrac{12}{52} = \dfrac{16}{52}$.

Check: There are sixteen aces and picture cards altogether in the pack.

Therefore, p (ace or picture) $= \dfrac{16}{52}$

Exercise 22.3

1.   The probability that a woman athlete will win a race is $\dfrac{1}{10}$.

The probability that she will be second is $\dfrac{1}{20}$.

The probability that she will be third is $\dfrac{1}{25}$.

What is the probability that she:

(a)   will be,
(b)   will not be in the first three places?

2.   The probability that a book may be published by a certain publisher is $\frac{1}{6}$, and by another $\frac{1}{4}$. What is the probability that it will be published by one of these two?

3.   An insurance company considers that the probabilities of rain, hail, snow, fog at a fete on a certain day are $\dfrac{1}{4}, \dfrac{1}{12}, \dfrac{1}{480}, \dfrac{1}{16}$ respectively. Assuming only one of these could happen:

(a)   What is the probability that one will happen?
(b)   What is the probability that it will be fine?

4.   The probabilities that a sealed envelope contains one, two, three, four or five tickets are $\dfrac{1}{2}, \dfrac{1}{4}, \dfrac{1}{8}, \dfrac{1}{16}, \dfrac{1}{32}$. Find the probability that it contains either more than five tickets or no tickets at all.

Probability Law III   Multiplication of probabilities

If A and B are two events and the probability of A occurring affects the probability of B occurring or vice versa, then the two events are **dependent**. If one event does *not* affect the other they are said to be independent.

Here we shall only consider the independent probabilities and we wish to find out the probability of both A and B occurring.

Example 22.6   What is the probability of drawing a black ace from a pack of playing cards?

Solution   There are two black aces in the pack. Therefore, the probability of drawing one of them is $\dfrac{2}{52}$.

We could answer the question another way, by asking what is the probability of drawing an ace which is also black?

The probability of drawing an ace $= \dfrac{4}{52}$.

The probability of drawing a black card $= \dfrac{1}{2}$ since half of the cards are black cards.

Hence the probability of drawing a card which is an ace *and* black is given by

$$p(\text{black ace}) = p(\text{ace}) \times p(\text{black card})$$

$$= \frac{4}{52} \times \frac{1}{2}$$

$$= \frac{2}{52} \quad \text{which we know is correct}$$

Example 22.7   A farmer takes a horse and a cow to market. He reckons he has a 50–50 chance of selling the horse but only a 25–75 chance of selling the cow from previous experience.

What is the probability that he will sell:

(i)   both;
(ii)   one or the other;
(iii)   neither?

Solution

(i)   Let selling the horse be represented by H.
        Let not selling the horse by represented by $\overline{H}$. (This is not the mean!)
        Let selling the cow be represented by C.
        Let not selling the cow be represented by $\overline{C}$.

Then the probabilities of H and $\overline{H}$ occurring are equal, i.e. 1:1. But the probabilities of C_s and $\overline{C}$ occurring are in the ratio of 1:3.

In other words $p(H) = \dfrac{1}{2}$ and $p(C) = \dfrac{1}{4}$.

If H occurs, C or $\overline{C}$ can also occur. But $\overline{C}$ is three times more likely than C.

A representation of this is as in Figure 22.2(a).

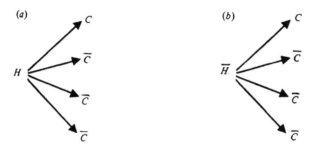

Figure 22.2

Similarly, if $\overline{H}$ occurs then C can occur or $\overline{C}$ can occur. But again $\overline{C}$ is three times more likely than C. This can be represented as in Figure 22.2(b). These diagrams illustrate the probable outcomes in the correct ratios.

Hence the combined outcomes are as follows:

(H, C)	$(\overline{H}, \overline{C})$
$(H, \overline{C})$	$(\overline{H}, \overline{C})$
$(H, \overline{C})$	$(\overline{H}, \overline{C})$
$(H, \overline{C})$	$(\overline{H}, \overline{C})$

We see that the probability of (H, C) occurring is one out of eight, i.e. $\dfrac{1}{8}$. This is obtained another way by using:

(i) $p(H \text{ and } C)$ $= p(H) \times p(C) = \dfrac{1}{2} \times \dfrac{1}{4} = \dfrac{1}{8}$

(ii) $p(H \text{ or } C \text{ but not both})$ $= p(H) \times p(\overline{C}) + p(\overline{H}) \times p(C)$

$$= \frac{1}{2} \times \frac{3}{4} + \frac{1}{2} \times \frac{1}{4} = \frac{3}{8} + \frac{1}{8} = \frac{1}{2}$$

(iii) $p(\overline{H} \text{ and } \overline{C})$ $= 1 - (\text{i}) - (\text{ii}) = 1 - \dfrac{1}{8} - \dfrac{1}{2} = \dfrac{3}{8}$

Exercise 22.4   Repeat Example 22.7 with $P(H) = \dfrac{1}{2}, \ P(C) = \dfrac{1}{3}$.

This example is illustrated in Figure 22.3.

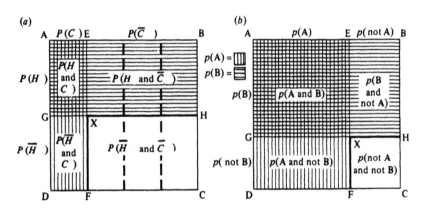

Figure 22.3

The representation of probabilities of independent outcomes

In Figure 22.3(a) a square ABCD of unit side is shown. It therefore has unit area.

A horizontal line GH is drawn dividing the square into two equal areas. Hence the rectangle ABHG and the rectangle GHCD are each of area ½ square unit. These two areas can be said to represent the equal probabilities H and $\overline{H}$ in Example 22.7.

Similarly if a vertical line EF is drawn at a distance ¼ unit from AD then the areas AEFD and EBCF can be said to represent the probabilities of C and $\overline{C}$.

Hence the events H and C occurring simultaneously is represented by the area AEXG (EF cuts GH at X).

i.e. $p(H \text{ and } C) = \dfrac{1}{8}$ since the whole area is 1 unit

Similarly in Figure 22.3(b):

if AEFD represents the probability of event A occurring, $p(A)$
EBCF represents the probability of event A not occurring, $p(\overline{A})$
ABHG represents the probability of event B occurring, $p(B)$
GHCD represents the probability of event B not occurring, $p(\overline{B})$

then AEXG represents the event of both A and B occurring, i.e. $p(A \text{ and } B)$

Hence $p(A \text{ and } B) = p(A) \times p(B)$.

Probability Law III may be stated as

If A and B are independent events, then $P(A \text{ and } B) = P(A) \times P(B)$

SUMMARY OF PROBABILITY LAWS

Law I

If the probability of an event occurring is p, the probability of it not occurring is $(1 - p)$.

Law II

If A and B are mutually exclusive events, then $P(A \text{ or } B) = P(A) + P(B)$.

Law III

If A and B are independent events, then $P(A \text{ and } B) = P(A) \times P(B)$.

Exercise 22.5

1. There are four stages of a rocket firing, 1, 2, 3 and 4. If the probabilities of success with these are $\dfrac{4}{5}, \dfrac{5}{6}, \dfrac{6}{7}, \dfrac{7}{8}$, respectively, what is the probability of total success?

2. In the men's pentathlon, an athlete has a three out of five chance of winning each of the first two events, but only a 50–50 chance of winning each of the last three events. What is his chance of winning all five?

PROBABILITY TREES

In a complicated probability problem it is often useful to draw a diagram illustrating the various outcomes, which can then be seen at a glance. Such a diagram is called a probability tree. We start with a very simple example.

Example 22.8

A man tosses a coin. If it is a head he draws a card from a pack. If it is a tail he tosses the coin again. Illustrate the various outcomes and find the probabilities of getting an ace, a picture, any other card, a head or a tail.

Solution

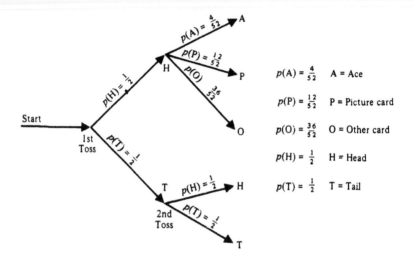

$p(A) = \frac{4}{52}$ A = Ace

$p(P) = \frac{12}{52}$ P = Picture card

$p(O) = \frac{36}{52}$ O = Other card

$p(H) = \frac{1}{2}$ H = Head

$p(T) = \frac{1}{2}$ T = Tail

Figure 22.4

Hence:

$$p(H \text{ followed by } A) = \frac{1}{2} \times \frac{4}{52} = \frac{2}{52} \qquad \text{TOTAL}$$

$$p(H \text{ followed by } P) = \frac{1}{2} \times \frac{12}{52} = \frac{6}{52} \qquad = \frac{2+6+18+13+13}{52}$$

$$p(H \text{ followed by } O) = \frac{1}{2} \times \frac{36}{52} = \frac{18}{52} \qquad = \frac{52}{52}$$

$p(T$ followed by a card)	$= 0$	
$p(T$ followed by $H)$	$= \dfrac{1}{2} \times \dfrac{1}{2} = \dfrac{1}{4}$	$= 1$
$p(T$ followed by $T)$	$= \dfrac{1}{2} \times \dfrac{1}{2} = \dfrac{1}{4}$	as expected

Example 22.9   A company bids for two contracts, A and B. The probability of getting A is 0.7. The probability of getting B depends on whether or not it gets A, and is 0.8 if it gets A, but only 0.4 if it does not get A. Draw a tree diagram to illustrate the probabilities of the outcomes.

(a)   What is the probability that it gets both contracts?

(b)   What is the probability of getting A but not B?

(c)   What is the probability of getting B but not A?

(d)   What is the probability of getting neither contract?

(e)   What is the probability of getting exactly one contract?

(f)   What is the probability of getting at least one contract?

Solution

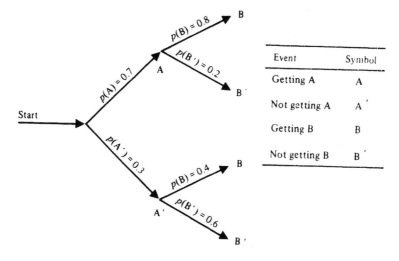

Figure 22.5

(a)   the probability of getting A and B $= 0.7 \times 0.8 = 0.56$

(b)   the probability of getting A but not B $= 0.7 \times 0.2 = 0.14$

(c)   the probability of not getting A but getting B $= 0.3 \times 0.4 = 0.12$

(d)   the probability of not getting A nor B $= 0.3 \times 0.6 = \underline{0.18}$

                                             Total $= 1.00$

(e)   the probability of getting exactly one contract
       $=$ the probability of getting either A or B but not both
       $= 0.14 + 0.12[(b) + (c)]$
       $= 0.26$

(f)   the probability of getting at least one contract
       $=$ the probability of getting either A or B or both
       $= 0.14 + 0.12 + 0.56 \,[(b) + (c) + (a)]$
       $= 0.82$ which of course is the same as $[1 - (d)]$

Exercise 22.6

1.   A die is thrown and if the result is an odd number a coin is tossed. If it is an even number no further action is taken. Draw a diagram showing the nine possible outcomes and calculate the probability of each.

2.   A male candidate has a 95% probability of passing exam A. If he fails the first time he is allowed to take it again, but not a third time. If he passes, he sits for exam B, which he can only take once. He has a 90% chance of passing B if he takes it. Draw a probability diagram showing the possibilities, and calculate the probability that he will pass both exams:

       (a)   the first time;
       (b)   eventually.

23

Probability Distributions

From the histogram to the continuous distribution

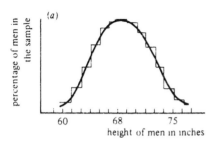 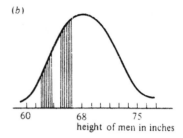

Figure 23.1

If we take a sample of men, say 1000, and measure their heights to the nearest inch, we should probably obtain a histogram similar to the one in Figure 23.1(a) in which only the tops of the columns are indicated. We could sketch a reasonably smooth curve through the midpoints of the tops of the columns as shown in Figure 23.1(b).

If we take a larger sample of men, say 10,000, we could take more accurate measurements, say to the nearest tenth of an inch, and obtain ten times as many columns and get an even more accurate picture of what is happening in the whole population. The 'zig-zags' would be much finer. In this way, by taking larger and larger samples and narrower and narrower ranges we would eventually build up a picture which would be entirely made up of lines of which we have shown only a few in Figure 23.1(b). The shape of the curve would alter slightly from our original small sample curve but it would be of the same type. This type of graph is called a *continuous distribution* for obvious reasons. The height of the curve at any point gives the proportion of men of that height, and the area under the curve between given limits represents the proportion of the population having heights within those limits.

We have already said that the histogram may be considered as columns of blocks, each block representing one member of the population in the distribution. In the same way, it may help you to view the continuous distribution as a heap of dots, like a heap of sand, each dot representing one member of the very large population being studied. The dots are piled highest in the centre where the frequency is greatest.

Many populations have a distribution pattern of this type with a single 'bulge' and fading away towards zero on either side. Mathematicians have fitted several useful equations to this type of curve. The most important is the normal distribution, which we shall meet later in this chapter.

We have discussed the way in which a histogram becomes a continuous curve when the numbers in the sample are increased and the class widths are decreased. Consider another example. Suppose records have been kept, over a long period, of the daily rainfall in centimetres at a certain place. The distribution of days when a certain rainfall was registered would look something like Figure 23.2.

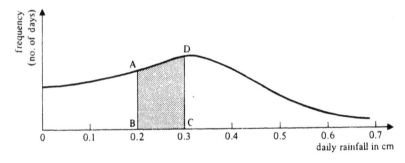

Figure 23.2

This is a typical continuous distribution. In the example given, the total number of days on which the rainfall was between 0.2 and 0.3 cm (say) is represented by the area ABCD. This area is the sum of a very large number of very narrow columns of a histogram. Hence, the proportion of the total number of days on which the rainfall was in this range is given by

(area ABCD) ÷ (total area under curve)

This is the probability that a day selected at random will have a rainfall in the range 0.2 to 0.3 cm. The vertical scale is chosen so that the total area under the curve is 1 unit. In this case, the required probability is simply given by the area under the curve between AB and CD, since dividing by unity makes no difference. We call this a *probability distribution*.

Note that the total area under the probability distribution curve must be 1 unit.

TYPES OF PROBABILITY DISTRIBUTIONS

We now take a look at three different probability distributions, the binomial distribution (which we already know something about), the Poisson distribution and the normal or Gaussian distribution which we will study in greater detail.

1. The Binomial Probability Distribution

This distribution is useful when 'fault-finding' in industry, cutting down faulty batches to the minimum. Industry tries to do this by 'sampling'. The reasons for sampling and testing these samples are fairly obvious.

1. It may be impossible to check every item produced (e.g. every tiny screw) or it may be too expensive.

2. It may be impossible for various other reasons; for example, testing the life of a piece of electrical equipment may mean running it until it breaks, which of course one cannot do with every item. Imagine testing every battery produced in a factory until it went flat! However, taking a sample need not be too expensive, but is certainly necessary in order to build up the firm's reputation for reliability, and to give an indication for what length of time a guarantee should be given.

Therefore sampling and testing are very necessary.

Example 23.1 In a certain town, records indicate that in a particular month it will probably be fine 3 days out of 4, i.e.

p = probability that it will be fine $= \frac{3}{4}$

∴ q = probability that it will rain $= \frac{1}{4}$

on any one day.

Find

(a) the probability that in a 5 day period it will stay fine on all 5 days;

(b) the probability that it will be fine at least 4 days out of the 5;

(c) the probability that it will be fine on 3 days and rain on 2;

(d) the probability that it will be fine on 2 days and rain on 3;

(e)   the probability that it will be fine on at least 1 day.

(f)   the probability that it will rain on at least 1 day

Solution   We use $(p + q)^n$ where $p = \frac{3}{4}$, $q = \frac{1}{4}$, $n = 5$:

$$(\tfrac{3}{4}+\tfrac{1}{4})^5 = (\tfrac{3}{4})^5 + 5(\tfrac{3}{4})^4(\tfrac{1}{4}) + \frac{5\cdot4}{2\cdot1}(\tfrac{3}{4})^3(\tfrac{1}{4})^2 + \frac{5\cdot4\cdot3}{3\cdot2\cdot1}(\tfrac{3}{4})^2(\tfrac{1}{4})^3; \text{ etc.}$$

(a)   is the first term $= (\tfrac{3}{4})^5 = 0.2373$

(b)   is the first term + second term $= 0.2373 + 0.3955 = 0.6328$

(c)   is the third term $= 0.2637$

(d)   is the fourth term $= 0.0879$

(e)   is found by calculating the probability of 5 wet days, i.e. $(\tfrac{1}{4})^5 = 9.7656 \times 10^{-4}$ and subtracting from 1.

    This gives 0.9990 to 4 d.p.

(f)   You should be able to adapt the answer to (e) to end up with 0.7627.

Example 23.2   A batch of manufactured items is known to contain 10% which are defective. What are the probabilities that, when a sample of four items is taken from the batch, four, three, two, one, zero items are defective? (Assume that the batch is large enough for the sample taken not to affect the proportion of defectives.)

Solution   We know $p(1 \text{ item defective}) = \dfrac{1}{10} = p \text{ (say)}$

$$n = 4$$

$$p(1 \text{ item satisfactory}) = \frac{9}{10} = q \text{ (say)}$$

$$(p+q)^4 = p^4 + 4p^3q + 6p^2q^2 + 4pq^3 + q^4$$

∴ $p(\text{all defective}) = p^4 = \dfrac{1}{10^4} = 0.000$

∴ $p(3 \text{ defective}) = 4p^3q = \dfrac{4.9}{10^4} = 0.0036$

$$p(2 \text{ defective}) = 6p^2q^2 = \frac{6.81}{10^4} = 0.0486$$

$$p(1 \text{ defective}) = pq^3 = \frac{4.729}{10^4} = 0.6561$$

$$p(2 \text{ defective}) = q^4 = \frac{9^4}{10^4} = 0.6561$$

Total of all probabilities = 1.0000

A histogram of these results is shown in Figure 23.3.

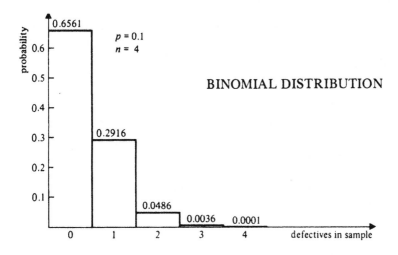

Figure 23.3

Notes on the histogram of the binomial distribution

1. If the width of the column is 1 unit, the area of the column represents the probabilities of zero, one, two, three, four defectives.

2. The total area of the graph is unit area; this will always be so if we include all possible outcomes.

3. This distribution is useful in such cases as the one quoted but its use is limited.

4. We can extend the idea to the case when a large sample is taken from a very large batch. The graph will approximate to a continuous curve as the width of the column decreases, but the area under the curve will still represent the total probability of all outcomes and will still be unit area.

5. The mean $= np$.

6. The variance $= npq$.

7. The standard deviation $= \sqrt{npq}$.

    (For proof of 5, 6 and 7, see any textbook on mathematical statistics.)

8. In practice binomial probabilities are rarely calculated directly. These probabilities have been tabulated for $n = 2$ to $n = 49$ by the National Bureau of Standards and can be read off directly.

Exercise 23.1

1. Calculate correct to 3 d.p. the binomial probabilities for $p = \frac{3}{4}$, $n = 8$.

    Calculate the mean and variance, i.e. np and npq.

2. A pack of cards is shuffled and cut five times in succession. What are the probabilities of getting an ace:

    (i)   four times;
    (ii) three times?

    (Leave answer in factor form.)

3. If 2/5 of the voters in a certain constituency favour Mr A, what is the probability that, of five voters interviewed at random, the majority will favour Mr B, assuming a 'straight' fight.

4. In the manufacture of screws, it is found that in the long term 1 in 20 are rejected as defective. Calculate the probability that twelve screws chosen at random will contain:

    (i)   two rejects;
    (ii) not more than two rejects;
    (iii) at least three rejects.

2.   The Poisson Distribution

We have seen that the binomial distribution is useful when we take a sample of definite size and count the number of times (a) something happens, and (b) something does not happen. But in many cases the problem does not involve just 'yes' and 'no', but could occur in a random way throughout time. We can count the number of times the lightning flashes during a storm, but we cannot count how many times it did not flash. We can count the number of cars arriving at a petrol station in a given period – we cannot say how many times they did not arrive. This type of probability does not fit a binomial distribution.

How do we overcome this? The answer to this problem is the Poisson distribution. It has been found by observation to fit cases such as the above with remarkable accuracy.

We know that the definition of e^x is

$$e^x = \lim_{n \to \infty} \left(1 + \frac{x}{n}\right)^n = 1 + x + \frac{x^2}{2!} + \frac{x^3}{3!} + \dots$$

Can we use this fact? The answer is 'yes'.

Imagine a time axis extending to infinity and divided up into small elements of time, δt, as in the figure below.

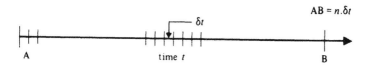

Figure 23.4

We can imagine our 'lightning flashes' slotting into some of these tiny elements of time of which there are n (say) where $n \to \infty$; the exponential expansion can be adapted for use as a probability distribution in cases like the above where there are a large number of very small time slots in which to fit random events.

In order for this expansion to be used as a probability distribution the sum of all the terms must equal unity, the total of all possible probabilities.

We derive the Poisson distribution as follows.

We have used p as the probability of success and q (= $1 - p$) as the probability of failure in discussing the binomial distribution, and n as the number of trials. The Poisson distribution is of the form assumed by the distribution when p is small and n is large (i.e. $p \to 0$ and $n \to \infty$) but the mean number of successes np is constant and equal to λ which is finite. In $(p + q)^n$ let $p = \dfrac{\lambda}{n}$.

The probabilities of 0, 1, 2, 3,... successes are

$$q^n, nq^{n-1}p; \frac{n(n-1)}{2!}q^{n-2}p^2; \frac{n(n-1)(n-2)}{3!}q^{n-3}p^3;...$$

by expanding $(q + p)^n$. Put $np = \lambda$.

These become

$$q^n, \lambda q^{n-1}; \frac{1-\dfrac{1}{n}}{2!}\lambda^2 q^{n-2}; \frac{\left(1-\dfrac{1}{n}\right)\left(1-\dfrac{2}{n}\right)}{3!}\lambda^3.q^{n-3};...$$

When $p \to 0$ and $n \to \infty$, $q \to 1$. These terms then become proportional to
$$1, \lambda, \frac{\lambda^2}{2!}, \frac{\lambda^3}{3!},....$$

This is the exponential series whose sum is e^λ.

However, the sum of the actual probabilities must be unity, since it is the limit of $(p + q)^n$ and $p + q = 1$.

Hence we obtain the actual probabilities by dividing by e^λ, i.e. multiplying by $e^{-\lambda}$. Thus the actual probabilities of 0, 1, 2, 3,... successes are

$$e^{-\lambda}, \lambda e^{-\lambda}, \frac{\lambda^2}{2!}e^{-\lambda}, \frac{\lambda^3}{3!}e^{-\lambda};... \text{ respectively}$$

Note that λ is the long-run average number of occurrences of the event in a given period. λ is also called the *expectation* (or expected number) of the occurrence of an event. It is found by observation over a long period.

Definition of the Poisson distribution

$$p(X = r) = e^{-\lambda} \cdot \frac{\lambda^r}{r!}$$

which we read as 'the probability of an event X occurring r times is given by $e^{-\lambda} \cdot \frac{\lambda^r}{r!}$, where $r = 0, 1, 2,...$'

($0! = 1$ by definition).

Example 23.3   Radio-active particles randomly strike a Geiger counter at a rate of 0.01 per second. Calculate the probabilities that zero, one, two, three or four particles will strike the counter in any 1 minute.

Solution

Expected rate of striking = 0.01 per second
                          = 0.6 per minute

This is the long-run average striking rate and must have been observed during thousands of hours of experimentation, i.e. $\lambda = 0.6$. Therefore probabilities of zero, one, two, three or four particles striking in 1 minute are the terms

$$e^{-0.6}, 0.6e^{-0.6}, \frac{(0.6)^2}{2!}e^{-0.6}, \frac{(0.6)^3}{3!}e^{-0.6} \text{ and } \frac{(0.6)^4}{4!}e^{-0.6}$$

$\therefore$ $p_0 = 0.5488, \quad p_1 = 0.3293, \quad p_2 = 0.0988, \quad p_3 = 0.0198, \quad p_4 = 0.0030$

For the Poisson distribution:

the mean $(\mu) = \lambda$;   the variance $(\sigma^2) = \lambda$;
the standard deviation $(\sigma) = \sqrt{\lambda}$

(The proof of this needs a more formal approach than is possible in a book of this type, but is quite straightforward. See any textbook on mathematical statistics.)

Exercise 23.2

1.   In a book of 400 pages, there are 40 errors. Find the probabilities of zero, one, two or three errors per page.

2.   With the information of question 1, in a similar book of 500 pages, how many pages approximately would you expect to contain:

      (i)   two errors or less;   (ii)   three errors or more?

3.   In a certain area, the number of violent storms occurring each year is a random variable whose probability distribution can be approximated closely with a Poisson distribution having a parameter $\lambda = 8$. Find the probability that in a given year fewer than six storms hit the area.

3.   The Normal or Gaussian Distribution

We have discussed the binomial distribution and the Poisson distribution, which enable us to deal with the occurrence of distinct events, e.g. the number of defective items in a sample of known size or the probability of breakdown or accidents in a factory in a given time. We now need a distribution pattern which will deal with quantities that are not integers. For example, suppose the length of a manufactured bar of steel should be 10 cm exactly. How do we estimate the probability of it being 0.001 cm too long or too short?

The distribution required here is the normal distribution, so called because it is a distribution which occurs frequently. The normal curve (Figure 23.5) is bell shaped and symmetrical about the mean μ.

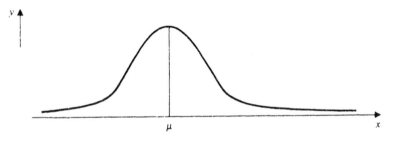

Figure 23.5

For example, if we take the heights of a large number of children *of a given age* and take very small class widths, the frequency distribution (i.e. the histogram) evens out to form a smooth curve, the most frequent height being, of course, μ, the mean value, tailing off on either side.

Mathematicians have worked on this type of curve for over 200 years and have shown that its equation is

$$y = \frac{1}{\sigma\sqrt{2\pi}}\, e^{-\frac{(x-\mu)^2}{2\sigma^2}}$$

Where μ is the mean value and σ the standard deviation. Discovery of the distribution is variously attributed to Gauss, De Moivre and Laplace.

STANDARD DEVIATION AND THE NORMAL CURVE

For a *normal* distribution, as illustrated in Figure 23.6, the standard deviation gives a good guide as to what proportion of the distribution falls within certain areas.

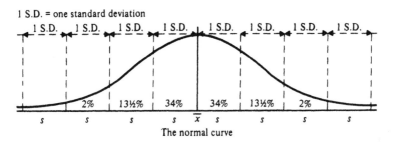

Figure 23.6

Dividing the area under the curve by standard deviations from the mean, it can be shown that:

(a) 68% of the distribution lies within one standard deviation either side of the mean (i.e. 34% on each side of the mean).

(b) 95% of the distribution lies within two standard deviations either side of the mean.

(c) 99% of the distribution lies within three standard deviations on either side of the mean.

This is a very useful guide when estimating what proportion of manufactured items may be rejected when certain limits are set to variation from some standard of measurement of the dimension concerned. This assumed that the dimension has a distribution of the above type. If we take any other distribution which is reasonably symmetrical about the mean and is 'unimodal', i.e. has one maximum point, we can often assume the above results for the normal distribution still to be approximately true.

THE STANDARD NORMAL DISTRIBUTION

The equation is very formidable and difficult to deal with. It can be easily transformed into a simpler form by moving the axes and changing the scale. When this has been done it is said to be in *standard form*, for which we have tables enabling us to read off the facts we need, given the mean μ and the standard deviation σ, of our particular distribution.

What do we mean by 'standardising'?

As an example consider the general parabola

$$y = ax^2 + bx + c$$

All parabolae are fundamentally the same. The difference between them is due to:

(a)   their position relative to the axes;
(b)   the scales used on the axes.

By suitable substitutions for x and y, and by changing the scale, the equation of the curve can be converted to the simple form $y = x^2$, which is much simpler to handle.
     In the same way we can convert any normal curve to standard form by:

1.   Moving the y axis so that it is at the centre of symmetry.

2.   Increasing or decreasing the scale so that the area under the curve is unit area.

Tables can then be used to give the probability of a quantity lying between given limits. This is represented by the area under the standard normal curve between these limits.

GRAPH OF THE STANDARD NORMAL CURVE

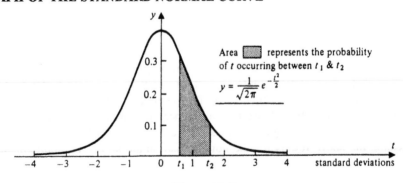

Area ▨ represents the probability of t occurring between t_1 & t_2

$$y = \frac{1}{\sqrt{2\pi}} e^{-\frac{t^2}{2}}$$

Figure 23.7

N.B.1   The total area under the curve between $-\infty$ and $+\infty$ is one unit and represents all the probabilities of t occurring.

N.B.2   The probability of an event occurring between t_1 and t_2 is given by the area under the curve between t_1 and t_2.

N.B.3   The scale on the t axis is in standard deviations.

To convert the normal distribution to standard form

The equation is
$$y = \frac{1}{\sigma\sqrt{2\pi}} e^{-\frac{(x-\mu)^2}{2\sigma^2}}$$

Let $\dfrac{x-\mu}{\sigma} = t$ in order to make the curve symmetrical about the y axis and take the units on the t axis in units of standard deviation, i.e. we are making $\sigma = 1$ unit.

Then
$$y = \frac{1}{\sqrt{2\pi}} e^{-\frac{t^2}{2}} \qquad \text{which is more manageable}$$

It can be shown that with y defined as above: $\displaystyle\int_{-\infty}^{\infty} y\,dt = 1$

(The proof of this is too difficult at this stage.) The $\dfrac{1}{\sqrt{2\pi}}$ factor ensures that the area under the curve is 1 unit.

N.B.1 $\displaystyle\int_{a}^{b} y\,dt$ (the area under the curve between a and b) is the probability of t being greater than or equal to a and less than or equal to b.

We say:
$$p(a \le t \le b) = \int_{a}^{b} y\,dt$$

N.B.2 These integrals have been calculated numerically by expanding $e^{-t^2/2}$ as a series, and are printed as tables that give the various probabilities.

N.B.3 Sometimes the tables give the values of the areas to the left and right of the y axis, i.e. the area is 0 when $t = 0$ and 0.5 when $t = \infty$ (Figure 23.8(a)). Sometimes the tables give the area from $-\infty$ to the value of t required, giving area 0.5 when $t = 0$ and 1 when $t = \infty$ (Figure 23.8(b)). In both cases, common sense gives the value of the area between two given t values.

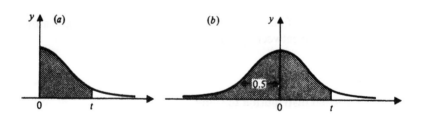

Figure 23.8

Example 23.4   From tables, evaluate the following (for $\mu = 0$, $\sigma = 1$):

(i) $p(0.7 \leq t \leq 0.9)$;
(ii) $p(|t| \leq 0.6)$;
(iii) $p(-0.7 \leq t \leq 0.9)$.

(Use the table (appendix) at the end of the book.)

Solution

(i)   From the table:

$$p(t = 0.7) - 0.2580 \tag{1}$$

$$p(t = 0.9) = 0.3159 \tag{2}$$

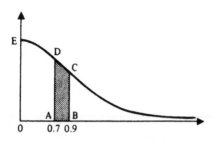

Figure 23.9

On the half-probability curve, this means that

(2) represents the area OBCE   (Figure 23.9)

Therefore   area required = shaded area ABCD = (2) – (1) = 0.0579

Hence $p(0.7 \le t \le 0.9) = 0.0579$

(ii)   Read $p(|t| \le 0.6)$ as the 'probability that mod t is less than or equal to 0.6'. This is the same as saying the 'numerical value of t is less than or equal to 0.6' or $-0.6 \le t \le +0.6$ or 't lies between -0.6 and $+0.6$ and includes -0.6 and $+0.6$'.

From tables:

$$p(t = 0.6) = 0.2257$$

$p(t = -0.6)$ is not given, but by symmetry is exactly the same (Figure 23.10).

Therefore $p(-0.6 \le t \le +0.6) = 2 \times 0.2257 = 0.4514$

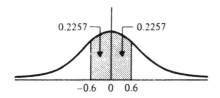

Figure 23.10

(iii) $p(-0.7 \le t \le 0.9)$ is represented by the shaded area (Figure 23.11)

$p(t = -0.7) = p(t = +0.7) = 0.2580$

$p(t = 0.9) = 0.3159$

so $p(-0.7 \le t \le 0.9) = 0.2580 + 0.3159 = 0.5739$

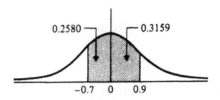

Figure 23.11

Example 23.5   For $\mu = 12$, $\sigma = 4$ in a normal distribution, find:

(i) $p(x \leq 14)$;
(ii) $p(x \leq 8)$;
(iii) $p(5.5 \leq x \leq 11)$.

Solution   In this case we have to convert to standard form first by putting

$$t = \frac{x - \mu}{\sigma} = \frac{x - 12}{4}$$

(i)   When $x = 14$, $t = \dfrac{14 - 12}{4} = \dfrac{1}{2}$

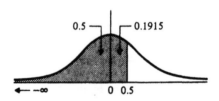

Figure 23.12

so $p(x \leq 14) = p(t \leq 0.5)$ in standard tables   (Figure 23.12)
             = (probability from $-\infty$ to 0) + (probability from 0 to 0.5)
             = 0.5 + 0.1915 = 0.6915

(ii)   When $x = 8$, $t = \dfrac{8-12}{4} = -1$.

Hence $p(x \le 8) = p(t \le 1)$

This is represented by the area from $-\infty$ to -1.

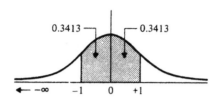

Figure 23.13

$$p(t = -1) = p(t = +1) = 0.3413$$

i.e. $p(t \le -1) = 0.5 - 0.3413$

$$= 0.1587$$

(iii)   When $x = 5.5$, $t = \dfrac{5.5-12}{4} = -1.625$

$$x = 11, \quad t = \dfrac{11-12}{4} = -0.25$$

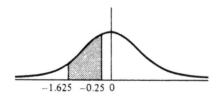

Figure 23.14

Required probability $= p(t \le 1.625) - p(t \le 0.25)$
$$= 0.4479* - 0.0987 \qquad \text{*(by interpolation)}$$
$$= 0.3492$$

Example 23.6   The amount of permissible impurity in a certain chemical is 3%. When batches of the chemical are tested, it is found that the impurity varies and is normally distributed about a mean of 2% with standard deviation of 0.5%.

(i)   What is the probability of getting a permissible batch?

(ii)   What percentage of batches are accepted in the long run?

Solution   Here $\mu = 2\% = 0.02$ $\sigma = 0.5\% = 0.005$

We require the probability of impurity to be less than or equal to 3%, i.e. ≤ 0.03.

Therefore $p(x \leq 0.03)$ is the required probability

Standardise:

Put $t = \dfrac{x - \mu}{\sigma} = \dfrac{x - 0.02}{0.005}$

When $x = 0.03$,

$$t = \frac{0.03 - 0.02}{0.005}$$

$$= \frac{0.01}{0.005} = 2$$

$$p(t \leq 2) = 0.5 + 0.4772 \qquad \text{from tables (Figure 23.15)}$$

$$= 0.9772$$

This is the probability that a batch chosen at random will be permissible.

Thus nearly 98% of all batches are acceptable in the long run.

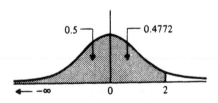

Figure 23.15

Exercise 23.3

1. Using a cumulative normal frequency distribution table find:

    (i) $p(t \le 0.3)$;
    (ii) $p(t \ge 0.3)$;
    (iii) $p(t \le 0.1)$;
    (iv) $p(0.1 \le t \le 0.3)$;
    (v) $p(t \ge 0.1)$;
    (vi) $p(|t| \le 0.1)$.

2. In a factory, electrical components are manufactured having a mean resistance of 50 ohms. The standard deviation is 2 ohms. If the distribution of these values is known by past experience to be normal, find how many rejects per 1000 there will be if a tolerance of ±6.6 ohms from the mean is allowable.

SAMPLING FROM NORMAL POPULATIONS

When we take a sample we are trying to learn something about the population from which it is drawn. If this sample is taken in an unbiased way, i.e. it is a random sample, we believe it gives a good idea of what is happening to the whole population, and the bigger the sample the truer the picture.

However, these 'spokespeople' for the whole population do not all carry the same authority. We need to know how reliable they are, i.e. whether they are 'well informed', or 'usually well informed', or merely saying 'rumour has it that...' to use the language of journalism.

Statisticians need to be discriminating about this. We realise that the sample mean and standard deviation will vary with the different samples taken – only slightly, but we want to know how slightly. This is why we have distinguished between $\bar{x}$ (the sample mean) and μ (the true mean).

CONFIDENCE INTERVALS AND SAMPLING

If we take several different samples of the same size, say ten samples each of size 100, we shall get ten slightly different values for the mean $\bar{x}$. Let us call these $\bar{x}_1, \bar{x}_2, \bar{x}_3, ..., \bar{x}_{10}$. Suppose $\bar{x}_1$ is the least and $\bar{x}_{10}$ is the greatest. The others lie somewhere in between. It is our hope that the true mean, μ, also lies somewhere in this range.

If we take many more samples we find that the $\bar{x}$s tend to cluster round their own mean. They are themselves normally distributed and this distribution is related to the distribution of the original population in the following way.

The Central Limit Theorem

If $\bar{x}$ is the mean of a random sample of size n from a normal population whose mean is μ and standard deviation is σ, then $\bar{x}$ itself is normally distributed with the same mean μ but with standard deviation $\dfrac{\sigma}{\sqrt{n}}(=\hat{\sigma})$, pronounced 'sigma cap' or 'sigma hat', and called the standard error of the mean.

Example 23.7　　　In a production process, wire is made into springs. The machine can accept the wire if the mean length (μ) is 4 cm and the variance (σ^2) is 0.0050. A sample of 50 springs is taken from each day's production and the process is under control if their mean length $\bar{x}$, which is normally distributed, does not vary more than 0.015 cm on either side of the mean value. We need to know the probability that the sample taken will fail this criterion even though the process is under control.

Let us illustrate this on the normal distribution curve:

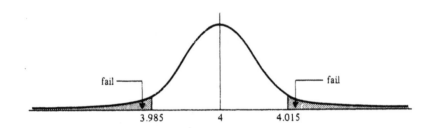

Figure 23.16

We have to find the probability that $\bar{x}$ falls outside the limits 3.985 to 4.015 cm when $\mu = 4$, $\sigma^2 = 0.005$.

Solution　　　Now

$$\hat{\sigma}^2 = \frac{\sigma^2}{n} = \frac{0.005}{50} = 0.0001$$

Therefore $\hat{\sigma} = 0.01$.

To standardise the distribution, let

$$t = \frac{\overline{x} - \mu}{\hat{\sigma}}$$

When $\qquad \overline{x} = 4.015, \ t = \dfrac{4.015 - 4}{0.01} = 1.5$

When $\qquad \overline{x} = 3.985, \ t = \dfrac{3.985 - 4}{0.01} = -1.5$

Hence if t lies outside the limits -1.5 to $+1.5$ the sample will fail. From the standardised curve using the table in the appendix, we see that the probability that the wire will fail the criterion (represented by the shaded areas) is

$2 \times 0.0668 = 0.1336$
$\qquad\quad = 13.36\%$

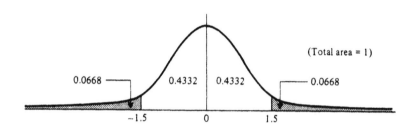

Figure 23.17

Here we have a *confidence interval* of 86.64%.

THE 95% CONFIDENCE INTERVAL

Suppose we take random samples of size n (where n is large) from a normally distributed population of mean μ and standard deviation σ. Suppose the mean of our samples is $\overline{x}$ and the standard deviation is $\hat{\sigma}$. Then we know that $\overline{x}$ is also normally distributed with mean μ and $\hat{\sigma} = \dfrac{\sigma}{\sqrt{n}}$. If we wish to investigate the range into which 95% of our $\overline{x}$ s will fall, we proceed as follows:

Required probability $95\% = 0.95$, i.e. 0.475 on either side of the vertical axis in the standardised normal curve (Figure 23.18).

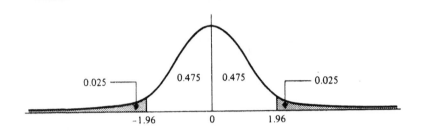

Figure 23.18

From the table (appendix) at the back of the book:

$$0.475 = p(t \le 1.96)$$

But the standardised tables apply only to $t = \dfrac{\bar{x} - \mu}{\hat{\sigma}} = \dfrac{\bar{x} - \mu}{\sigma / \sqrt{n}}$.

Hence $\dfrac{\bar{x} - \mu}{\sigma / \sqrt{n}}$ must lie between -1.96 and $+1.96$ for 95% certainty.

This gives $-1.96 \le \dfrac{\bar{x} - \mu}{\sigma / \sqrt{n}} \le +1.96$.

Thus $$-1.96 \frac{\sigma}{\sqrt{n}} \le \bar{x} - \mu \le +1.96 \frac{\sigma}{\sqrt{4}}$$

i.e. $$-1.96 \frac{\sigma}{\sqrt{n}} + \mu \le \bar{x} \le \mu + 1.96 \frac{\sigma}{\sqrt{4}}$$

or $$\bar{x} - 1.96 \frac{\sigma}{\sqrt{n}} \le \mu \le \bar{x} + 1.96 \frac{\sigma}{\sqrt{4}}$$

This interval $\left[\bar{x} - \dfrac{1.96\sigma}{\sqrt{n}}, \bar{x} + \dfrac{1.96\sigma}{\sqrt{n}} \right]$ is called the 95% confidence interval for μ.

What does this really mean? It is saying that if we take a lot of samples all of the same size n, work out the mean $\bar{x}$ for each sample and substitute in the given inequality, then in 95% of the cases the true value μ will lie within the limits obtained.

We use the table (appendix) to work out other confidence intervals in the same way. Diagrammatically, Figure 23.19 is transformed into Figure 23.20.

In general, in order to obtain a $(1 - \theta)$ confidence interval for the true mean μ, where θ is the probability of no confidence (5% or 0.05 in the case of 95% confidence), we organise the work as follows:

1. Express the confidence interval as a decimal (e.g. 99% = 0.99).

2. Divide this by 2, since the table gives only half the picture (e.g. ½ × 0.99 = 0.4950).

3. Use this figure (0.4950) to find the corresponding value of t from the table (here, $t = 2.575$).

4. Substitute this value of t in the interval

$$\left[\bar{x} - t\frac{\sigma}{\sqrt{n}}, \ \bar{x} + t\frac{\sigma}{\sqrt{n}} \right]$$

5. When $\bar{x}$, σ and n are known, this gives the required confidence interval. To simplify problems we quote some of the values commonly used in the following table. (The values of t are obtained from the table in the appendix.)

6. For normal distributions (only), where σ is unknown it may be replaced by s, where s is the standard deviation of the *sample*.

Confidence interval req'd	$(1 - \theta)$	$p = \frac{1}{2}(1 - \theta)$	t
99.9%	0.999	0.4995	3.295
99%	0.990	0.4950	2.575
98%	0.980	0.4900	2.326
95%	0.950	0.4750	1.960
90%	0.900	0.4500	1.645

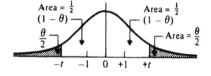

Figure 23.19

transforms into

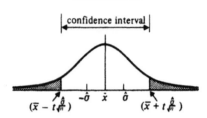

Figure 23.20

Exercise 23.4

1.   Use the table of special confidence intervals just quoted to answer the following:

    (a)   What is the 95% confidence interval for $\bar{x} = 49.9$, $\sigma = 2.02$, $n = 100$?

    (b)   What is the 99% confidence interval for the same values of $\bar{x}$, σ and n?

2.   If in measuring the specific gravity of a substance a mean of 2.705 is found from sixteen experiments, with standard deviation of 0.029, find a 95% confidence interval for the actual specific gravity of the substance, assuming the sample is a random sample from a normal population.

3.   At a local firm producing light bulbs, experiments were conducted to find the average lifetime of a certain type of bulb. Taking a random sample of 22 bulbs the average was 722 hours with standard deviation of 54 hours. Taking the lifetime of the population of light bulbs as being normally distributed find a 98% confidence interval for the true average lifetime of the bulbs.

4.   At a certain large store a sample of 100 accounts outstanding was taken and the mean was found to be £58.14 with a standard deviation of £15.30. Find the 95% confidence interval for the actual mean size of accounts outstanding at this store, assuming that these accounts are normally distributed around the mean.

The cumulative normal distribution function $\quad y = p(t) = \dfrac{1}{\sqrt{(2)\pi}} \displaystyle\int_0^t e^{-\frac{1}{2}t^2}\,dt$

t	0.00	0.01	0.02	0.03	0.04	0.05	0.06	0.07	0.08	0.09
0.0	.0000	.0040	.0080	.0120	.0160	.0199	.0239	.0279	.0319	.0359
0.1	.0398	.0438	.0478	.0517	.0557	.0596	.0636	.0675	.0714	.0753
0.2	.0793	.0832	.0871	.0910	.0948	.0987	.1026	.1064	.1103	.1141
0.3	.1179	.1217	.1255	.1293	.1331	.1368	.1406	.1443	.1480	.1517
0.4	.1554	.1591	.1628	.1664	.1700	.1736	.1772	.1808	.1844	.1879
0.5	.1915	.1950	.1985	.2019	.2054	.2088	.2123	.2157	.2190	.2224
0.6	.2257	.2291	.2324	.2357	.2389	.2422	.2454	.2486	.2517	.2549
0.7	.2580	.2611	.2642	.2673	.2704	.2734	.2764	.2794	.2823	.2852
0.8	.2881	.2910	.2939	.2967	.2995	.3023	.3051	.3078	.3106	.3133
0.9	.3159	.3186	.3212	.3238	.3264	.3289	.3315	.3340	.3365	.3389
1.0	.3413	.3438	.3461	.3485	.3508	.3531	.3554	.3577	.3599	.3621
1.1	.3643	.3665	.3686	.3708	.3729	.3749	.3770	.3790	.3810	.3830
1.2	.3849	.3869	.3888	.3907	.3925	.3944	.3962	.3980	.3997	.4015
1.3	.4032	.4049	.4066	.4082	.4099	.4115	.4131	.4147	.4162	.4177
1.4	.4192	.4207	.4222	.4236	.4251	.4265	.4279	.4292	.4306	.4319
1.5	.4332	.4345	.4357	.4370	.4382	.4394	.4406	.4418	.4429	.4441
1.6	.4452	.4463	.4474	.4484	.4495	.4505	.4515	.4525	.4535	.4545
1.7	.4554	.4564	.4573	.4582	.4591	.4599	.4608	.4616	.4625	.4633
1.8	.4641	.4649	.4656	.4664	.4671	.4678	.4686	.4693	.4699	.4706
1.9	.4713	.4719	.4726	.4732	.4738	.4744	.4750	.4756	.4761	.4767
2.0	.4773	.4778	.4783	.4788	.4793	.4798	.4803	.4808	.4812	.4817
2.1	.4821	.4826	.4830	.4834	.4838	.4842	.4846	.4850	.4854	.4857
2.2	.4861	.4864	.4868	.4871	.4875	.4878	.4881	.4884	.4887	.4890
2.3	.4893	.4896	.4898	.4901	.4904	.4906	.4909	.4911	.4913	.4916
2.4	.4918	.4920	.4922	.4925	.4927	.4929	.4931	.4932	.4934	.4936
2.5	.4938	.4940	.4941	.4943	.4945	.4946	.4948	.4949	.4951	.4952
2.6	.4953	.4955	.4956	.4957	.4959	.4960	.4961	.4962	.4963	.4964
2.7	.4965	.4966	.4967	.4968	.4969	.4970	.4971	.4972	.4973	.4974
2.8	.4974	.4975	.4976	.4977	.4977	.4978	.4979	.4979	.4980	.4981
2.9	.4981	.4982	.4983	.4983	.4984	.4984	.4985	.4985	.4986	.4986
3.0	.4987	.4987	.4987	.4988	.4988	.4989	.4989	.4989	.4989	.4990
3.1	.4990	.4991	.4991	.4991	.4992	.4992	.4992	.4992	.4993	.4993
3.2	.4993	.4993	.4994	.4994	.4994	.4994	.4994	.4995	.4995	.4995
3.3	.4995	.4995	.4996	.4996	.4996	.4996	.4996	.4996	.4996	.4997
3.4	.4997	.4997	.4997	.4997	.4997	.4997	.4997	.4997	.4997	.4998
3.5	.4998	.4998	.4998	.4998	.4998	.4998	.4998	.4998	.4998	.4998
3.6	.4998	.4998	.4999	.4999	.4999	.4999	.4999	.4999	.4999	.4999
3.7	.4999	.4999	.4999	.4999	.4999	.4999	.4999	.4999	.4999	.4999
3.8	.4999	.4999	.4999	.4999	.4999	.4999	.4999	.4999	.4999	.5000
3.9	.5000	.5000	.5000	.5000	.5000	.5000	.5000	.5000	.5000	.5000
t	0.00	0.01	0.02	0.03	0.04	0.05	0.06	0.07	0.08	0.09

We are indebted to the following for permission to reproduce copyright material: John Wiley & Sons, Inc. for table (Appendix 9) from *Reliability Technology* by A. E. Green and A. J. Bourne.

Answers

Chapter 0

Exercise 0.1

With hierarchy of operations: **1.** 7 **2.** 22 **3.** 6 **4.** 0 **5.** 56 **6.** 15
Without hierarchy of operations: **1.** 7 **2.** 24 **3.** 2 **4.** 0 **5.** 68 **6.** 7

Exercise 0.2

1. 6.7765 **2.** 2.124374 **3.** 2.86122 **4.** 0.07623593

Exercise 0.3

Calculation	Estimate of the answer	Calculator display (to 4 decimal places)
$\dfrac{81.23}{4.1+3.5}$	$\dfrac{80}{8}=10$	10.6882
$\dfrac{427}{11.2-3.14}$	$\dfrac{400}{11-3}=\dfrac{400}{8}=50$	52.9777
$\dfrac{6.71-2.04}{11.62+4.12}$	$\dfrac{7-2}{12+4}=\dfrac{5}{16}=0.3$	0.2967
$7.26 \times (6.91-1.32)$	$7 \times (7-1)=42$	40.5834
$\dfrac{1.42+6.17-0.32}{1.71-09.82}$	$\dfrac{1.5+6-0.5}{2-1}=7$	8.1685
$\dfrac{1}{4.86-2.49}$	$\dfrac{1}{2.5}=0.4$	0.4219
$\dfrac{3}{7}+\dfrac{2}{4}$	$\dfrac{1}{2}+\dfrac{1}{2}=1$	0.9286
$\dfrac{17.2}{1.4}-\dfrac{3.61}{0.42}$	$13-9=4$	3.6905
$\dfrac{17.7}{3.02 \times (9.61-4.26)}$	$\dfrac{18}{3(10-4)}=\dfrac{18}{18}=1$	1.0955

Exercise 0.4

1. 5 **2.** 14 **3.** 3.7417 **4.** 4.0988 **5.** 2.7928 **6.** 1.1619
7. 38.2322 **8.** 0.2615 **9.** 1 **10.** 121 **11.** 2.56 **12.** 0.04
13. 0.49 **14.** 1.7424 **15.** 16.4917 **16.** 0.0047 **17.** 1.072 **18.** 0.7147
19. 3.7457 **20.** 3.7457 **21.** 5.6486 **22.** 1.3075

Chapter 1

Exercise 1.1

1. (a) three is less than four (b) five is greater than one
   (c) eleven is less than thirteen
2. (a) $2 < 6$ (b) $1 < 15$ (c) $20 > 9$

Exercise 1.4

1. 2,3,5,7,11,13,17,19,23,29,31,37,41,43,47,53,59,61,67,71
2. Because all other even numbers have 2 as a factor.
3. (i) $2^3 \times 3^2 \times 5$ (ii) $2 \times 3^2 \times 5^2$ (iii) $3^3 \times 5 \times 7^2$

Exercise 1.5

1. 2 **2.** 2 **3.** 37 **4.** 35 **5.** 30 **6.** co-prime

Exercise 1.6

1. 60 **2.** 1260 **3.** 1260 **4.** 2184 **5.** 11,858 **6.** 55,770

Exercise 1.7

5. 19 **6.** 31 **7.** 47 **8.** 86 **9.** 93 **10.** 83

Exercise 1.8

1. 5 **2.** 10 **3.** 2 **4.** 157 **5.** 43 **6.** 50

Exercise 1.9

2. No, because there is no number which can be subtracted from any other
   number, *in any order*, without changing it.

Exercise 1.10

4. No. We cannot find a number which combines by division with any number, *in any order*, without changing it.

Exercise 1.11

1. 38 **2.** 9 **3.** 52 **4.** 8 **5.** 33.25 **6.** 17 **7.** 76 **8.** 73 **9.** 11 **10.** 22 **11.** 78 **12.** 30 **13.** Division by zero is not allowed.

Chapter 2

Exercise 2.1

1. (a) 23 (b) 6 (c) 3 (d) 0 (e) −12 (f) − 4 (g) −9 (h) 18
2. (a) 4 (b) 5 (c) 21 (d) −1 (e) −20 (f) 18 (g) −2

Exercise 2.2

1. −3.5 **2.** −35 **3.** 35 **4.** 35 **5.** −60 **6.** 1 **7.** −1 **8.** 9 **9.** −64 **10.** 360 **11.** −4 **12.** −4 **13.** −1 **14.** 1 **15.** $\frac{1}{2}$ **16.** −12 **17.** $-\frac{4}{7}$ **18.** −8

Exercise 2.3

1. $\frac{19}{12}$ **2.** $\frac{41}{28}$ **3.** $\frac{5}{36}$ **4.** $\frac{1}{26}$ **5.** $-\frac{1}{20}$ **6.** $\frac{67}{88}$ **7.** $-\frac{63}{48}$ **8.** $\frac{31}{12}$
9. $\frac{50}{7}$ **10.** $\frac{147}{16}$ **11.** $\frac{79}{6}$ **12.** $\frac{1802}{35}$

Exercise 2.4

1. $\frac{21}{55}$ **2.** $-\frac{10}{27}$ **3.** $-\frac{11}{4}$ **4.** 1 **5.** $\frac{25}{36}$ **6.** $\frac{64}{125}$ **7.** $\frac{4}{9}$ **8.** $-\frac{4}{9}$
9. $-\frac{27}{125}$ **10.** −3 **11.** $-\frac{525}{24}$

Exercise 2.5

1. $\frac{4}{3}$ **2.** −10 **3.** $\frac{12}{5}$ **4.** $\frac{145}{84}$ **5.** $\frac{35}{36}$ **6.** $\frac{5}{7}$ **7.** $\frac{3}{4}$ **8.** $\frac{133}{18}$

Exercise 2.6

1. $\frac{31}{48}$ **2.** $\frac{15}{4}$ **3.** $-\frac{50}{33}$ **4.** $\frac{624}{665}$ **5.** $\frac{13}{32}$

Exercise 2.7

1. 1 , no 2. no, closed 3. yes 4. 12 , no 5. $\frac{9}{10}$, no
6. 5 , not closed under multiplication.
7. it could mean either $\frac{3}{4}$ divided by 5 $(=\frac{3}{20})$ or 3 divided by $\frac{4}{5}$ $(=\frac{15}{4})$.

Chapter 3

Exercise 3.1

1. (a) $4y$, -5 (b) $5xy$, $7xy$ (c) -15 (d) -7
   (e) $-10ab$ (according to our 'rule of thumb' for letters)
2. (a) $P = S - C$ (b) $e = c - t$ (c) percentage error $= \dfrac{100e}{t}$

   (d) $A = 3.142r^2$ (e) $a = \dfrac{v^2}{r}$ (f) $f = \dfrac{u+v}{uv}$ (g) $c = f + 5v$

   (h) $v = u + at$ (i) $F = 1.8C + 32$ (j) $C = \dfrac{5(F-32)}{9}$

Exercise 3.2

1. (a) 3 (b) 4 (c) -7 (d) -11 (e) 10 (f) -10 (g) 0 (h) 13
   (i) 0 (j) 74 (k) -56 (l) -16 (m) -1 (n) -1 (o) $-\frac{28}{3}$ (p) 0
2. (a) $\frac{5}{6}$ (b) $\frac{7}{12}$ (c) $\frac{3}{4}$ (d) $\frac{1}{6}$ (e) $\frac{1}{12}$ (f) $-\frac{1}{12}$ (g) $\frac{13}{12}$ (h) 2
   (i) $\frac{1}{2}$ (j) $\frac{61}{144}$ (k) $\frac{1}{3}$ (l) $\frac{7}{24}$ (m) $\frac{10}{3}$ (n) $\frac{4}{9}$ (o) $\frac{1}{5}$ (p) $\frac{4}{15}$

Exercise 3.3

1. 18.84 2. 120 3. 314 4. 200 5. 395.64 6. 1.167 7. 20
8. 0.57735 $(=\frac{1}{\sqrt{3}})$ 9. 1 10. 35 m/s 11. $B = \dfrac{RQ}{NA}$, 0.015 tesla

Exercise 3.4

1. $19x$ 2. $2x$ 3. $3x$ 4. $-9x$ 5. $-7x$ 6. $3x$ 7. $6y$ 8. $10n$ 9. $2x^2$
10. $9pq$ 11. $-4x + y + 2z$ 12. $8x + x^2 + 2x^3$ 13. $6xy$ 14. $8xy^2 + x^2y$
15. $ab^3 - 4a^2b^2 + 7a^2b$

Exercise 3.5

1. $12xy$ 2. $30pq$ 3. $-abc$ 4. $4xy$ 5. $42xyz$ 6. $\dfrac{15ab}{c}$ 7. $-\dfrac{7a}{5b}$

8. $\dfrac{2y}{z}$ 9. $\dfrac{3}{7}$ 10. $22m^3n^3$ 11. $\dfrac{3x^2y^2z^3}{2}$ 12. $-\dfrac{2}{7r}$

Chapter 4

Exercise 4.2

2. (a) a^8 (b) a^5 (c) a^9 (d) a^3b^6 (e) $8a^4$ (f) $15a^3b^3$
   (g) $36x^2$ (h) $36x^2$ (i) $216x^3$

3. (a) a^4 (b) a^6 (c) a^5 (d) a^4b^4 (e) a^3 (f) $\dfrac{35yz^3}{22x}$

   (g) $\dfrac{1}{5}$ (h) $\dfrac{1}{3xyz}$ (i) $\dfrac{7x^3}{2y}$

Exercise 4.4

1. (a) $9x^6$ (b) x^3y^6 (c) $x^4y^8z^{12}$ (d) $256a^4b^4c^4$
   (e) $32a^{10}b^{15}c^{20}$ (f) x^{24}

2. (a) $\dfrac{a^8}{b^{12}}$ (b) $\dfrac{27a^6}{64b^9}$ (c) $\dfrac{1}{a^3b^6c^3}$ (d) $\dfrac{18xy}{abc}$

Exercise 4.5

1. $8a^3$ 2. $9a^2b^2$ 3. a^7b^5 4. $\dfrac{a}{b}$ 5., 6., 7., 8. not possible

9. $27a^4$ 10. $12a^2$

Exercise 4.6

1. (a) $8\sqrt{2}+2\sqrt{3}-6$ (b) $1+3\sqrt{2}-8\sqrt{3}$
   (c) $7\sqrt{5}-7\sqrt{2}$ (d) $2\sqrt{6}-3\sqrt{2}$
2. (a) $7+6\sqrt{2}-\sqrt{3}-4\sqrt{5}$ (b) $8-2\sqrt{3}+7\sqrt{5}$
3. (a) $3\sqrt{3}+9\sqrt{2}-\sqrt{6}-6$ (b) $9\sqrt{5}-5\sqrt{6}+4\sqrt{10}-10$

Exercise 4.7

1. 10   2. 9   3. $6x$ 4. $6y$ 5.   4   6. $9\sqrt{3}$ 7. 18   8. 245   9. $392\sqrt{7}$

10. $5\sqrt{3}$ **11.** $4\sqrt{5}$ **12.** $\sqrt{10}$ **13.** $2\sqrt{3}$ **14.** $5\sqrt{x}$ **15.** $4\sqrt{x}$

Exercise 4.8

1. $6\sqrt{6}$ **2.** $15\sqrt{6}$ **3.** $6\sqrt{30}$ **4.** $4\sqrt{3}$ **5.** $2\sqrt{2}$ **6.** $\sqrt{7}$
7. $5\sqrt{2}$ **8.** $3\sqrt{y}$ **9.** $8\sqrt{yz}$

Exercise 4.9

1. $5\sqrt{2}$ **2.** $2\sqrt{10}$ **3.** $10\sqrt{3}$ **4.** $9\sqrt{3}$ **5.** $2\sqrt{7}$ **6.** $2x\sqrt{2x}$
7. $4\sqrt{3}$ **8.** $12\sqrt{5}$ **9.** $30\sqrt{2}$ **10.** $20\sqrt{3}$ **11.** $6\sqrt{5}$ **12.** $35\sqrt{2}$
13. $\sqrt{6}$ **14.** $10\sqrt{2}$ **15.** $56\sqrt{2}$

Exercise 4.10

1. $\dfrac{3\sqrt{2}}{2}$ **2.** $\dfrac{\sqrt{2}}{4}$ **3.** $\dfrac{5\sqrt{3}}{3}$ **4.** $\dfrac{\sqrt{2}}{4}$ **5.** $3\sqrt{6}$ **6.** $\dfrac{\sqrt{10}}{5}$ **7.** $\dfrac{\sqrt{3}}{24}$ **8.** $\dfrac{2\sqrt{10}}{15}$

Chapter 5

Exercise 5.1

1. (a) $3x + 6$ (b) $2a + 2b$ (c) $12x + 21y$ (d) $15m - 20n$ (e) $7x - 21y$
 (f) $-x - y$ (g) $-2x + 3y$ (h) $-3x + 4$ (i) $-6f - 4g$ (j) $12xy + 28x$
 (k) $5xya + 10xyb - 15xyc$ (l) $6x^3y^3 - 9\,x^2y^2$ (m) $3a + 3a^2 + 3a^3$
 (n) $-12x^2yab^2 + 21x^2ybcd$
2. (a) $4x^2 + 21y$ (b) $15p - 4q$ (c) $5a^3 + 8$ (d) $2x^2 - 3x^3$
 (e) $a^2 - 3ab + 2b^2$ (f) 0 (g) $4 - x - y + z$ (h) $4x + 15$
3. (a) $7x + 10$ (b) $7x + 14$ (c) $4x - 2$ (d) $-5x + 34y$
 (e) $10x - 19y + 17z$ (f) $7x^2 + 12x$ (g) $-57x$ (h) $6x^3 + 19x^2 - 12x$

Exercise 5.2

2. (a) $xy - 3x + 4y - 12$ (b) $fg - 6f - 7g + 42$
 (c) $2xy + 12x - 3y - 18$ (d) $2yz + 28z - 25y - 35$
 (e) $2mp - mq + 4np - 2nq$ (f) $a^2x - a^2y + b^2x - b^2y$
 (g) $10ax + 10bx + 15ay + 15by$ (h) $-2bp + 2ap - 2bq + 2aq$
 (i) $2ab^2c - 3a^2b + 8bc^2 - 12ac$ (j) $a^2 + 3a - 18$
 (k) $x^2 - 9x + 20$ (l) $a^2 + 3ab + 2b^2$

Exercise 5.3

1.(a) x^2+3x+2 (b) x^2+5x+6 (c) $x^2+7x+12$ (d) $6x^2+7x+2$
(e) $10x^2-27x-28$ (f) x^2+x-2 (g) x^2-3x+2 (h) $6x^2-17x+5$
(i) $6x^2-16x+8$ (j) $12x^2+4x-21$

2. (a) $x^2+xy-6y^2$ (b) $6p^2-7pq-3q^2$ (c) $6y^2-13yz+6z^2$
(d) $6x^2-xy-12y^2$ (e) $2a^2b^2-abcd-3c^2d^2$ (f) x^4+7x^2+12
(g) $3x^4+4x^2y^2-4y^4$ (h) $a^2+2ab+b^2$ (i) $a^2-2ab+b^2$ (j) a^2-b^2

Exercise 5.4

1. x^2+2x+1 2. $x^2-10x+25$ 3. $4y^2+12y+9$ 4. $9z^2-42z+49$
5. $a^2+2ab+b^2$ 6. $m^2+6mn+9n^2$ 7. $x^4+2x^2y^2+y^4$
8. $a^2b^2-2abcd+c^2d^2$ 9. $4x^2-9$ 10. $4x^2-1$
11. $25a^2-4b^2$ 12. $9x^2-36$

Exercise 5.5

1. 1, 6, 15, 20, 15, 6, 1 ; seven terms
    1, 7, 21, 35, 35, 21, 7. 1 ; eight terms

2.

```
                    1
                 1     1
              1     2     1
           1     3     3     1
        1     4     6     4     1
     1     5    10    10     5     1
  1     6    15    20    15     6    1
1     7    21    35    35    21    7    1
1  8   26   56   70   56   26   8   1
1  9   36   82  126  126  82   36  9   1      ; power 9
1  10  45  118  208  252  208  118  45  10  1  ; power 10
```

Exercise 5.6

1. $1+6x+15x^2+20x^3+\ldots$ 2. $1-6x+15x^2-20x^3+\ldots$
3. $1+8x+28x^2+56x^3+\ldots$ 4. $1-8x+28x^2-56x^3+\ldots$
5. $1+8x+24x^2+32x^3+\ldots$ 6. $1-6x+\frac{27}{2}x^2-\frac{27}{2}x^3+\ldots$
7. $3^5\left(1+\frac{5}{3}x+\frac{10}{9}x^2+\frac{10}{27}x^3+\ldots\right)$
8. (i) $\dfrac{7.6.5}{3.2}(2x)^3$ (ii) $\dfrac{11.10.9.8}{4.3.2.1}(-x)^4$

Exercise 5.7

1. (a) $x^5+10x^4+40x^3+80x^2+80x+32$
   (b) $625-500x+150x^2-20x^3+x^4$
2. $3^8-16\times3^7x+122\times3^6x^2$
3. $1024+5120x+11{,}520x^2+15{,}360x^3$
4. (a) 14   (b)   7,185,024
5.   6000
6. (a) $1125x^8$ (b) $-1792x^3$ (c) $-63x^4$

Chapter 6

Exercise 6.1

1. x^2y 2. xy^2z^2 3. $4xy$ 4. y 5. xy^2z^2 6. ab^2 7. $3a^2c^2$ 8. 5

Exercise 6.2

1. $2(x+2)$ 2. $5(2x+5)$ 3. $4(2x+3y)$ 4. $4(x-y)$
5. $a(b-3c)$ 6. $6(x-1)$ 7. $x(1-3x)$ 8. $3(2x-y^2)$
9. $5x(1-2y)$ 10. $4y(1-4y)$ 11. $2a^2(4a+7)$ 12. $a^2b(1-5ab^4)$
13. $py(y+1)$ 14. $2\pi r(r+h)$ 15. $5(x-2y+4z)$ 16. $x(p+q+r)$
17. $8x^2y^3(y-2x+3y^2)$ 18. $3xy(3x+2y-1)$ 19. $5ab(1+2ab-3a^2b^2)$
20. $-pqr(1+2q+3r)$ 21. $-11xyz(1+2xz^2-6y^2z)$

Exercise 6.3

1. (a) $(x+2)(y+6)$ (b) $(x-3)(y+1)$ (c) $(x+1)(y-5)$
   (d) $(x-1)(y-2)$ (e) $(x+3)(y+5)$ (f) $(x-3)(y-5)$
   (g) $(x+2)(y+3)$ (h) $(x-1)(y-5)$

2. (a) $(a+b)(x+y)$ (b) $(x+y)(p-q)$ (c) $(ax+z)(ax+z)$
   (d) $(2x+y)(r-2s)$ (e) $2(a-b)(2x+3y)$ (f) $(q-1)(p^2q-xy)$
   (g) $(x+y)(y-1)$ (h) $(w+z)(z-2)$ (i) $(b-3a)(x+2y)$
   (j) $(x-4y)(x-y)$ (k) $(a^2+b^2)(4+a)$ (l) $(ax+b)(x+1)$

Exercise 6.4

1. yes 2. yes 3. no 4. no 5. no 6. 0 7. no 8. yes 9. no

Exercise 6.5

1. $(x+3)(x+2)$ 2. $(x+3)(x+1)$ 3. $(x+4)(x+2)$ 4. $(x+4)(x+6)$
5. $(x+6)(x+9)$ 6. $(x+3)(x+17)$ 7. $(x-2)(x-1)$ 8. $(x-2)(x-8)$
9. $(x-3)(x-1)$ 10. $(x-4)(x-8)$ 11. $(x+5)(x-4)$ 12. $(x-8)(x+5)$
13. $(x-5)(x-2)$ 14. $(x-20)(x+3)$ 15. $(y+2)(y+3)$ 16. $(y+1)(y+4)$
17. $3(a-3)(a+1)$ 18. $(a+4)(a-1)$ 19. $(p-2)(p-4)$ 20. $2(p-3)^2$
21. $(z+4)(z-2)$ 22. $(z-4)(z+3)$ 23. $5(x-3y)(x+2y)$
24. $(x+7y)(x-y)$ 25. $(x-9y)(x-y)$ 26. $(x-5y)(x-2y)$
27. $(p+7q)(p-2q)$ 28. $(p-5q)(p+3q)$ 29. $(h-5k)(h-3k)$
30. $(h+7k)(h+2k)$

Exercise 6.6

1. $(3x+1)(x+3)$ 2. $(3x+1)(2x+1)$ 3. $(3x+7)(x+5)$
4. $(3x+2)(2x+5)$ 5. $(3x-1)(2x+3)$ 6. $(5x-2)(x+7)$
7. $(4x+1)(5x-1)$ 8. $(3x+1)(2x+3)$ 9. $(3+2x)(x-2)$
10. $(7-3x)(x-2)$ 11. $(3-2x)(2x+5)$ 12. $(4+3x)(5-2x)$
13. $(3x+2y)(2x+y)$ 14. $(3x+4y)(2x-3y)$ 15. $(9x-2)(4x+3)$
16. $(3x+2y)(2x-5y)$ 17. $(3x-y)(3x+4y)$ 18. $(1+5xy)(1-4xy)$
19. $(5xy-4)(3xy+2)$ 20. $(2x-5y)^2$ 21. $(2p-9q)(p+q)$
22. $(3b-16c)(b+c)$ 23. $(x-7)(3-x)$ 24. $(12+x)(7-x)$

Exercise 6.7

1. (a) $(x-1)(x+1)$ (b) $(x-6)(x+6)$ (c) $(x-7)(x+7)$
   (d) $(x-y)(x+y)$ (e) $(2x-3)(2x+3)$ (f) $(2x-5y)(2x+5y)$
   (g) $(x-\sqrt{5})(x+\sqrt{5})$ (h) $(2x-8)(2x+8)$ (i) $(9-x)(9+x)$
   (j) $(10-3x)(10+3x)$ (k) $(\sqrt{3}x-\sqrt{5})(\sqrt{3}x+\sqrt{5})$
   (l) $(x-y)(x+y)(x^2+y^2)$
2. (a) $(x-1)^2$ (b) $(x-2)^2$ (c) $(x-3)^2$ (d) $(3x+1)^2$
   (e) $(3x+2y)^2$ (f) $-(x-3)^2$ (g) $2(x+8)^2$ (h) $(4x+5)^2$
   (i) $-(4x+3)^2$ (j) $3(x-5)^2$ (k) $5(x+4)^2$ (l) $(x^2-2)^2$

Chapter 7

Exercise 7.1

1. 4 2. 0.25 3. 3 4. $\frac{23}{8}$ 5. $\frac{8}{23}$ 6. 15 7. 9 8. 20 9. $\frac{4}{9}$ 10. 6 11. 5
12. 20 13. 6 14. −4 15. −4 16. $\frac{5}{7}$ 17. $\frac{6}{5}$ 18. 1.5 19. 5 20. 9 21. 4

Exercise 7.2

1. 7 2. 4 3. 2 4. 5 5. 3 6. 4 7. 21 8. 15 9. 21 10. 3 11. 1
12. $\frac{18}{4}=\frac{9}{2}$ 13. 3 14. 8 15. 2 16. 7 17. 5 18. 6 19. 4 20. 3
21. 0 22. 2 23. 3 24. 11

Exercise 7.3

1. 4 2. 26 3. 19 4. 24 5. 5 6. −8 7. 5 8. 1 9. 4 10. 9 11. 3 12. 3

Exercise 7.4

1. 2 2. $\frac{6}{5}$ 3. 11 4. $\frac{24}{23}$ 5. −17 6. 17 7. $-\frac{1}{5}$ 8. −1 9. 1.5 10. 17
11. 1.2 12. $\frac{2}{3}$ 13. $-\frac{22}{3}$ 14. $\frac{7}{8}$ 15. 1.5 16. 6

Exercise 7.5

1. 6 , 18 2. 24 miles 3. 5 mph 4. 4 , 10 5. 3 , 36 6. 34 , 51
7. 6 , 14 8. 80% 9. 80 10. £3.90 11. 16 cm , 9 cm 12. 180 sq. cm
13. $45p$, $75p$ 14. 140 s A wins 15. £5000 16. £30,000

Exercise 7.6

2. (a) $-1 \leq x \leq 1$ **(b)** $3 < x < 7$ **(c)** empty set **(d)** empty set

Exercise 7.7

1. $10 < 12$, $-4 < -2$ **2.** $21 < 35$, $\frac{3}{7} < \frac{5}{7}$ **3.** $-7 < -2$, $3 < 8$
4. $10 > -15$, $\frac{2}{5} > -\frac{3}{5}$ **5.** $-8 > -12$, $-4 > -8$ **6.** $12 < 20$, $3 < 5$

Exercise 7.8

1. $x > 3$ **2.** $x < -4$ **3.** $x < 3$ **4.** $x > 6$ **5.** $x > 7$ **6.** $x > 4$ **7.** $x \geq \frac{4}{3}$
8. $x \geq 3$ **9.** $x < -3$ **10.** $x < \frac{6}{7}$ **11.** $x < 4$ **12.** $x \geq -\frac{6}{7}$ **13.** $x > 6.5$
14. $x < -5$ **15.** $x \leq 6$ **16.** $x \leq 1.5$ **17.** $x < 4$ **18.** $x < \frac{9}{5}$ **19.** $x < -\frac{1}{2}$
20. $x < \frac{3}{2}$ **21.** $-\frac{3}{2} < x < \frac{3}{2}$ **22.** $4 \leq x \leq 8$ **23.** $-1 \leq x \leq \frac{1}{3}$
24. $-2 \leq x < 5$

Exercise 7.9

1. $x > 8$ or $x < 2$ **2.** $3 < x < 11$ **3.** $x \geq 6$ or $x \leq -12$ **4.** $1 < x < 4$
5. $1.5 < x < 2.5$ **6.** $-2 < x < 5$ **7.** $x \leq 2$ or $x \geq 3$ **8.** $x < \frac{1}{3}$ or $x > \frac{7}{3}$
9. $x < \frac{1}{3}$ or $x > \frac{5}{3}$ **10.** $-\frac{5}{6} \leq x \leq \frac{13}{6}$ **11.** $x \geq 1$ or $x \leq -2$
12. $x > \frac{34}{5}$ or $x < -4$

Chapter 8

Exercise 8.1

1. $v = \frac{8}{3}$ **2.** $l = \frac{105}{11}$ **3.** 0.11 to 2 d.p. **4.** $x = -\frac{13}{12}$ **5.** 13.06
6. $h = 9.924$ **7.** 14.57 **8.** 1.989

Exercise 8.2

1. – 8. (See answers to Ex. 8.1.) **9.** $r = \dfrac{A}{\pi d}$ **10.** $r = \dfrac{v^2}{\sqrt{3g}}$ **11.** $t = \sqrt{\dfrac{2x}{g}}$

12. $a = \dfrac{2(s - ut)}{t^2}$ **13.** $\dfrac{4\pi^2 l}{T^2}$ **14.** $u = v - st$ **15.** $m = \dfrac{Fr}{gr + v^2}$

16. $t = \sqrt{\dfrac{1-x}{1+x}}$ **17.** $\dfrac{y - x^2 y}{x^2}$ **18.** $y = \dfrac{x^2 z}{(1 - x^2)}$ **19.** $x = \dfrac{(4y + 3)}{(5 - y)}$

20. $r = \dfrac{(V-1)R}{V}$ **21.** $R = \dfrac{Vr}{(V-1)}$ **22.** $a = \dfrac{bc}{(2c-b)}$ **23.** $a = \dfrac{(D^2-d^2)f}{(D^2+d^2)}$

24. $a = \dfrac{2xk}{2k-xv^2}$ **25.** $r = \dfrac{uv(\mu-1)}{(u-v)}$ **26.** $\mu = 1 + \dfrac{r(u-v)}{uv}$

Exercise 8.3

1. $y^2 = 4ax$ **2.** $y = 1 + \dfrac{1}{x-1} = \dfrac{x}{x-1}$ **3.** $q = \dfrac{1}{p-h} + \dfrac{1}{h} = \dfrac{p}{h(p-h)}$

4. $v^2 = u^2 + 2as$ **5.** $l = \dfrac{I}{mgh}$, $h = \dfrac{I}{ml}$ **6.** $v = \dfrac{2-c}{2(1-c)}$

Chapter 9

Exercise 9.1

1. $x = 0$ or $x = 7$ **2.** $x = 0$ or $x = -4$ **3.** $x = 0$ or $x = 2$ **4.** $x = 0$ or $x = 3$
5. $x = 0$ or $c = -4$ **6.** $x = 0$ or $x = \frac{5}{3}$ **7.** $x = \pm 7$ **8.** $x = \pm\sqrt{17}$
9. $x = \pm 5$ **10.** $x = \pm\sqrt{26}$ **11.** $x = \pm 7$ **12.** $x = \pm y$ **13.** $x = \pm\frac{3}{2}$
14. $x = \pm 4$ **15.** $x = \pm\frac{5}{2}y$ **16.** $x = \pm\frac{10}{3}$ **17.** imaginary **18.** imaginary
19. imaginary **20.** imaginary **21.** imaginary

Exercise 9.2

1. $x = -2$ or -3 **2.** $x = -1$ or -3 **3.** $x = -2$ or -4 **4.** $x = -4$ or -6
5. $x = -6$ or -9 **6.** $x = -3$ or -17 **7.** $x = 1$ or 3 **8.** $x = 4$ or 8
9. $x = 8$ or -5 **10.** $x = 2$ or 5 **11.** $x = 5$ or 12 **12.** $y = -2$ or -3
13. $y = -1$ or -4 **14.** $a = 3$ or -1 **15.** $a = 1$ or -4 **16.** $p = 3$ only
17. $z = 4$ or -3 **18.** $x = 3y$ or $-2y$ **19.** $x = 1$ only **20.** $x = 2$ only
21. $x = 3$ only **22.** $x = y$ or $-7y$ **23.** $x = y$ or $9y$ **24.** $p = 2q$ or $-7q$
25. $h = -2k$ or $-7k$ **26.** $x = -\frac{1}{3}$ or $x = -3$ **27.** $x = -\frac{1}{2}$ or $x = -\frac{1}{3}$

Exercise 9.3

1. (a) $x^2 + 12x + 6^2 = (x+6)^2$ **(b)** $x^2 - 18x + (-9)^2 = (x-9)^2$
   (c) $x^2 + 7x + \left(\frac{7}{2}\right)^2 = (x+\frac{7}{2})^2$ **(d)** $x^2 - 5x + \left(-\frac{5}{2}\right)^2 = (x-\frac{5}{2})^2$
   (e) $3(x^2 + 2x + 1^2) = 3(x+1)^2$ **(f)** $2x^2 + \frac{5}{2}x + \left(\frac{5}{4}\right)^2 = (x+\frac{5}{4})^2$

2. (a) 1 , $x = 7$ or $x = 8$ **(b)** $\sqrt{5}$, $x = \dfrac{1 \pm \sqrt{5}}{2}$

(c) 9 , $x = -2$ or $x = -5$ **(d)** -12 , complex roots

(e) 12 , $x = \dfrac{16 \pm \sqrt{12}}{2} = 8 \pm \sqrt{3}$ **(f)** -4 , complex roots

(g) 36 , $x = 1$ or $x = -\frac{1}{5}$ **(h)** 41 , $x = \dfrac{3 \pm \sqrt{41}}{8}$

(i) -71 , complex roots   **(j)** 17 , $x = \dfrac{7 \pm \sqrt{17}}{4}$

(k) -31 , complex roots   **(l)** 61 , $x = \dfrac{9 \pm \sqrt{61}}{10}$

(m) 121 , $x = 1$ or $x = -\frac{7}{4}$ **(n)** 73 , $x = \dfrac{-5 \pm \sqrt{23}}{6}$

Exercise 9.4

1. length 10, width 16 **2.** 9 mph , 10 mph **3.** 32 mph   **4.** 37 and 39
5. 10 , 11 and 12   **6.** $3 + \sqrt{15}$ **7.** either 15 by 40 or 20 by 30
8. (a) 15 **(b)** 21
9. (a) eight sides **(b)** ten sides   **(c)** n is non-integer; hence no polygon has 40 diagonals.
10. (a) 0.55 s (on way up) and 1.45 s (coming back down)
       (b) 1 second (only one root; therefore 5 metres is highest point)
       (c) complex roots; therefore never gets to 6 metres high
11. ten guests (including guests of honour)

Chapter 10

Exercise 10.1

1. $a = 2$, $b = 1$ **2.** $a = 5.75$, $b = 1.25$ **3.** $a = 1$, $b = 1$
4. $a = 1$, $b = -0.75$ **5.** $a = 2.8$, $b = 1.2$ **6.** $a = -1$, $b = 4$
7. $x = 1.4$, $y = -0.2$ **8.** $y = 4$, $z = 1$ **9.** $p = -1$, $q = 2$

Exercise 10.2

1. $x = -2$, $y = 3$ **2.** $x = -1$, $y = 2$ **3.** $s = 6.5$, $t = 3.6$
4. $a = \frac{2}{3}$, $b = -4$ **5.** $x = \frac{1}{3}$, $y = -\frac{1}{2}$ **6.** $x = 19$, $y = -17$

Exercise 10.3

1. $x = 2$, $y = 3$ or $x = 3$, $y = 2$ 2. $x = 4$, $y = 3$ or $x = -1.4$, $y = 4.8$
3. $x = 8$, $y = 2.5$ or $x = -2.5$, $y = 8$ 4. $x = 2$, $y = 1$ or $x = \frac{74}{35}$, $y = \frac{29}{35}$
5. $x = -8.5$, $y = 9.5$ or $x = 9$, $y = -8$ 6. $x = 4$, $y = 3$ or $x = 20$, $y = 35$

Exercise 10.4

1. 15 , 20   2. 7 , 13   3. seven 'tenners' , eleven 'fivers'
4. Ferry = £ 211 , cost /night = £ 27 , cost of fourteen day break £589
5. 100 miles from London at 2 p.m.
6. Xtra costs £1.60 per pint and Yukkybrew £1.70 per pint.

Chapter 11

Exercise 11.1

1. (a) 1   (b) 1   (c) 1   (d) 1   (e) $\frac{1}{6}$ (f) 3   (g) 1.25 (h) -1
2. (a) $\frac{1}{4}$ (b) $\frac{1}{9}$ (c) 1   (d) 4   (e) $\frac{1}{8}$ (f) $\frac{1}{27}$ (g) -1 (h) 8
3. (a) 2   (b) 10   (c) 25   (d) 343 (e) 16   (f) 216 (g) $\frac{81}{16}$ (h) $\frac{9}{16}$

Exercise 11.2

1. (a) $x^{1/5}$ (b) $x^{2/3}$ (c) $x^{4/7}$ (d) $x^{6/2} = x^3$
2. (a) $3^{1/2}$ (b) $100^{1/2}$ (c) $2^{1/3}$ (d) $16^{1/5}$ (e) $4^{1/6}$ (f) $49^{1/7}$ (g) $81^{1/9}$
   (h) $11^{1/11}$ (i) $17^{3/2}$ (j) $17^{2/3}$ (k) $5^{6/2} = 5^3$ (l) $13^{8/4} = 13^2$
3. (a) $\sqrt{4}$ (b) $\sqrt{5}$ (c) $\sqrt[3]{7}$ (d) $\sqrt[4]{11}$ (e) $\sqrt{64}$ (f) $\sqrt[3]{64}$ (g) $\sqrt[5]{\frac{1}{6}}$ (h) $\sqrt{\frac{3}{4}}$
4. (a) 2 (b) 5 (c) 16 (d) 512 (e) 4 (f) 8 (g) $\frac{1}{9}$ (h) $\frac{1}{32}$
5. (a) $\frac{4}{7}$ (b) $\frac{5}{6}$ (c) $\frac{4}{9}$ (d) $x^{1/5}$ (e) 2.4 (f) 1.5 (g) 1 (h) 5

Exercise 11.3

1. $\frac{1}{3}$ 2. $\frac{1}{3}$ 3. $\frac{1}{3}$ 4. $\frac{1}{5}$ 5. $\frac{1}{8}$ 6. $\frac{1}{81}$ 7. 16 8. 512 9. $\frac{1000}{27}$

Exercise 11.4

1. (a) $\log_3 81 = 4$ (b) $\log_2 8 = 3$ (c) $\log_7 2401 = 4$ (d) $\log_{10} 2 = 0.301$

   (e) $\log_3\left(\frac{1}{27}\right) = -3$ (f) $\log_{1/2} 4 = -2$ (g) $\log_4 32 = \frac{5}{2}$

   (h) $\log_{216}\left(\frac{1}{36}\right) = -\frac{2}{3}$ (i) $\log_7 1 = 0$ (j) $\log_{43} 1 = 0$

   (k) $\log_{17} 17 = 1$ (l) $\log_{0.5} 0.5 = 1$

2. (a) $16 = 4^2$ (b) $16^{1/2} = 4$ (c) $2^5 = 32$ (d) $\left(\frac{1}{2}\right)^{-3} = 8$ (e) $\frac{1}{4} = 4^{-1}$

   (f) $10^0 = 1$ (g) $3^0 =$ (h) $10^2 = 100$ (i) $100^{1/2} = 10$

   (j) $27^{2/3} = 9$ (k) $10^1 = 10$ (l) $17^1 = 17$

3. 6 , $\frac{1}{6}$ 4. 2 , $\frac{1}{2}$ 5. $\frac{3}{2}$, $\frac{2}{3}$ 6. (a) 3 (b) $\frac{1}{2}$ (c) -1 (d) $-\frac{3}{2}$

7. (a) 5 (b) 2 (c) -2 (d) $\frac{1}{4}$ (e) 1 (f) 3 (g) 10 (h) 25

Exercise 11.5

3. (a) $\log 6 = \log 2 + \log 3 = 0.7781$

   (b) $\log 1.5 = \log\left(\frac{3}{2}\right) = \log 3 - \log 2 = 0.1761$

   (c) $\log 8 = \log 2^3 = 3\log 2 = 0.903$ (d) 0.9542 (e) 1.3010

   (f) -0.1761 (g) $\log 5 = \log\left(\frac{10}{2}\right) = \log 10 - \log 2 = 0.6990$ (h) 1.1761

4. (a) $\log x + \log y$ (b) $\log x + \log y + \log z$

   (c) $2\log x + 3\log y + 4\log z$ (d) $\frac{1}{2}\log x + \frac{1}{2}\log y$ (e) $\log x - \log y$

   (f) $\log x + \log z - \log y$ (g) $\log x - \log z - \log y$ (h) $-\log x$

   (i) $\frac{1}{2}\log x - \frac{1}{2}\log y$ (j) $1 - 2\log x$ (k) $3\log x -$ (l) $1 - \frac{1}{2}\log x$

5. (a) $\log\left(\frac{25}{343}\right)$ (b) $\log 6$ (c) $\log\sqrt{2}$ (d) $\log\left(\frac{343}{4}\right)$ (e) $\log_a xy^2$

   (f) $\log_a \dfrac{x^{1/2} y^4}{z^3}$ (g) $\log_a \dfrac{x}{\sqrt{y}}$ (h) $\log_a \dfrac{x^2 y^6}{z^3}$

Chapter 12

Exercise 12.1

1.   independent variable : Fahrenheit scale , F
       dependent variable : Celsius scale , C

2.   independent variables : time , rate of interest
       dependent variable : interest

3.   independent variable : radius . R
       dependent variable : circumference , C

Exercise 12.2

1. **(a)** $f(0) = -3$, $f(1) = -2$, $f(5) = 122$, $f(-4) = -67$, $f(16) = 4093$
   $g(0) = 0$, $g(1) = -1$, $g(-4)$ does not exist , $g(16) = -4$
   (b) range of $f(x)$ is $-3 \leq f(x) \leq 15.622$
   range of $g(x)$ is $-5 \leq g(x) \leq 0$

2. **(a)** $\{-3, 0, 3, 6\}$ **(b)** $-1 \leq f(x) \leq 4$
   (c) $\{1, \frac{1}{2}, \frac{1}{3}, \frac{1}{4}, \frac{1}{5}\}$ **(d)** $2 \leq h(u) \leq 17$

3. $g(f(2)) = 9$ $f(g(2)) = 15$
4. $g(f(3)) = -18$ $f(g(3)) = 36$
5. $g(f(1)) = 7$ $f(g(1)) = \sqrt{7}$
   $g(f(2)) = 4\sqrt{2} + 3$ $f(g(2)) = \sqrt{11}$
   $g(f(3)) = 4\sqrt{3} + 3$ $f(g(3)) = \sqrt{15}$
   $g(f(4)) = 11$ $f(g(4)) = \sqrt{19}$
   $g(f(5)) = 4\sqrt{5} + 3$ $f(g(5)) = \sqrt{23}$

6. $g(f(x)) = g(x^2) = \dfrac{1}{x^2}$ $f(g(x)) = f\left(\dfrac{1}{x}\right) = \dfrac{1}{x^2}$

7. $g(f(x)) = g(x^2) = \sqrt{x^2} = x$ $f(g(x)) = f(\sqrt{x}) = (\sqrt{x})^2 = x$

Exercise 12.3

In the following , the numbers given are gradient followed by y intercept.
2. **(a)** 2 , 1 **(b)** 2 , −1 **(c)** −2 , 1 **(d)** −2 , −1 **(e)** −1 , 2
(f) $\frac{3}{2}$, $\frac{5}{2}$ **(g)** −3 , −6 **(h)** $-\frac{3}{4}$. −3 **(i)** 0 , 3 **(j)** infinity , doesn't
cross.

If the gradient is positive , graphs slope 'upwards' from left to right.
If the gradient is negative , graphs slope 'downwards' from left to right.

Chapter 13

Exercise 13.1

1. $v = 26 + 3t$, $u = 26$, acceleration $= 3$
2. **(i)** unsound reading for $t = 50$; **(ii)** $a = 0.34$, $b = 91.5$; **(iii)** 25°
3. $a = 0.16$; $b = 40$; $t = 50$; $R = 54.4$ (all approximate)

Exercise 13.4

1. (a) Plot $\dfrac{1}{y}$ against x ; a = gradient , b = intercept on $\dfrac{1}{y}$ axis.

(b) Plot y^2 against x ; a = gradient , b = intercept on vertical axis.

(c) Rearrange as $\dfrac{1}{y} = -\left(\dfrac{a}{b}\right)x + \dfrac{1}{b}$ and plot $\dfrac{1}{y}$ against x.

$\dfrac{1}{b}$ = intercept on vertical axis; hence $b - \left(\dfrac{a}{b}\right)$ = gradient ; hence a

(d) Rearrange as $\dfrac{y}{x} = ax + b$ and plot $\dfrac{y}{x}$ against x

a = gradient , b = intercept.

(e) Rearrange as $\dfrac{y}{x} = a - bx^2$.

(f) One possibility is to rearrange as $\dfrac{1}{xy} = \left(\dfrac{1}{b}\right)x - \dfrac{a}{b}$.

2. $a = 1.8$ approx. , $b = 0.1$ approx.
3. $a = 0.5$ approx. , $b = 0.2$ approx.
4. Wrong results are for $x = 30$ and $x = 70$; $a = 2$, $b = -5$.

Chapter 14

Exercise 14.1

1. 2 **2.** 1 **3.** -6 **4.** $2a$ **5.** $\frac{10}{3}$

Exercise 14.2

1. $5x^4$ **2.** $9x^2$ **3.** 5 **4.** -1 **5.** x^3 **6.** $\dfrac{3}{\sqrt{x}}$ **7.** $\dfrac{1}{3}x^{-2/3}$ **8.** $4x$ **9.** $-\dfrac{2}{x^2}$

10. $-\dfrac{1}{2x^2}$ **11.** $1-\dfrac{1}{x^2}$ **12.** $-\dfrac{3}{x^2}-\dfrac{1}{2}x^{-1/2}$ **13.** x^5 **14.** x^6 **15.** x^7

16. x^n **17.** $3x^2+\dfrac{4}{x^2}$ **18.** $-\dfrac{1}{x^2}+\dfrac{20}{x^6}$ **19.** $2x+4$ **20.** $6x+2$

21. $2a^2x-4ax^3$ **22.** $2ax + b$

Chapter 15

Exercise 15.1

1. (i) 44 m/s   (ii)   12 m/s^2 (iii) $\frac{11}{3}$ s (iv) $\frac{242}{3}$ m
2. (i) $v = 0$; $a = 6$; $t = 1$ or 2 (ii) $v = -3$; $a = 0$; $t = 1$ or 3
   (iii) $v = 4$; $a = -4$; $t = \sqrt{2}$ or 0

Exercise 15.2

1. $(2, 4)$; max 2. $\left(-\frac{1}{2}, \frac{3}{4}\right)$; min 3. $\left(\frac{3}{2}, -\frac{5}{4}\right)$; min
4. $(-1, -2)$; min , $(1, 2)$; max 5. $(2, 4)$; max, $(0, 0)$; min
6. $(2, 25)$; max 7. $N = 600$, £36,000   8. $\dfrac{wl^2}{8}$ 9. $\dfrac{E^2}{4R}$
10. 1.5 11. 0.4 12. (i) $A = x^2 + 4xy$ (ii) $4 = x^2 y$ (iii) 12

Exercise 15.3

1. $9(3x-2)^2$ 2. $-8(5-2x)^3$ 3. $-5(3-x)^4$ 4. $18(4+3x)^5$
5. $2(x^3 - 2x + 1)(3x^2 - 2)$ 6. $3(3x^2 + 5x - 1)^2(6x+5)$ 7. $\dfrac{2}{\sqrt{4x-1}}$
8. $\dfrac{-1}{\sqrt{1-2x}}$ 9. $\dfrac{a}{2\sqrt{ax+b}}$ 10. $-\dfrac{a}{3\sqrt[3]{(1-ax)^2}}$ 11. $\dfrac{1}{(1-x)^2}$
12. $\dfrac{-3}{(2+3x)^2}$ 13 $\dfrac{-3}{(3x+4)^2}$ 14. $\dfrac{-12x^3}{(3x^4-7)^2}$ 15. $\dfrac{-3x^2}{(x^3-a^3)^2}$

Exercise 15.4

1. $2x+5$ 2. $3(x^2+5x+1)^2(2x+5)$ 3. $\frac{1}{2}(x+3)^{-\frac{1}{2}}$ 4. $(2x+3)^{-\frac{1}{2}}$
5. $30x+7$ 6. $\dfrac{-t}{\sqrt{1-t^2}}$ 7. $\frac{1}{3}(x+1)^{-\frac{2}{3}}$ 8. $\frac{4}{3}x(2x^2+1)^{-\frac{2}{3}}$
9. $4x^3 - 2x - 2$ 10. $3(x+1)^4(6x+1)$ 11. $4x(4x+1)^2(10x+1)$
12. $2(x+1)(4x^2+2x+7)$ 13. $\frac{1}{2}(x^2-3x+2)x^{-\frac{1}{2}} + x^{\frac{1}{2}}(2x-3)$
14. $\dfrac{-2}{(x-1)^2}$ 15. $\dfrac{6x^2-8x-12}{(3x-2)^2}$ 16. $\dfrac{2a}{(a-x)^2}$ 17. $\dfrac{-1}{4t^{\frac{3}{2}}}$
18. $\dfrac{-2x^2+4x}{(x^2-x+1)^2}$ 19. $\dfrac{2}{(1-t^2)^{\frac{3}{2}}}$ 20. $\dfrac{2t}{(1+t)^3}$ 21 $-\dfrac{a}{2}(ax+b)^{-\frac{3}{2}}$

Chapter 16

Exercise 16.1

1. All answers are $+ c$.

(a) $\dfrac{x^7}{7}$ (b) $\dfrac{x^{11}}{11}$ (c) $\dfrac{x^{10}}{10}$ (d) $\dfrac{x^3}{3}$ (e) $\dfrac{x^2}{2}$

(f) x (g) constant (h) $\dfrac{x^{2.4}}{2.4}$ (i) $-\dfrac{x^{0.4}}{0.4}$ (j) $2x^{1/2}$

2. (a) $-\dfrac{1}{2x^2}$ (b) $-\dfrac{1}{3x^3}$ (c) $-\dfrac{1}{4x^4}$ (d) $-\dfrac{1}{5x^5}$

3. (a) $\dfrac{2}{3}x^{3/2}$ (b) $2x^{1/2}$ (c) $-\dfrac{5}{x}$ (d) $\dfrac{9}{5}x^{5/3}$

4. (a) $\dfrac{x^3}{3}+\dfrac{x^2}{2}$ (b) $\dfrac{x^4}{4}-a^3x$ (c) $\dfrac{ax^3}{3}+\dfrac{bx^2}{2}+cx$

5. (a) x^3+x^2-x (b) $5x-\dfrac{x^4}{4}+\dfrac{2x^3}{3}-\dfrac{x^2}{2}$ (c) $\dfrac{10x^{3/2}}{3}+\dfrac{4x^{5/2}}{5}$

Exercise 16.2

1. $10\frac{5}{12}$, yes , a combination of $15\frac{3}{4}$ and $-5\frac{1}{3}$. This does not, however , give the correct value for the area between the curve and the x axis.

2. $\int_{-1}^{2} ydx = -4.5;$ therefore the area $= 4.5$

3. $\int_{1}^{3} ydx = 14$ **4.** $\dfrac{2}{3}$ **5.** $\dfrac{5}{12}$ and $\dfrac{8}{3}$

Exercise 16.3

1. $\dfrac{56\pi}{3}$ **2.** $\dfrac{326\pi}{3}$ **3.** 625π **4.** $\dfrac{\pi}{30}$

Chapter 17

Exercise 17.1

1. (a) 0.5 (b) 0.5446 (c) 1.1504 (d) 0.9988 (e) 0.9617 (f) 1.1319

Exercise 17.2

1. *The following diagrams are not to scale.*

(a)
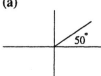

(b)

(c)
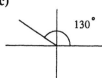

(d)
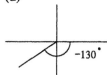

(e)

(f)

(g)
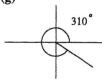

(h)

(i)

(j)
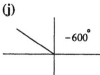

(k)

on positive
y axis

(l)
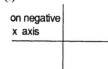

on negative
x axis

2. (a)–(h) 50° (i) 60° (j) 60° (k) 90° (l) 180°
   (a) and (h) are equivalent   (b) and (g) are equivalent
   (c) and (f) are equivalent   (d) and (e) are equivalent.

Exercise 17.3

These diagrams are not to scale.

1. 1.1918
tan 50° = 1.1918
tan 230° = tan 50°

2. −0.9063
sin 65° = 0.9063
sin −115° = − sin 65°

3. 0.9205
cos 23° = 0.9205
cos 743° = cos 23°

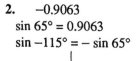

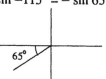

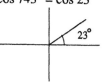

4.

0.0698

sin 4° = 0.0698

sin −544° = sin 4°

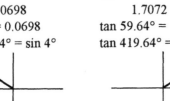

5.

1.7072

tan 59.64° = 1.7072

tan 419.64° = tan 59.64

59.64°

6.

0.3859

cos 67.3° = 0.3859

cos −67.3° = cos 67.3°

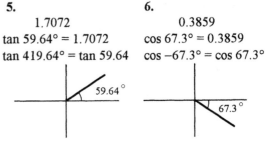

67.3°

Exercise 17.4

1. If θ is the required angle then, to 1 d.p. :

   $\theta = 180n - 15.8$ n odd $\theta = 180n + 15.8$ n even

   angles are 15.8°, 164.2°, 375.8°, 524.2°.

2. $\theta = 360n \pm 74.3$; angles are 74.3°, 285.7°, 434.3°, 645.7°.

3. $\theta = 180n + 15.1$; angles are 15.1°, 195.1°, 375.1°, 555.1°.

4. **(a)** −159.3°, −20.7°, 200.7°, 339.3°

   (b) −262.3°, −82.3°, 97.7°, 277.7°

   (c) no solutions

5. **(a)** 22.6°, 157.4°, 202.6°, 337.4°

   (b) 8.9°, 81.1°, 188.9°, 261.1°

   (c) 18.8°, 78.8°, 138.8°, 198.8°, 258.8°, 348.8°

6. Rearrange as $\tan\theta = 0.75$, giving $\theta = 180n + 36.9$.

Exercise 17.5

1. **(a)** 2.5593 **(b)** 1.2208 **(c)** −0.3640 **(d)** −1.2868 **(e)** −3.4203 **(f)** −0.0175

2. $\theta = 360n \pm 78.5$; 78.5°, 281.5°

3. $\theta = 180n - 0.38$ n odd, $\theta = 180n + 0.38$ n even ; 0.38°, 179.62°

4. $\theta = 180n + 52.6°$; 52.6°, 232.6°

5. **(a)** ±33.6°, ±326.4°

   (b) −326.4°, −9.8°, 189.8°, 350.2°

   (c) −355.95°, −265.95°, −175.95°, −85.95°, 4.05°, 94.05°, 184.05°, 274.05°

   (d) ±43.2°, ±136.8°, ±223.2°, ±316.8°

Exercise 17.6

3. **(a)** −169.9°, −10.1°, 190.1°, 349.9°   **(b)** ±60°, ±180°, ±300°

   (c) ±45°, ±135°, ±225°, ±315°   **(d)** 0°, ±228.2°, ±360°

   (e) −321.8°, −218.2°, 38.2°, 141.8°   **(f)** 9.7° 80.3°, 189.7°, 260.3°

4. (a) $\theta = 180n - 13$ n odd   , $\theta = 180n + 13$ n even

  (b) $\theta = 360n \pm 41.4$

  (c)   no real roots

  (d) $\theta = 180n - 14$, $\theta = 180n + 45$

5. $\cos A = -\frac{4}{5}$, $\tan A = -\frac{3}{4}$

6. $\sin A = -\frac{7}{25}$, $\cos A = -\frac{24}{25}$, $\cot A = \frac{24}{7}$, $\operatorname{cosec} A = -\frac{25}{7}$, $\sec A = -\frac{25}{24}$

7. $\sin A = -\frac{8}{17}$, $\tan A = -\frac{8}{15}$, $\cos A = \frac{15}{17}$, $\operatorname{cosec} A = -\frac{17}{8}$, $\cot A = -\frac{15}{8}$

Chapter 18

Exercise 18.1

1. (a) 0.1745 radians   **(b)** 0.2269 radians   **(c)** 1.0472 radians

  (d) 2.0071 radians   **(e)** 7.1384 radians   **(f)** −12.4267 radians

2. (a) 17.1887°   **(b)** 28.6479°   **(c)** 120.3211°

  (d) 269.2902°   **(e)** 359.8175°   **(f)** −63.0254°

3.

Degrees	30	45	60	90	120	135	150	180
Radians	$\dfrac{\pi}{6}$	$\dfrac{\pi}{4}$	$\dfrac{\pi}{3}$	$\dfrac{\pi}{2}$	$\dfrac{2\pi}{3}$	$\dfrac{3\pi}{4}$	$\dfrac{5\pi}{6}$	π
Degrees	210	225	240	270	300	315	330	360
Radians	$\dfrac{7\pi}{6}$	$\dfrac{5\pi}{4}$	$\dfrac{4\pi}{3}$	$\dfrac{3\pi}{2}$	$\dfrac{5\pi}{3}$	$\dfrac{7\pi}{4}$	$\dfrac{11\pi}{6}$	2π

4. (a) $\dfrac{\pi}{36}$ **(b)** $\dfrac{\pi}{18}$ **(c)** $\dfrac{\pi}{15}$ **(d)** $\dfrac{\pi}{12}$ **(e)** $\dfrac{\pi}{8}$

  (f) $\dfrac{2\pi}{15}$ **(g)** $\dfrac{\pi}{5}$ **(h)** $\dfrac{2\pi}{5}$ **(i)** $\dfrac{5\pi}{2}$ **(j)** 4π

Exercise 18.3

1. $-3\cos x + 4\sin x$ **2.** $8\cos 8x$ **3.** $-\frac{1}{2}\sin 3x$ **4.** $2\cos(2x+5)$

5. $\sin(\frac{\pi}{6} - x)$ **6.** $9\cos(3x-2)$ **7.** $4\cos\frac{2}{3}x$ **8.** $2\pi n \cos 2\pi n x$

9. $-2\cos x \sin x$ **10.** $2\cos\frac{1}{2}x$ **11.** $\dfrac{\cos x}{2\sqrt{\sin x}}$ **12.** $\cos x - \frac{1}{2}\cos 3x$

13. $\sqrt{x}\cos x + \dfrac{\sin x}{2\sqrt{x}} = \dfrac{2x\cos x + \sin x}{2\sqrt{x}}$ **14.** $2\cos 2x \cos x - \sin 2x \sin x$

15. $4x\cos 3x - 6x^2 \sin 3x$ **16.** $-b\cos ax \sin bx - a\sin ax \cos bx$

17. $2x\sin x + x^2\cos x$ **18.** $\dfrac{x\cos x - \sin x}{x^2}$ **19.** $\dfrac{2\cos x}{(1-\sin x)^2}$

20. $-\csc^2 x$ **21.** $-\cot x \csc x$

Chapter 19

Exercise 19.1

1. (a) $3e^{3x}$ (b) $4e^{4x-1}$ (c) $6e^{2x}$ (d) e^{4x+2} (e) $-\frac{1}{2}e^{-x/2}$

   (f) $4xe^{2x^2}$ (g) $e^x - e^{-x}$ (h) $e^{\sin x}\cos x$

2. (a) $e^x(6x+3x^2)$ (b) $e^{ax}(1+ax)$

   (c) $e^x(\cos x + \sin x)$ (d) $e^{5x}(\sec^2 x + 5\tan x)$

3. (a) $\dfrac{4}{e^{2x}+e^{-2x}+2}$ (b) $e^{-x}(2-x^2)$

4. All answers are $+ c$.

   (a) $-e^{-x}$ (b) $\frac{1}{3}e^{3x}$ (c) $\frac{1}{4}e^{4x-1}$ (d) $\frac{3}{2}e^{2x}$ (e) $\frac{1}{16}e^{4x+2}$

   (f) $-2e^{-x/2}$ (g) $e^x - e^{-x}$ (h) $\frac{1}{4}e^{2x^2}$

Exercise 19.2

1. (i) 6623 (ii) 2 days   2. (i) $N = N_0 e^{-\lambda t}$ (ii) when $t = \frac{1}{\lambda}\ln 2$

Chapter 20

Exercise 20.1

1. $q = 2.6p^{0.8}$, $a = 2.6$ approx. , $b = 0.8$ approx.

2. $y = 4.2e^{-1.5x}$

3. $T = 2e^{1.8t}$, $T_0 = 2$ approx. , $\mu = 0.3$ approx.

Exercise 20.2

1. (a) $\dfrac{1}{x+6}$ (b) $\dfrac{1}{x-11}$ (c) $-\dfrac{1}{5-x} = \dfrac{1}{x-5}$

   (d) $\dfrac{5}{5x+1}$ (e) $\dfrac{4}{4x-3}$ (f) $\dfrac{1}{7x-3}$

2. All answers are $+ c$.
   (a) $\ln(x+3)$ (b) $\ln(x-7)$ (c) $-\ln(3-x)$
   (d) $\frac{1}{3}\ln(3x+4)$ (e) $\frac{1}{3}\ln(3x-4)$ (f) $-\frac{1}{8}\ln(11-8x)$

Exercise 20.3

1. (a) $\dfrac{1}{x}$ (b) $\dfrac{4}{4x-1}$ (c) $\dfrac{4}{x}$ (d) $\dfrac{\cos x}{\sin x} = \cot x$

   (e) $\dfrac{-\sin x}{\cos x} = -\tan x$ (f) $\dfrac{2a}{a^2-x^2}$ (g) $\dfrac{e^x - e^{-x}}{e^x + e^{-x}}$ (h) $\frac{1}{2}\cot x$

2. (a) $x^2(1+3\ln x)$ (b) $1+\ln 2x$

   (c) $a\ln bx\cos ax + \dfrac{1}{x}\sin ax$ (d) $\dfrac{1}{x}(\ln ax + \ln bx)$

3. All answers are $+ c$.
   (a) $\ln(1+x^2)$ (b) $\ln(2+2x^3)$ (c) $\ln(x^2-3x)$
   (d) $-\ln\cos x$
4. (a) 0.4055 (b) -2.0793 (c) 0.5893 (d) 0.2903
   (e) 4.9034 (f) 0.1371 (g) 9.543

Chapter 21

Exercise 21.1

1. 47.1
2. (a) 47.1
   (b) Even though the marks are different, the total marks are the same. Hence the means are the same.

Exercise 21.2

1. (a) 3.5 (b) (i) none (ii) none
   Equal additions in positions symmetrical about the mean, make no difference to the mean.
2. 4.54

Exercise 21.3

1. $\bar{x} = 33.9$, $s = 1.51$ 2. $\bar{x} = 162.6$, $s = 23.2$

Chapter 22

Exercise 22.1

1. $2^4 = 16$, yes
2. There are 16 possible outcomes:

$$H\,H\,H\,H \; ; \; H\,H\,H\,T \; ; \; H\,H\,T\,H \; ; \; H\,H\,T\,T$$
$$H\,T\,H\,H \; ; \; H\,T\,H\,T \; ; \; H\,T\,T\,H \; ; \; H\,T\,T\,T$$
$$T\,H\,H\,H \; ; \; T\,H\,H\,T \; ; \; T\,H\,T\,H \; ; \; T\,H\,T\,T$$
$$T\,T\,H\,H \; ; \; T\,T\,H\,T \; ; \; T\,T\,T\,H \; ; \; T\,T\,T\,T$$

(a) $\frac{1}{16}$ (b) $\frac{4}{16} = 4\left(\frac{1}{2}\right)^3 \times \left(\frac{1}{2}\right)$ (c) $\frac{6}{16} = \frac{4.3}{2.1}\left(\frac{1}{2}\right)^2 \times \left(\frac{1}{2}\right)^2$ (d) $\frac{4}{16}$ (e) $\frac{1}{16}$

3. $\left(\frac{1}{2}+\frac{1}{2}\right)^4 = \left(\frac{1}{2}\right)^4 + 4\left(\frac{1}{2}\right)^3\left(\frac{1}{2}\right) + \frac{4.3}{2.1}\left(\frac{1}{2}\right)^2\left(\frac{1}{2}\right)^2 + \frac{4.3.2}{3.2.1}\left(\frac{1}{2}\right)\left(\frac{1}{2}\right)^3 + \left(\frac{1}{2}\right)^4$, yes

4. (a) $\frac{18}{100}$ (b) $\frac{81}{100}$ (c) $\frac{1}{100}$ since $P(A) = \frac{9}{10}$, $P(B) = \frac{1}{10}$

Exercise 22.2

1. (a) $\frac{1}{2^{10}}$ (b) $\frac{1023}{1024} = 1 - \frac{1}{2^{10}}$ 2. $\left(\frac{1}{6}\right)^3$, $1 - \left(\frac{1}{6}\right)^3$

Exercise 22.3

1. (a) $\frac{19}{100}$ (b) $\frac{81}{100}$ 2. $\frac{5}{12}$ 3. (a) $\frac{191}{480}$ (b) $\frac{289}{480}$ 4. $\frac{1}{32}$

Exercise 22.4

1. (i) $\frac{1}{6}$ (ii) $\frac{1}{2}$ (iii) $\frac{1}{3}$

Exercise 22.5

1. $\frac{1}{2}$ 2. $\frac{9}{200}$

Exercise 22.6

1. $p(1, H) = p(3, H) = p(5, H) = p(1, T) = p(3, T) = p(5, T) = \frac{1}{12}$

$p(2) = p(4) = p(6) = \frac{1}{6}$

2. (a) $\frac{171}{200}$ (b) $\frac{171}{200} + \frac{171}{4000} = \frac{171 \times 21}{4000} = \frac{3591}{4000}$

Chapter 23

Exercise 23.1

1. 0.100 , 0.267 , 0.311 , 0.208 , 0.086 , 0.023 , 0.004 , 0.000(4) , 0.000
$np = 6$, $npq = 1.5$
2. (i) $\dfrac{60}{13^5}$ (ii) $\dfrac{1440}{13^5}$ 3. $\dfrac{2133}{3125}$
4. (i) 0.0988 , (ii) 0.9805 , (iii) 0.0195

Exercise 23.2

1. 0.6703 , 0.2681 , 0.0536 , 0.0072
2. (i) about 496 (ii) about 4 3. 0.1912

Exercise 23.3

1. (i) 0.6179 (ii) 0.3821 (iii) 0.5398 (iv) 0.0781 (v) 0.4602 (vi) 0.0796
2. 1 in 1000 $(p(33) = 0.4995)$

Exercise 23.4

1. (a) 49.50 – 50.30 (b) 49.38 – 50.42
2. $2.690 < \mu < 2.720$
3. $693.91 < \mu < 750.09$ hours
4. £55.14 $< \mu <$ £61.14

Index

A
absolute value, 150
acceleration, 267
addition and subtraction
    of algebraic terms, 70
    of directed numbers, 87
    of fractions, 44
    of surds, 83
addition, 11
addition formulae, 336
algebra, 55
arithmetic mean, 378
arithmetic progression, 161
associated acute angle, 311
associative law
    of addition, 12
    of multiplication, 15
assumed mean, 381
average, 378

B
binary operator, 11
binomial distribution, 417
binomial series, 99
brackets, 12, 61

C
calculator, 1
calculus, 245
cancelling common factors, 27
central limit theorem, 434
chain rule, 280
chance, 400
closed, 11

co-domain, 211
co-prime, 19
coding, 381
coefficients, 59
common denominators, 45
common logarithms, 206
commutative law
    of addition, 12
    of multiplication, 14
complex roots, 180
composite functions, 214
compound interest, 347
confidence intervals, 433
constant of integration, 287
constants, 58
construction of equations, 140
continuous distribution, 415
continuous function, 271
continuous random variable, 386
cosecant of an angle, 323
cosine of an angle, 302
cotangent of an angle, 323
cube, 16
cubic function, 226

D
data, 377
decimal fractions, 28
definite integrals, 291
denominator, 26
dependent variable, 211
derivative, 252
difference of two squares, 118
differential calculus, 249

differential coefficient, 252
differential equation, 358
differentiation
    of a constant, 260
    of a product, 282
    of a quotient, 284
    of exponential functions, 352
    of logarithmic functions, 355, 373
    of sums and differences, 262
    of trigonometric functions, 339
directed numbers, 25
discontinuous function, 270
discrete, 386
discrete random variable, 386
discriminant, 179
distributive law
    of division, 28
    of multiplication, 20
division, 26
    of fractions, 50
    by zero, 30
domain, 211

E
element of a set, 9
elimination, 166
equation, 12, 61, 121
equiprobable events, 402, 405
equivalent angles, 307
equivalent fractions, 45
even numbers, 16
explicit formulae, 154
exponent, 17, 60
exponential function, 352
exponentiation, 17
expressions, 59
extrapolate, 243

F
factorising quadratic expressions, 111
factors
    algebraic, 106
    numerical, 15

formula, 56
formulae, 56
    manipulation of, 154
    subject of, 154
    substitution in, 154
    transposition of, 156
fractional indices, 199
frequency, 385
frequency distribution, 385
frequency polygon, 391
function, 211
function of a function rule, 282

G
gaussian distribution, 424
gradient
    of a curve, 250
    of a line, 220
graph of a function, 211
graph of logarithm function, 355
graphs of exponential functions, 355
graphs of trigonometric functions, 314
grouping, 386

H
HCF
    numerical, 18
    of algebraic terms, 107
histogram, 390
hypotenuse, 303

I
identity element
    for addition, 24
    for multiplication, 16
improper fractions, 27
indefinite integrals, 292
independent variable, 211
indeterminate, 248
index, 17, 60
indices, 60
    laws of, 196
    the three rules, 75

   things we cannot do with, 82
inequalities, 11, 121, 143
infinity, 247
integer, 10
integration, 286
   as the inverse of differentiation, 286
   as the limit of a sum, 290
   to find areas, 291
intervals, 143
inverse, 22
irrational numbers, 32

L
law of natural growth, 345
laws of logarithms, 207
laws of probability, 402, 405
LCM, 19
like terms, 70
limit, 245
linear equations, 123
linear function, 216
local maximum, 270
local minimum, 270
logarithmic graph paper, 366
logarithmic scale, 366
logarithms, 196, 205
lowest terms, 27

M
major trigonometric ratios, 304
mean deviation, 395
measures of central tendency, 379
measures of spread, 393
median, 378
minor trigonometric ratios, 323
mode, 378
modelling, 230
modulus, 150
multiple, 15
multiplication, 14
   of fractions, 47
multiplication and division
   of algebraic terms, 71

   of surds, 84
mutually exclusive, 404

N
natural logarithms, 355
natural number, 10
negative gradient, 259
negative indices, 197
negative integers, 24
normal distribution, 424
number line,10
numbers, 9
numerator,26

O
odd numbers, 16
order
   numerical, 10
   of precedence, 21, 23

P
parameter, 166
Pascal's triangle, 100
perfect square, 119
pi, 33
point of inflection, 274
poisson distribution, 421
polynomial function, 229
population, 384
positive integers, 25
possibility, 400
power, 16, 60
prime factorisation, 16
prime number, 15
probability, 400
probability distribution, 415
probability trees, 412
products
   algebraic, 90
   of binomial expressions, 93
proper fractions, 27
Pythagoras, 328
Pythagorean identities, 330

Q
quadrant diagram, 305
quadratic equations, 171
    problems leading to, 181
    solution by factorising, 174
    solution by formula, 176
quadratic expression, 95
quadratic function, 222

R
radian, 304, 333
raising to a power, 17
range, 211
rational numbers, 26
rationalising denominators, 88
real number, 9
reciprocals, 28
recurring decimals, 34
roots, 123
rotating vector, 305

S
sample, 384
sampling, 384
    from normal populations, 433
secant of an angle, 323
set, 9
similar triangles, 302
simple interest, 346
simultaneous equations, 186
    one equation not linear, 191
    problems leading to, 192
    solution by elimination, 187
    solution by substitution, 189
sine of an angle, 302
small angles, 338
solids of revolution, 296
solution, 123
solving equations, 123
speed, 266
square, 16
square root, 32

standard deviation, 396
standard normal distribution, 426
stationary values, 269
straight line law, 232
subset, 9
substitution, 64
subtraction, 22
surds, 34, 83
symbols, 9

T
tally, 386
tangent of an angle, 302
terminating decimals, 34
terms, 58
tests for maxima and minima, 272
Theorem of Pythagoras, 328
trigonometry, 302
turning point
    maximum, 270
    minimum, 270

U
unlike terms, 70

V
variables, 58
variance, 396
velocity, 266

W
working mean, 381

Y
y intercept, 218

Z
zero, 241
zero gradient, 259
zero index, 81, 197
zero products, 171

Printed and bound by CPI Group (UK) Ltd, Croydon, CR0 4YY

03/10/2024

01040437-0014